Radio-Frequency Digital-to-Analog Converters

Radio-Frequency Digital-to-Analog Converters
Implementation in Nanoscale CMOS

Morteza S. Alavi

Jaimin Mehta

Robert Bogdan Staszewski

AMSTERDAM • BOSTON • HEIDELBERG • LONDON
NEW YORK • OXFORD • PARIS • SAN DIEGO
SAN FRANCISCO • SINGAPORE • SYDNEY • TOKYO
Academic Press is an imprint of Elsevier

Academic Press is an imprint of Elsevier
125 London Wall, London EC2Y 5AS, United Kingdom
525 B Street, Suite 1800, San Diego, CA 92101-4495, United States
50 Hampshire Street, 5th Floor, Cambridge, MA 02139, United States
The Boulevard, Langford Lane, Kidlington, Oxford OX5 1GB, United Kingdom

© 2017 Elsevier Ltd. All rights reserved.

No part of this publication may be reproduced or transmitted in any form or by any means, electronic or mechanical, including photocopying, recording, or any information storage and retrieval system, without permission in writing from the publisher. Details on how to seek permission, further information about the Publisher's permissions policies and our arrangements with organizations such as the Copyright Clearance Center and the Copyright Licensing Agency, can be found at our website: www.elsevier.com/permissions.

This book and the individual contributions contained in it are protected under copyright by the Publisher (other than as may be noted herein).

Notices

Knowledge and best practice in this field are constantly changing. As new research and experience broaden our understanding, changes in research methods, professional practices, or medical treatment may become necessary.

Practitioners and researchers must always rely on their own experience and knowledge in evaluating and using any information, methods, compounds, or experiments described herein. In using such information or methods they should be mindful of their own safety and the safety of others, including parties for whom they have a professional responsibility.

To the fullest extent of the law, neither the Publisher nor the authors, contributors, or editors, assume any liability for any injury and/or damage to persons or property as a matter of products liability, negligence or otherwise, or from any use or operation of any methods, products, instructions, or ideas contained in the material herein.

Library of Congress Cataloging-in-Publication Data
A catalog record for this book is available from the Library of Congress

British Library Cataloguing-in-Publication Data
A catalogue record for this book is available from the British Library

ISBN 978-0-12-802263-4

For information on all Academic Press publications
visit our website at https://www.elsevier.com/

Publisher: Joe Hayton
Acquisition Editor: Tim Pitts
Editorial Project Manager: Charlotte Kent
Production Project Manager: Nicky Carter
Cover Designer: Victoria Pearson

Typeset by SPi Global, India

Contents

Preface xi
Acknowledgment xiii
Acronyms xv

CHAPTER 1 Introduction 1
    1.1 The Conventional RF Radio 2
    1.2 Motivation 4
    1.3 The Book Objectives 6
          1.3.1 System Simulation of WCDMA Baseband Data 6
          1.3.2 Some Important Figures-of-Merit in RF Transmitters 8
    1.4 Analog Versus Digital RF Transmitters 9
    1.5 Analog-Intensive RF Transmitters 9
    1.6 Digitally Intensive RF Transmitters 12
    1.7 New Paradigm of RF Design in Nanometer-Scale CMOS 13
    1.8 All-Digital Polar Transmitter 14
    1.9 All-Digital I/Q Transmitter 16
    1.10 Conclusion 18
    1.11 Book Outline 19

CHAPTER 2 Digital Polar Transmitter Architecture 21
    2.1 Introduction to Narrowband Polar Transmitters 21
          2.1.1 Motivation 21
          2.1.2 Contrast With Conventional Analog Approaches 24
    2.2 Overview of the RFDAC-Based Polar Transmitter
         Architecture 25
          2.2.1 Overview of the DPA 28
          2.2.2 DCO Operating Frequency and CKV Clock 29
    2.3 Details of Phase Modulation 30
    2.4 Design Challenges for the Small-Signal Polar Transmitter 32
          2.4.1 DCO Phase Noise 32
          2.4.2 DCO Pulling and Pushing 32
          2.4.3 VGA/DPA Nonlinearity 34
          2.4.4 Amplitude-Phase Path Delay Mismatch 35
          2.4.5 Amplitude-Phase Path Transfer Function
                Mismatch 37
          2.4.6 Amplitude Path DC Offset 37
          2.4.7 Amplitude Path Dynamic Range Limitations 38
          2.4.8 Amplitude Path DAC Mismatches 39
          2.4.9 Additional Distortions in the DPA (DAC/Mixer) 40

CHAPTER 3   Digital Baseband of the Polar Transmitter 41
  3.1 Overview of the TX Digital Baseband 41
    3.1.1 Pulse Shaping Filter 43
    3.1.2 Resampler 47
    3.1.3 CORDIC 49
  3.2 Predistortion Module 51
    3.2.1 Overview 51
    3.2.2 Principle of Operation 55
    3.2.3 Analysis of the Quantization Effects 64
  3.3 Predistortion Self-Calibration 66
    3.3.1 Effect of Temperature Variations on Predistortion 69
  3.4 Interpolative Filter 69
  3.5 Polar Bandwidth Expansion 75

CHAPTER 4   RF Front-End (RFDAC) of the Polar Transmitter 77
  4.1 Overview of the RF Front-End 78
  4.2 $\Sigma\Delta$ Amplitude Modulation 80
    4.2.1 $\Sigma\Delta$ Overview 80
    4.2.2 Digital Design 81
    4.2.3 Transfer Function and Spectrum 82
  4.3 Digital Pre-PA 83
    4.3.1 Overview of DPA Functionality 83
    4.3.2 Analysis of DPA Quantization Noise 84
    4.3.3 DPA Structural Design 91
  4.4 DPA Transistor Mismatches 92
    4.4.1 Amplitude Mismatch 92
    4.4.2 Phase Mismatch 93
  4.5 Key Categories of Mismatches and DEM 96
    4.5.1 Key Categories 96
    4.5.2 Simulation-Based Specifications 101
    4.5.3 Dynamic Element Matching 102
    4.5.4 Measurement Results 104
  4.6 Clock Delay Alignment 106
    4.6.1 Explanation of the Problem 106
    4.6.2 Self-Calibration and Compensation Mechanism 108
  4.7 Analysis of Parasitic Coupling 112
    4.7.1 Possible Coupling Paths 112
    4.7.2 A Novel Method of Characterizing $\Sigma\Delta$ Parasitic
          Coupling Using Idle-Tones 113
    4.7.3 Relationship Between Idle Tones and $\Sigma\Delta$ Parasitic
          Coupling 114

Contents vii

CHAPTER 5 Simulation and Measurement Results of the Polar Transmitter 117
- 5.1 Simulation Results 117
- 5.2 Measurement Results 118
  - 5.2.1 Predistortion 118
  - 5.2.2 Transmitter Close-In Performance 118
  - 5.2.3 Transmitter Wideband Noise Performance 123
  - 5.2.4 Performance Comparison 125
- 5.3 Conclusion 125

CHAPTER 6 Idea of All-Digital I/Q Modulator 129
- 6.1 Concept of Digital I/Q Transmitter 129
- 6.2 Orthogonal Summing Operation of RFDAC 131
- 6.3 Conclusion 141

CHAPTER 7 Orthogonal Summation: A 2 × 3-Bit All-Digital I/Q RFDAC 143
- 7.1 Circuit Building Blocks of Digital I/Q Modulator 144
  - 7.1.1 Digitally Controlled Oscillator 145
  - 7.1.2 Divide-By-Two Circuit 146
  - 7.1.3 25% Duty Cycle Generator 146
  - 7.1.4 Sign Bit Circuit 147
  - 7.1.5 Implicit Mixer Circuit 148
  - 7.1.6 2 × 3-Bit I/Q Switch Array Circuits 148
- 7.2 Measurement Results 154
- 7.3 Conclusion 159

CHAPTER 8 Toward High-Resolution RFDAC: The System Design Perspective 161
- 8.1 System Design Considerations 162
- 8.2 Conclusion 169

CHAPTER 9 Differential I/Q DPA and Power-Combining Network 173
- 9.1 Idealized Power Combiner With Different DRACs 174
- 9.2 A Differential I/Q Class-E Based Power Combiner 179
- 9.3 Efficiency of I/Q RFDAC 187
- 9.4 Effect of Rise/Fall Time and Duty Cycle 190
- 9.5 Efficiency and Noise at Back-Off Levels 191
- 9.6 Design an Efficient Balun for Power Combiner 193
- 9.7 Conclusion 200

CHAPTER 10 A Wideband 2 × 13-Bit All-Digital I/Q RFDAC 201
10.1 Clock Input Transformer 202
10.2 High-Speed Rail-to-Rail Differential Dividers 203
10.3 Complementary Quadrature Sign Bit 204
10.4 Differential Quadrature 25% Duty Cycle Generator 205
10.5 Floorplanning of 2 × 13-Bit DRAC 207
10.6 Thermometer Encoders of 3-to-7 and 4-to-15 209
10.7 DRAC Unit Cell: MSB and LSB 210
10.8 MSB/LSB Selection Choices 214
10.9 Digital I/Q Calibration and DPD Techniques 216
  10.9.1 IQ Image and Leakage Suppression 217
  10.9.2 DPD Based on AM-AM and AM-PM Profiles 219
  10.9.3 DPD Based on I/Q Code Mapping 220
  10.9.4 DPD Required Memory and Time 224
  10.9.5 DPD Effectiveness Against the Temperature
    and Aging 225
  10.9.6 Verification of DPD I/Q Code Mapping 226
10.10 Conclusion 228

CHAPTER 11 Measurement Results of the 2 × 13-Bit I/Q RFDAC 229
11.1 Measurement Setup 230
11.2 Static Measurement Results 231
11.3 Dynamic Measurement Results 234
  11.3.1 LO Leakage and IQ Image Suppression of I/Q
    RFDAC 234
  11.3.2 The RFDAC's Linearity Using AM-AM/AM-PM
    Profiles 236
  11.3.3 The RFDAC's Linearity Using Constellation
    Mapping 238
11.4 Conclusion 243

CHAPTER 12 Future of RFDAC 245
12.1 The Outcome 245
12.2 Some Suggestions for Future Developments 247
12.3 Future Trends 249

APPENDIX A Appendix for the Polar Transmitter 251
A.1 EDGE Modulation 251
  A.1.1 Symbol Mapping and Rotation 252
  A.1.2 Pulse Shaping Filter and Modulation 255

A.2 RF System Specifications for the EDGE Transmitter 256
A.3 Details of the Simulation Model 259
      A.3.1 Digital Amplitude and Phase Data Generation 260
      A.3.2 RF Front-End Model 260

APPENDIX B Appendix for I/Q RFDAC 263
B.1 Universal Asynchronous Receiver/Transmitter 263
B.2 Matching Network Equations 263
B.3 AM-AM/AM-PM Relationship 266
B.4 DPD Bandwidth Expansion 267

References 269
Index 281

Preface

The work on this RFDAC idea can be traced all the way back to 2001 when our small group at Texas Instruments (TI) was busy inventing new digital methods of realizing the traditional analog-intensive RF transceiver functions. The design of our first digital RF processor (DRP) for Bluetooth in 130 nm CMOS was coming to an end but we still needed to obtain some form of RF power regulation so that the transmitter (TX) would not produce too little or too much RF power. Hence, we added a few other transistors parallel to the main transistor of the switched-mode (thus, digital!) power amplifier operating in class-E mode (theoretically, up to 100% efficient!). Little did we know that those extra transistors would become extremely handy when another group from TI-Israel asked us to add amplitude modulation (AM) to the original frequency-modulated (FM) Bluetooth signal for the new extended data rate mode. The fully digital vector modulation (AM combined with FM) could not be made any much simpler: We just expanded the number of transistors from 8 to 256 to get the sufficient AM resolution in addition to the fine frequency resolution from an all-digital phase-locked loop (ADPLL), which was invented a couple of years earlier in the same DRP group as part of my part-time PhD research. That SoC went into high-volume production in 90 nm and then in 45 nm CMOS, while we got busy at applying these newly invented techniques to cellular standards. Enter Jaimin Mehta, who joined my TX DRP group to work out precisely all the system, architectural, and circuits issues as well as those related to calibration and compensation for high-volume production of cellular radios. Given the pioneering work and enormous amount of research issues solved throughout this cycle, Jaimin was able to turn his full-time industrial work into a PhD thesis guided by my old advisor from nearby University of Texas at Dallas, Professor Poras T. Balsara. Chapters 2–5 and Appendix A are the outcome of that industrial work.

After I left the US industry for academia in Europe in 2009, I had plenty of time to think about future applications of RFDACs. And the future looked unmistakably broadband. It was clear that the polar topology would come to a crashing halt due to its $>10\times$ bandwidth expansion. Enter Morteza Alavi, he was then a new bright-eyed PhD student at Delft University of Technology, extremely hungry for new knowledge and eager to prove himself with almost impossible tasks. He took upon himself to adapt the polar RFDAC approach to the in-phase/quadrature (I/Q) topology, which would entirely avoid that pesky bandwidth expansion. In the next four PhD years, he worked furiously at solving all the issues and the net result is that we can now *directly* go from 2×13 I/Q bits of the digital baseband signal to 150 MHz wideband 23 dBm of output RF power with 1024-QAM constellation. Chapters 6–12 and Appendix B are the outcome of his research work.

My coauthors and I strongly believe in the huge potential of RFDACs not only in commercial applications but also as an avenue for further academic research. Our belief seems to be shared with our collaborators and industrial partners. We would like to sincerely thank all our collaborators and supporters, as well as critics, at

TI and Delft University of Technology who have helped to turn the idea of digital RF modulation into a reality. I would like to conclude with a statement from my dear TI mentor, Dr. Bill Krenik: "The best is yet to come."

Robert Bogdan Staszewski
(on behalf of coauthors, Morteza S. Alavi and Jaimin Mehta)
Dublin
October 2016

Acknowledgment

The authors acknowledge help from the following PhD students in finding typos and minor mistakes during final proofreading: Milad Piri, Feifei Zhang, and Kai Xu.

Acronyms

1D	one-dimensional
2D	two-dimensional
ADC	analog-to-digital converter
ADPLL	all-digital PLL
AM	amplitude modulation
balun	balanced-unbalanced
BER	bit error rate
CML	current-mode logic
CMOS	complementary metal-oxide-semiconductor
CORDIC	coordinate rotation digital computer
DAC	digital-to-analog converter
DAT	distributed active transformer
DC	direct current
DCO	digitally controlled oscillator
DNL	differential nonlinearity
DPA	digitally controlled power amplifier
DPD	digital predistortion
DRAC	digital-to-RF-amplitude converter
DSP	digital signal processing
EDGE	enhanced data rates for GSM evolution
ESD	electrostatic discharge
EVM	error vector magnitude
FFT	fast Fourier transform
FIR	finite impulse response
GSM	global system for mobile communications
I	in-phase
IC	integrated circuit
IQ	I/Q vector
ISI	inter symbol interference
LO	local oscillator
LPF	low-pass filter
LSB	least significant bit
LTE	long-term evolution
MSB	most significant bit
NF	noise figure
OFDM	orthogonal frequency-division multiplexing
OSR	over sampling ratio
PA	power amplifier
PAPR	peak-to-average power ratio
PLL	phase-locked loop
PM	phase modulation

PSD	power spectral density
PVT	process, voltage, and temperature
Q	quadrature-phase
QAM	quadrature amplitude modulation
QPSK	quadrature phase-shift keying
RF	radio frequency
RF-DAC	RF digital-to-analog converter
RMS	root mean square
RRC	root raised cosine
RX	receiver
SAW	surface acoustic wave
SDT	software-defined-transmitter
SNR	signal-to-noise ratio
SRAM	static random access memory
SRC	sample-rate converter
TDD	time-division duplexing
TX	transmitter
UART	universal asynchronous receiver/transmitter
VCO	voltage controlled oscillator
VSA	vector signal analyzer
VSWR	voltage standing wave ratio
WCDMA	wideband code division multiple access
WLAN	wireless local area network
ZOH	zero-order-hold

CHAPTER

Introduction

1

CHAPTER OUTLINE

- 1.1 The Conventional RF Radio 2
- 1.2 Motivation 4
- 1.3 The Book Objectives 6
  - 1.3.1 System Simulation of WCDMA Baseband Data 6
  - 1.3.2 Some Important Figures-of-Merit in RF Transmitters 8
- 1.4 Analog Versus Digital RF Transmitters 9
- 1.5 Analog-Intensive RF Transmitters 9
- 1.6 Digitally Intensive RF Transmitters 12
- 1.7 New Paradigm of RF Design in Nanometer-Scale CMOS 13
- 1.8 All-Digital Polar Transmitter 14
- 1.9 All-Digital I/Q Transmitter 16
- 1.10 Conclusion 18
- 1.11 Book Outline 19

Consumer electronic devices such as smartphones, tablets, and laptops are constantly evaluated against three key criteria: low-cost, high-power efficiency, and support of multimode/multiband communication standards such as Wi-Fi or wireless LAN (IEEE 802.11) [1], Bluetooth [2], GNSS,[1] second-generation (2G) cellular using GSM, third-generation (3G) cellular using WCDMA [3], and fourth-generation (4G) cellular using either of WiMAX or 3GPP LTE [4, 5]. These gadget devices comprise a myriad of IC chips to perform an extensive number of distinct functions such as multimedia streaming and gaming as well as supporting the aforementioned communication standards. As an example of contemporary gadget devices, Fig. 1.1 illustrates the mainboard of a smartphone, for example, the iPhone 5. It consists of an application processor (AP) unit, subscriber identification module (SIM) card slot,

---

[1] GNSS is the abbreviation of global navigation satellite system. It includes American's GPS, Russian's GLONASS, European Union's Galileo, and China's Beidou navigation system.

CHAPTER 1 Introduction

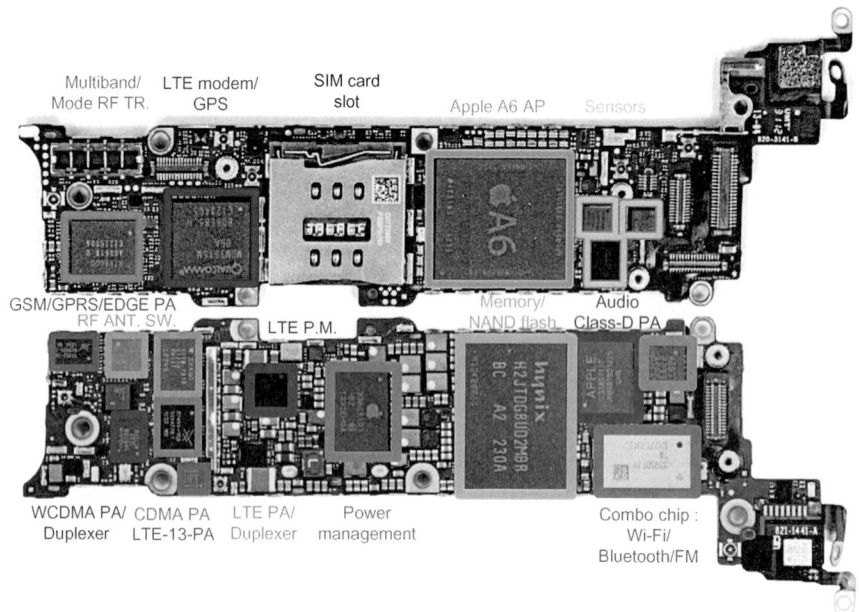

FIG. 1.1

The front and rear mainboard of iPhone-5.

Courtesy of Apple Inc.

NAND flash memory, power management unit, Class-D audio amplifier, and, most significantly, a number of RF transceiver modules that support today's universal communication standards such as GSM, CDMA, Wi-Fi/Bluetooth/FM, GPS, and LTE in combination with its power management unit. Over the past two decades, there have been tremendous efforts to design RF radios that will afford an opportunity to address the low-power, low-cost, and extremely power-efficient demands and, yet, they have also employed inventive transceiver architectures.

1.1 THE CONVENTIONAL RF RADIO

As depicted in Fig. 1.2A, a conventional RF transceiver consists of a baseband DSP unit, TX, and RX [6, 7]. The transmitter performs in the following manner: The anticipated transmitted information such as voice, video, or digital data like text/images are initially digitally processed, then encoded, and subsequently applied to a DAC in order to convert the digital data to their corresponding analog counterparts. Due to the fact that these analog signals comprise unwelcome noise and spectral replicas, the transmitter utilizes a low-pass filter (LPF) to reduce those undesirable artifacts. The filtered analog signals are subsequently mixed with an RF LO utilizing an

1.1 The conventional RF radio

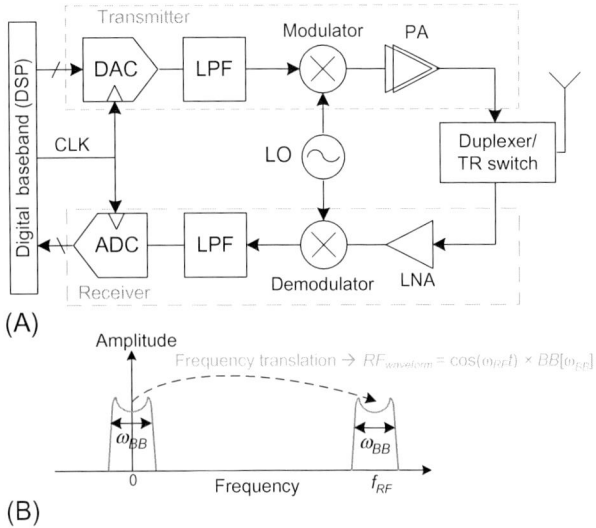

FIG. 1.2

(A) A conventional RF transceiver, (B) frequency-domain illustration of modulating the baseband signal.

upconverting mixer and, consequently, upconverted to their designated RF transmit frequency band. Otherwise stated, as illustrated in Fig. 1.2B, the LO signal translates the low-frequency baseband analog signals with the bandwidth of ω_{BB} into their equivalent high-frequency RF representation. In fact, the LO clock modulates the baseband signal. Thus, the upconverting RF mixer is referred to as an RF modulator. A PA is succeedingly employed to efficiently boost the RF signal. The subsequent RF amplified signal is then applied to either a duplexer or transmit/receive (T/R) switch[2] in order to be delivered to the transmitting antenna.

In the receiver, however, its sensitivity is ameliorated utilizing a low-noise amplifier (LNA). The subsequent signal is then down converted to baseband frequency utilizing a down-converter mixer, which is also referred to as a demodulator. By employing a few LPF circuits and then an ADC, the received signal is digitized for further digital baseband-processing operation.

Among transceiver blocks, the RF transmitter is considered to be the most power consuming block of the entire radio, thus being a hindrance for extending the battery lifetime of portable wireless devices. This is due to the fact that it comprises power hungry building blocks such as DAC, the upconverting mixer, and, most importantly, the RF PA. For example, in the aforementioned smartphone, as indicated in Fig. 1.1,

---

[2] If the radio uses frequency-division duplexing (FDD), it employs a duplexer. Otherwise, in the TDD communication system, it requires a T/R switch.

Table 1.1 Worldwide Smartphone Sales in 4Q13 (kUnits)

Company	4Q13 Units	4Q13 Market Share (%)	4Q12 Units	4Q12 Market Share (%)
Samsung	83,317.2	29.5	64,496.3	31.1
Apple	50,224.4	17.8	43,457.4	20.9
Huawei	16,057.1	5.7	8666.4	4.2
Lenovo	12,892.2	4.6	7904.2	3.8
LG	12,822.9	4.5	8038.8	3.9
Others	106,937.9	37.9	75,099.3	36.2
Total	282,251.7	100	207,662.4	100

Gartner (February 2014).

there are at least six different RF-PA modules[3] to support multiband/multimode communication standards.

Consequently, the power amplifiers are the most "power-hungry" building blocks of any portable device due to the high-power RF signal generation that is required by the corresponding communication standard to continuously ensure an impeccable transmitting operation. For example, the corresponding PA in the GSM standard should provide 2 W RF power. Considering a power efficiency as high as 40%, the battery must drain 5 W to generate such RF power which reduces the battery life of the portable device [8–10].

It is worth mentioning that, currently, there are over 6 billion cellular phone users and, in particular, the number of smartphone users is increasing exponentially. Table 1.1 depicts the third quarter of 2012–2013 global smartphone market. As is indicated, smartphone production increases approximately 45% every year. With the progressive number of wireless devices and the relative shortage of bandwidth availability, it will become increasingly difficult to transmit power-efficient, undistorted RF signals that do not interfere with other users.

1.2 MOTIVATION

Considering the aforementioned design challenges of the conventional RF transmitter in Section 1.1, intensive research has recently been directed toward the realization of digitally intensive and all-digital RF transmitters that provide high-output power with high efficiency while concurrently being highly reconfigurable [11–21]. This was also in response to the incredible advancements of the mainstream CMOS technology in both processing speed and circuit density as well as the relentless coercion to reduce total solution costs through integration of RF, analog, and digital circuitry. Since the digital baseband part of a wireless communication channel has

---

[3] 1-GSM/GPRS/EDGE; 2-WCDMA; 3-CDMA; 4-LTE; 5-LTE band 13; 6-Wi-Fi/Bluetooth/FM.

traditionally been implemented in the most advanced CMOS technology available at a given time for mass production, the need for single-chip CMOS integration has forced immutable changes to the way RF circuits are fundamentally designed. In this low-voltage nanometer-scale CMOS environment, the high-performance RF circuits must exploit the time-domain design paradigm and rely significantly on digitally assisted techniques. In other words, contemporary designers are analog thinkers; they tend to cogitate in terms of continuous voltages and currents. On the contrary, digital designers think in terms of discrete-time, discrete-value, and, most recently, time-domain operations.

Scientist researchers believe that the development of the concepts behind direct digital or all-digital RF transmitters is in their infancy and innumerable untapped applications and standards are still available. In the current Information Era where more and more data is wirelessly transmitted, it is essential that this information is transmitted in a clear and distortion-free manner, while exploiting as little energy as possible. The innovative concept should involve a method and a system for digitally generating wideband radio frequency (RF) signals that can be transmitted with (for narrowband) and without (for wideband) the requirement of the current state-of-the-art polar topology, which has only been successfully applied to narrowband modulation schemes [14].[4] Otherwise stated, the wireless LAN (WLAN) as well as the currently running wideband modulation standards such as the 3G cellular employing WCDMA and the 4G cellular using 3GPP LTE cannot exploit the current state-of-the-art digital polar RF technology [14] due to the wide bandwidth requirements. Solving the wide bandwidth modulation issues will afford an opportunity to introduce a universal all-digital RF transmitter. It will decrease the cost of production for new and existing IC customers and also will benefit the end users. Concurrently, it will reduce the IC area, which signifies increased space for other applications. It will also lower power consumption, thereby resulting in a longer battery lifetime. To summarize, the main advantages of utilizing the new direct digital transmitter architecture are:

1. The circuit integration level will be higher, thus the wireless portable devices can be smaller, lighter, and have nicer designs through smaller form factors. From a technical perspective, employing a direct digital transmitter eliminates the bulky DAC, analog LPFs, and analog mixers.
2. Due to the elimination of several power-hungry analog components, the power efficiency of the RF transmitter will be higher, therefore, it subsequently increases battery life or leads to smaller battery size.
3. The new topology can afford more sophisticated reconfigurability in order to handle various standards and even modulation methods that have yet to be invented.

---

[4] As will be explained in the following chapters, although recent publications have indicated that polar architecture can manage up to 20 MHz [18–21], the utilized polar structure is still very complicated. Moreover, it could not handle wider band signals such as 802.11n/ac with the modulation bandwidth of 40/160 MHz.

1.3 THE BOOK OBJECTIVES

Based on the previous explanations in Sections 1.1 and 1.2, the underlying objective of this book is, therefore, to implement an innovative, fully integrated, all-digital RF transmitter, which should be power-efficient and, yet, must manage a wideband complex-modulated signal in order to support multimode/multiband communication standards. Moreover, it should generate adequate RF power, minimal out-of-band spectra while the generated far-out noise of the all-digital RF transmitter must be low enough, that is, does not desensitize the companion receiver circuit blocks. Due to the zero-order-hold operation in the digital-to-RF interface, the direct digital RF transmitter, however, inherently creates spectral replicas at multiples of the baseband upsampling frequency away from the carrier frequency. In contrast, in the conventional analog RF transmitter, the spectral replicas have been eliminated due to utilizing the analog continuous-time filters directly following the DAC (see Fig. 1.2A). Moreover, the quantization noise induced by the limited effective baseband code resolution of the digital transmitter worsens out-of-band spectra. This anomaly also does not exist in the analog counterparts. To gain more insight into these phenomena, the following subsection will be conducive.

1.3.1 SYSTEM SIMULATION OF WCDMA BASEBAND DATA

As discussed previously, although the direct digital transmitter does not necessitate the bulky analog LPF, it requires another type of filtering based on a discrete-time digital approach. Consider a WCDMA baseband signal in which its data rate varies between 7.5 kb/s and 960 kb/s. Fig. 1.3 illustrates the filtering process

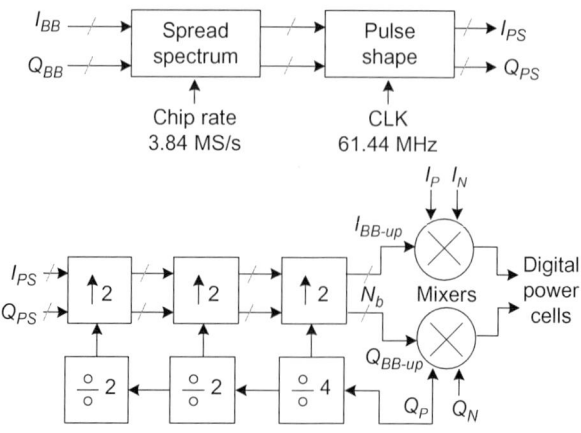

FIG. 1.3

Upsampling and interpolation process of WCDMA baseband signals.

whereby the spread spectrum filter is first applied to I_{BB} and Q_{BB} samples at a chip rate of 3.84 MHz. Subsequently, in order to constrain the occupied RF band, the resulting signals get pulse-shaped. The pulse-shaped oversampling rate is preferably 16; therefore, the baseband data rate will increase to 61.44 Mb/s. If these samples (I_{PS} and Q_{PS} shown in Fig. 1.3) are applied directly to the mixer of an all-digital transmitter, then the resulting signal will produce spectral images at every 61.44 MHz from the upconverted spectrum, which are very difficult to filter out. Therefore, it necessitates interpolation of these signals to higher data rates. Moreover, with the use of upsampling, the quantization noise spreads over a wider operational frequency range, which also improves the dynamic range or resolution. With the employment of three cascaded FIR filter-based interpolators, each with an upsampling factor of $L = 2$, the resulting signal sampling rate is 535 MHz. Consequently, the images repeat every 535 MHz from the desired spectrum, which are now easier to filter out. It should be mentioned that, as depicted in Fig. 1.3, the FIR clocks are synchronized to the carrier clocks using clock frequency dividers which are a divide-by-4, and consecutive divide-by-2.

The simulation results of Fig. 1.3 are depicted in Fig. 1.4, which also indicates that, by increasing the resolution (N_b) of the all-digital transmitter, the quantization noise is decreased and, thus, the dynamic range of the WCDMA signal improves. Each additional bit corresponds to a 6 dB improvement in the dynamic range. The far-out spectrum of Fig. 1.4 demonstrates the sampling image replica of the original signal, which is located 535 MHz away. This reveals that the baseband signals[5] are ultimately upsampled by a factor of 128 (16 × 8) to spread the quantization noise, thus lowering its spectral density.

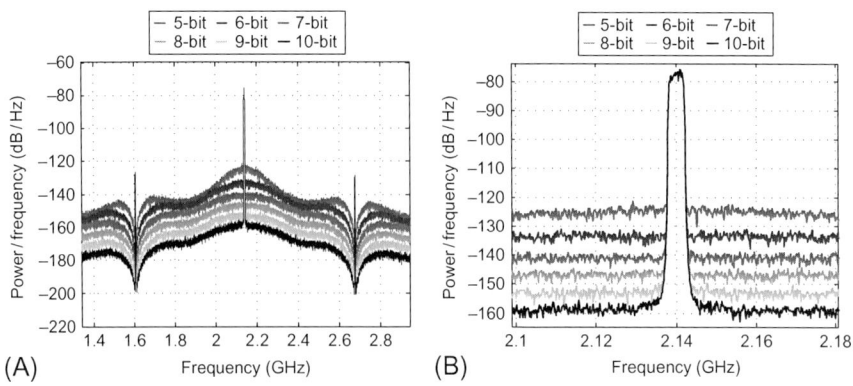

FIG. 1.4

Simulated WCDMA RF output spectrum for different resolutions of all-digital transmitter: (A) far-out; (B) close-in range.

---

[5] Note that the WCDMA chip rate is 3.84 MS/s.

In conclusion, an all-digital RF transmitter at the equal RF output power level compared to its analog counterpart creates more noisy out-of-band spectrum.

1.3.2 SOME IMPORTANT FIGURES-OF-MERIT IN RF TRANSMITTERS

The RF transmitter can be evaluated by employing a number of figures-of-merit and they will be discussed as follows. First, the drain efficiency of the transmitter is defined as:

$$\eta_{drain} = \frac{P_{RFout}}{P_{DC\text{-}PA}} = \frac{V_{RFout} \times I_{RFout}}{V_{DC\text{-}PA} \times I_{DC\text{-}PA}} \quad (1.1)$$

where P_{RFout} is the generated RF output power at the 50 Ω load, and $P_{DC\text{-}PA}$ is the corresponding DC power consumption of the PA. Moreover, V_{RFout} and I_{RFout} are RMS output voltage and current, respectively. Second, the system efficiency of the transmitter is defined as:

$$\eta_{system} = \frac{P_{RFout}}{P_{DC\text{-}Total}} = \frac{P_{RFout}}{P_{DC\text{-}PA} + P_{DC\text{-}Driver}} \quad (1.2)$$

where $P_{DC\text{-}Total}$ is the related DC power consumption of the entire RF transmitter, which includes the DC power consumption of PA and the upconverting clock driver circuit. The RF transmitter dynamic performance is subsequently divided into in-band performance and out-of-band spectral purity. The EVM is conventionally utilized for the evaluation of the in-band performance and calculates the difference between the symbol sampling points of the measured and reference waveforms. The EVM result is defined as the square root of the ratio of the mean error vector power to the mean reference power and is expressed as a percentage [3]:

$$EVM_{RMS} = \frac{\sqrt{\sum_{i=1}^{N_{pt}} \frac{1}{N_{itr}} \times (IQ_{ideal}(i) - IQ_{real}(i))^2}}{\sqrt{\sum_{i=1}^{N_{pt}} \frac{1}{N_{itr}} \times (IQ_{ideal}(i))^2}} \quad (1.3)$$

where N_{itr} is the number of iterations for each point, and N_{pt} is the number of constellation points. Moreover, $IQ_{ideal}(i)$ and $IQ_{real}(i)$ are "ith" ideal (reference) and real (measured) constellation points, respectively. Additionally, the out-of-band spectral purity can be evaluated as an adjacent channel power ratio (ACPR)[6] and far-out noise performance at the corresponding receiver frequency band of interest. ACPR is utilized to measure the nonlinear distortion in the transmitted signal. Moreover, ACPR in combination with the modulation scheme determines the maximum allowable nonlinearity of the related RF transmitter. ACPR is defined as follows [22, 23]:

---

[6]Sometimes ACPR is expressed as adjacent channel leakage ratio (ACLR).

$$ACPR_{adj} = \frac{P_{adj}}{P_{main}} \qquad (1.4)$$

where P_{adj} is the total adjacent right/left channel and P_{main} is the total RF power within the main transmit channel. Note that, instead of ACPR, the third-order intermodulation product (IM_3) can also be employed [22, 24]. Furthermore, as discussed previously, the far-out noise performance of the RF transmitter strongly depends on quantization noise and baseband upsampling rate. It is noteworthy that the far-out noise at the receiver frequency band of interest[7] must be less than its related thermal noise of -173.83 dBm/Hz.[8] Thus, as will be explained in more detail in Chapter 8, the baseband code resolution as well as the baseband upsampling rate should be suitably selected so as not to deteriorate the out-of-channel spectrum as well as the far-out noise performance.

1.4 ANALOG VERSUS DIGITAL RF TRANSMITTERS

The following sections give an overview of the various existing types of RFIC-TX architectures that are required to determine their advantages and disadvantages in order to develop an innovative approach to devise a fully integrated all-digital RF transmitter. Though this has already been explained in many ways in the existing literature [24–26], it is still beneficial to reiterate their distinctions. Section 1.5 discusses analog-intensive I/Q or Cartesian as well as polar RF transmitters. In Section 1.6, the need for digital-intensive RF transmitters is examined. Section 1.7 explains the new paradigm of RF transmitter design utilizing the nanoscale CMOS process technology. In Section 1.8, an all-digital polar transmitter is explored while Section 1.9 unveils all-digital Cartesian transmitters. Finally, the conclusion of the chapter is drawn in Section 1.10.

1.5 ANALOG-INTENSIVE RF TRANSMITTERS

Beginning around the 1990s, virtually all RFIC-TXs have been analog intensive and based on an architecture similar to that depicted in Fig. 1.5. At the transmitter back-end, the baseband user information symbols become pulse-shaped to obtain two orthogonal components of complex-number digital samples, that is, I and Q that are constrained to the allocated frequency channel.

They are then converted into an analog continuous-time domain through a DAC with a typical ZOH function. The LPF following the DAC subsequently

---

[7] This condition is applicable in full-duplex systems.
[8] This noise power level is obtained assuming conjugate matching at the input of the receiver, 50 Ω input load, and temperature at 27°C [24, 25]. Note that, the NF and SNR of the receiver should also be considered which will be discussed in more detail in Chapter 8.

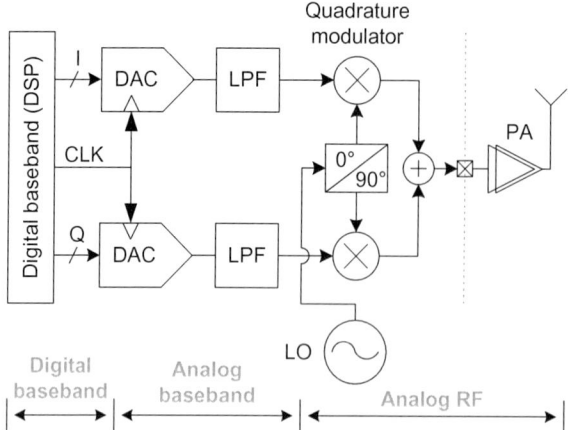

FIG. 1.5

Traditional analog-intensive I/Q RF transmitter.

filters out the switching harmonics. Thus obtained analog baseband signal is then upconverted (frequency translated) into RF through a quadrature modulator. The resulting (typically external) PA increases the RF power level at the antenna to that of what is required by the wireless standard, which could be as high as 2 W for a GSM handset. The frequency synthesizer-based LO performs the frequency translation. It is typically realized as a charge-pump PLL with $\Sigma\Delta$ dithering of the divider modulus to realize the fractional-N frequency division ratio. The complete architecture and monolithic circuit design techniques of the conventional transmitter exhibited in Fig. 1.5 have been described extensively in an innumerable amount of literature, particularly in the latest editions of text books [24, 25]. These architectures have been successfully employed in integrated CMOS transmitters [27–34] for over a decade (since the late 1990s). Unfortunately, their beneficial lifetime is gradually coming to an end [35] to make room for analog polar and more digitally intensive architectures. In summary, due to linear summation of I and Q signal paths, this architecture can manage large modulation bandwidth. On the other hand, it comprises power-hungry as well as bulky components, which make this transmitter deficient.

An alternative to the I/Q topology of Fig. 1.5 is the polar realization exhibited in Fig. 1.6 in which the two uncorrelated, that is, orthogonal, components (alternative to the I and Q components) are amplitude ρ and phase θ:

$$\rho = \sqrt{I^2 + Q^2}$$
$$\theta = \tan^{-1}\left(\frac{Q}{I}\right) \quad (1.5)$$

The complex-envelope signal is $S = \rho \exp(j\theta)$. The phase modulation could be performed by a direct or indirect frequency modulation of a PLL. The amplitude

1.5 Analog-intensive RF transmitters

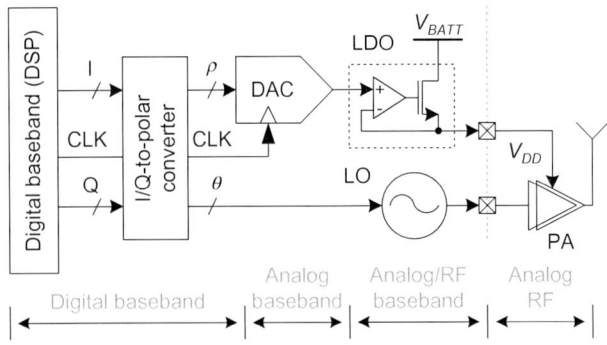

FIG. 1.6

Traditional analog-intensive polar RF transmitter.

modulation could be performed by a low drop-out (LDO) voltage regulator modulating the V_{DD} supply of a high-efficiency switched-mode PA, such as a class-E PA. The analog polar TX architecture is a more recent development that addresses the inherent poor power efficiency and noise issues of the I/Q architecture. Notwithstanding these benefits, polar topologies exhibit certain serious disadvantages:

1. Their phase and amplitude paths exploit heterogeneous circuits whose delays must be accurately aligned to avoid spectral distortion at the final recombining stage, yielding spectral re-growth.
2. The required instantaneous bandwidth is significantly larger than in the I/Q approach. Otherwise stated, the composite modulator bandwidth is typically limited due to the bandwidth constrains in the practical DC-to-DC switchers or linear converters.

The latter aspect can be understood by examining the nonlinear operations that the amplitude ρ and phase θ experience during conversion from the I/Q representation (Eq. 1.5). The resulting signal bandwidth expansion can be observed in Fig. 1.7. This figure compares the baseband bandwidth (ω_{BB}) expansion of an example WCDMA I/Q modulator with its polar counterpart. Note that the bandwidth expansion is due to the fact that ρ and θ enhance the even ($2\omega_{BB}$, $4\omega_{BB}$, $6\omega_{BB}$, ...) and odd ($3\omega_{BB}$, $5\omega_{BB}$, $7\omega_{BB}$, ...) harmonics of baseband signal, respectively.[9]

Over the past several years, such analog-intensive polar transmitters (both small-signal, that is, at the transceiver IC level, and large signal, that is, encompassing PA) have been touted for their reconfigurability, implementational, and performance benefits more so than their traditional Cartesian counterparts that are based on

---

[9] Note that according to Eq. (1.5) the envelope signal is an even function (square function) which entails the bandwidth expansion due to even harmonics. On the other hand, the phase signal is an odd function (arctangent function), which causes the bandwidth expansion because of odd harmonics.

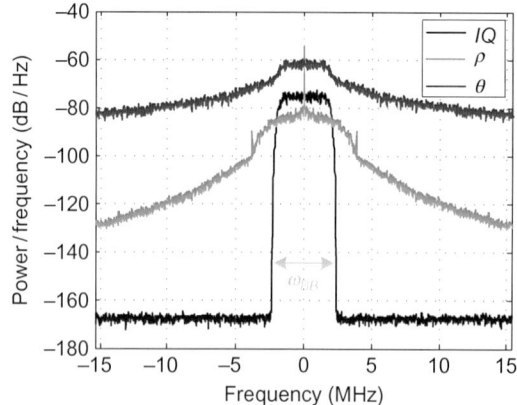

FIG. 1.7

Polar modulator bandwidth expansion of the WCDMA signal.

an I/Q upconversion mixer. A number of recent publications have demonstrated their superiority with highly integrated silicon realizations [36–42] but only for the *narrowband* (200 kHz allocated bandwidth) modulation standard, Enhanced Data rates for GSM Evolution (EDGE), of the 2G cellular. The polar architecture migration to *wideband* modulation standards, such as 3G (WCDMA—allocated bandwidth of 5 MHz), 4G (3GPP LTE, WiMAX—allocated bandwidth of up to 20 MHz), and other evolving wideband wireless standards (e.g., 802.11n/ac) continues to be a daunting task and, thus far, there has only been one very recent silicon demonstrator that produces low-efficiency and low-RF output power (3 dBm) discussed in the open literature [43]. The effort, however, appears to continue with another theoretical proposal of the polar topology for WCDMA [44] and one 90-nm CMOS demonstration of the RF front-end block components of the WCDMA digital polar transmitter [45].

1.6 DIGITALLY INTENSIVE RF TRANSMITTERS

As mentioned above, the digital approach to design RF circuits and architectures is taking over in industry. The primary contributors to this sea-like transformation are the ever-improving cost advantages and processing capabilities of the CMOS technology that have been occurring at regular intervals at the pace according to the so-called Moore's law. Basically, with every CMOS process technology advancement node (i.e., from 90-nm to 65-nm, then to 40-nm, and then to 28-nm, and so on) occurring every 18–24 months, the digital gate density, being a measure of the digital processing capability, doubles (i.e., gate area scaling factor of $0.5\times$). Simultaneously, the basic gate delay, being a measure of the digital processing speed, improves linearly (i.e., gate delay scaling factor of $0.7\times$). Likewise, the cost of fabricated silicon per unit area remains approximately the same at its high-volume production maturity stage. Indeed, over the last decade, the cost of silicon charged by IC fabs has remained constant. The main implication of this is that a cost of a given digital function, such as a GSM detector or a digital audio decoder, can be cut in half every 18–24 months when transitioned to an upgraded CMOS technology. At the same time, the circuits consume proportionately less power and are faster [26].

Unfortunately, these astonishing benefits of digital scaling are not shared by traditional RF circuits. Additionally, the strict application of the architectures (Figs. 1.5 and 1.6) to the advanced CMOS process node might actually result in a larger silicon area, inadequate RF performance, and higher consumed power. The constant scaling of the CMOS technology has had an unfortunate effect on the linear capabilities of analog transistors. To maintain reliability of scaled-down MOS devices, the V_{DD} supply voltage continues to decrease, while the threshold voltage V_t remains almost constant (to maintain the low level of leakage current). This negatively affects the available voltage margin when the transistors are intended to operate as current sources. Moreover, the implant pockets that were added for the benefit of digital operation have drastically degraded the MOS channel dynamic resistance r_{ds}, thus severely reducing the quality of MOS current sources and the maximum available voltage self-gain

$$A_{v\text{-intrinsic}} = g_m \cdot r_{ds} \qquad (1.6)$$

where g_m is the transconductance gain of a transistor. Furthermore, due to the thin gate dielectric becoming increasingly thinner, large high-density capacitors realized as MOS switches are becoming unacceptably leaky. This prevents an efficient implementation of low-frequency baseband filters and charge pump PLL loop filters.

1.7 NEW PARADIGM OF RF DESIGN IN NANOMETER-SCALE CMOS

An early attempt at designing RF circuits in advanced CMOS has revealed a new paradigm:

In a deep-submicron CMOS process, time-domain resolution of a digital signal edge transition is superior to voltage resolution of analog signals [11].

On a pragmatic level, this indicates that a successful design approach in this environment would exploit this paradigm by emphasizing the following:

1. fast switching characteristics or high f_T (20 ps and 250 GHz in 40-nm CMOS process, respectively) of MOS transistors: high-speed clocks and/or fine control of timing transitions,
2. high density of digital logic (1 Mgates/mm^2) and also static random access memory (4 Mb/mm^2) results in extremely inexpensive digital functions and assistant software,

3. ultra-low equivalent power-dissipation capacitance C_{pd} of digital gates leading to both low switching power consumption

$$P_T = f \cdot C_{pd} \cdot V_{DD}^2 \tag{1.7}$$

as well as potentially low coupling power into sensitive analog blocks, and
4. small device geometries and precise device matching made possible by the fine lithography to create high-quality analog and RF data converters;

while avoiding the following:

1. biasing currents that are commonly exploited in analog designs,
2. reliance on voltage resolution with continuously decreasing supply voltages and increasing noise and interferer levels, and
3. nonstandard devices that are not required for memory and digital circuits, which constitute the majority of the silicon die area.

Despite the early misconceptions that the digitalization of RF would somehow produce more phase noise, spurs, and distortion, the resulting digitally intensive architecture is inclined to be, overall, more robust by actually producing lower phase noise and spurious degradation of the transmitter chain and a lower noise figure of the receiver chain in face of millions of active logic gates on the same silicon die, as repeatedly substantiated in subsequent publications [13, 14, 17, 19, 46–49]. Additionally, the new digital TX architecture would be highly reconfigurable with analog blocks that are controlled by software to guarantee the best achievable performance and parametric yield. An additional benefit would be an effortless migration from one process node to the next without significant modifications.

1.8 ALL-DIGITAL POLAR TRANSMITTER

A transmitter architecture that is amenable to the digital nanoscale CMOS technology is depicted in Fig. 1.8 [13, 14]. This digital polar transmitter has found its way into commercial products. Note that, this digital-to-RF architectural transformation is influenced by the ever-improving cost advantages and process capabilities of CMOS technology, and various successful implementations have been brought to fruition [14]. The work of Chowdhury et al. [19] and, most recently, Ye et al. [21, 49] proposed wideband polar modulators, which achieve high output power and high efficiency.

The LO in Fig. 1.8 is realized as an ADPLL that produces a phase or frequency-modulated digital clock carrier employing the two-point digital modulation scheme. The clock is fed into a digital-to-RF-amplitude converter (DRAC) that produces an RF output of which the envelope is substantially proportional to the amplitude control word (ACW) or ρ. Hence, the architecture is termed as a digital polar TX.

The DRAC is realized as an all-digital RF power generation circuit as illustrated in Fig. 1.9A [50]. It is also referred to as a DPA. Note that compared to the traditional

1.8 All-digital polar transmitter

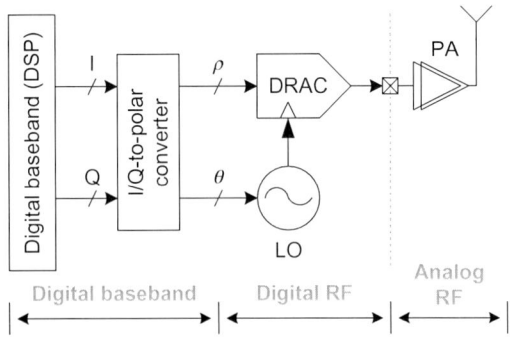

FIG. 1.8
All-digital polar transmitter.

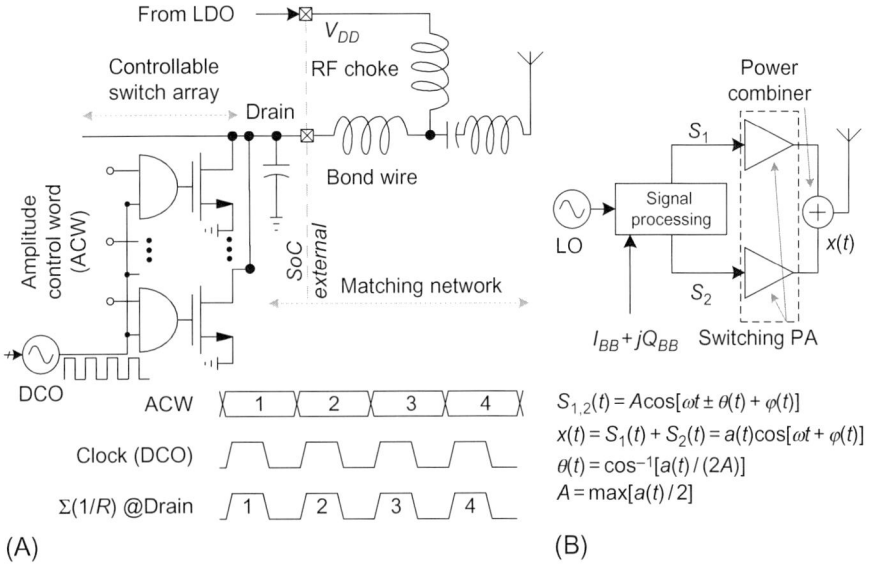

FIG. 1.9
(A) DRAC in digital polar TX [50]. (B) Digital outphasing TX [51].

power amplifier, which amplifies the analog/RF input signal, in this context, the input signal of the DPA is an internal digital clock, therefore, measuring the *amplification* gain of the DPA seems a bit problematic. The DRAC has proven its exploitation by the amplitude modulation in more advanced modulation schemes, such as the extended data rate (EDR) mode of Bluetooth, EDGE, and WCDMA.

The DRAC operates as a near-class-E RF PA and is driven by the square wave output of a DCO in the ADPLL. The rise-/fall-time of digital signals, including clocks, is typically 40–60 ps in 65 nm low-power CMOS process, which makes the trapezoidal shape almost square for cellular band signals. A large number of core NMOS transistors are employed as on/off switches, each with a certain conductance $1/R$, and are followed by a matching network that interfaces with an antenna or an external PA. The number of active switches, and thus the total conductance $\sum(1/R)$, is digitally controlled and establishes the instantaneous amplitude of the output RF envelope. The RF output power is incited by coherently moving the resonating energy through the LC tank between the load and the switches. The supply V_{DD} replenishes the energy lost to the load and internally. The RF amplitude is contingent upon the relationship between the total switch conductance and the conductance of the matching network.

The class-E PA operation is attempted in order to achieve the maximum output power (all the switches are active), where the highest achievable efficiency is of the utmost importance. Note that there could be different methods of conversions from digital to RF amplitude as well as classes of operation. For example, class-D PA [18, 52] would utilize two switching devices operating complementary and connected to the supply/ground and the matching network. Fine amplitude resolution is achieved through high-speed $\Sigma\Delta$ transistor switch dithering. The timing diagram of Fig. 1.9A [50] assumes that the data changes with every RF clock cycle, which is reasonable considering the high-speed dithering. In practice, the integer ACW signals would change every certain number of DCO cycles. Despite the high speed of digital logic operation, the overall power consumption of the transmitter architecture is reasonably low.

As previously indicated, the significant advantage of the polar architecture is its high-power efficiency. However, due to the nonlinear signal processing, there is a significant bandwidth expansion that makes it unusable for very wideband signals. This has prompted researchers to consider alternative architectures. The outphasing architecture [53], which avoids the amplitude modulation at the component level, is not only highly efficient at maximum RF output power, but also efficiently operates at back-off power levels. Moreover, a few research groups have recently demonstrated various digital implementations [51, 54–56]. The digital transmitter depicted in Fig. 1.9B [51] produces 26 dBm peak RF power while its drain efficiency is 35%. However, the outphasing architecture still experiences bandwidth expansion issues due to the nonlinear signal transformation which might limit its usefulness for very wideband signals.

1.9 ALL-DIGITAL I/Q TRANSMITTER

As mentioned in Section 1.5, due to the bandwidth expansion of ρ and θ, which could be as much as 10× the original I/Q signal bandwidth, it might be problematic to apply the digital polar TX architecture to the wideband modulation, especially the most recent 3GPP LTE cellular and 802.11n/ac wireless connectivity standards. The required bandwidths would exploit approximately hundreds of MHz. Even though

1.9 All-digital I/Q transmitter

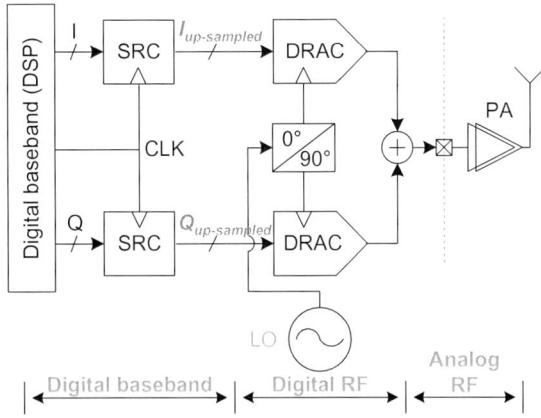

FIG. 1.10

Digital I/Q TX.

Kavousian et al. [17] introduced a wideband digital polar PA, the efficiency was inadequate. In addition, Chowdhury et al. [19, 48] and Ye et al. [49] proposed wideband polar modulators, which achieve high-output power and high efficiency. Nonetheless, they require a complicated baseband processing of ρ and θ signals. Consequently, the digitally intensive I/Q architecture has been introduced [16, 57–66] to maintain the digital RF approach while addressing the bandwidth expansion issue of the polar topology. A typical architecture is illustrated in Fig. 1.10. Compared to Fig. 1.5, a new optional circuit, SRC, is included to convert the lower-rate baseband I/Q signal (processed at integer multiple of the symbol rate) into a much higher rate that is necessary to spread the quantization noise, thus lowering its spectral density. Note that this optional SRC circuit could also be incorporated in the digital polar TX of Fig. 1.8 as a component of the I/Q-to-polar converter.

The operation of the digital I/Q modulator is as follows. The I and Q digital samples drive their respected DRAC converters that produce two RF signal components in which their amplitudes are ideally proportional to the respective I/Q digital inputs. The two amplitude-modulated RF components are subsequently combined to produce the desired composite RF output. The output is then either directly transmitted to the antenna or through a high-performance power amplifier. Note that, the phase of the subsequent RF signal is implicitly the result of the arctangent function of "DRAC Q RF output voltage" to "DRAC I RF output voltage." This is due to the fact that the phase of the DRAC Q output is 90° delayed with respect to that of the I output.

This digital I/Q architecture does not appear to compare favorably with the polar topology of Fig. 1.8. The digital I/Q architecture seems increasingly complex with extra circuitry contributing noise and creating signal distortion. The frequency modulation of the LO in Fig. 1.10 could be satisfactorily accomplished through the familiar two-point modulation scheme of the ADPLL. The phase component is thus under the closed-loop feedback, which reduces the noise and distortion content falling within the ADPLL loop bandwidth. Closing the loop around the I/Q modulator RF output is typically much more difficult. The traditional issues of the timing misalignment in the analog polar architecture are no longer an issue with the digital approach. The digital discrete-time operation is, by construct, clock-cycle accurate while the modern technology can clearly support sampling rates even at the GHz range, as well as ultra-fast settling of the DPA conversion circuits (with the speed governed by f_T). Consequently, the circuits of digital architectures can ensure fine time accuracy which is constant and not subject to processes and environmental changes. The DRAC could be implemented as a digitally controlled RF-modulated current source [16, 58]. In this manner, the addition of the two DRAC output components could be as uncomplicated as connecting them electrically. For the I/Q signal orthogonality to maintain, these two current sources must be ideal such that one signal path output does not influence the operation of the other. This indicates the need to resort to current source impedance boost techniques, such as cascoding. Unfortunately, stacking of the MOS transistors in a cascode structure as a current source is difficult in the modern low-voltage technologies and further produces an excessive amount of leakage and noise. In order to mitigate the aforementioned issues, consequently, there is a need for a digital TX architecture that is capable of supporting advanced wideband wireless modulation standards but which also avoids the intrinsic bandwidth expansion issues of the polar topology and the severe noise issues of the conventional digital I/Q architectures.

1.10 CONCLUSION

In Section 1.4, four types of RF transmitter architectures were briefly described. The analog I/Q modulators are the most straightforward and widely employed RF transmitters. They are later replaced by analog polar counterparts to address their poor power efficiency and noise performance. On the other hand, in the analog polar RF transmitters, their related amplitude and phase signals must be aligned or spectral regrowth is inevitable. Utilizing digital intensive polar RF transmitters mitigates the latter alignment issue. Nonetheless, polar transmitters suffer from an additional issue that is related to their nonlinear conversion of in-phase and quadrature-phase signals into the amplitude and phase representation. Therefore, the polar RF transmitters are not able to manage very large baseband bandwidth of the most stringent communication standards, therefore, reusing I/Q modulators based on digitally intensive implementation appears to be a reasonable approach to resolve this issue. The digital I/Q RF transmitters, however, suffer again from inadequate power efficiency. Moreover, the combination of in-phase and quadrature-phase paths must be orthogonal to produce an undistorted-upconverted-modulated RF signal. In Chapter 6, a new digital I/Q modulator that alleviates those extreme issues will be uncovered.

1.11 BOOK OUTLINE

This book is organized as follows:

Although the presented RFDAC-based approach can be adapted to be used for any cellular or noncellular standard, EDGE is first used as a benchmark in Chapters 2–5 to design and implement the *narrowband* architecture and validate the design using silicon implementation. Chapter 2 provides an overview of the presented transmitter architecture along with the details of the peripheral modules of a cellular SoC. It also provides a list of key challenges encountered during the work. The transmitter datapath is loosely broken down into two separate sections. The first section covers all the modules running below a clock rate of 200 MHz, implemented in a traditional VLSI flow. These modules are covered in Chapter 3. The modules running above 200 MHz, containing some digital modules designed using VLSI flow, as well as some analog modules with fully digital control, are covered in Chapter 4. It also covers the analysis of some of the parasitic effects encountered due to the high-speed digital nature of the design. Simulation and silicon measurement results are presented throughout Chapters 2–4 for the individual modules. Key simulation and silicon measurement results covering the performance of the complete transmitter are summarized in Chapter 5. Appendix A.3 explains the details of the simulation model used for the design of the presented transmitter.

In Chapter 6, a new all-digital I/Q RF modulator is introduced, which is a basis for the *wideband* RFDAC. This concept suggests that utilizing an upconverting RF clock with a 25% duty cycle ensures the orthogonal summation of I_{path} and Q_{path}. It is clarified that electric summing of I and Q digital unit array switches are the most appropriate I/Q orthogonal summation approach. Moreover, in order to cover all four quadrants of the constellation diagram, differential quadrature upconverting RF clocks should be used. In Chapter 7, a 2×3-bit all-digital I/Q (i.e., Cartesian) RF transmit modulator is implemented. The radio-frequency digital-to-analog converter (RFDAC) functions according to the concept of orthogonal summing and it is based on a time-division duplexing (TDD) manner of an orthogonal I/Q addition whereby a simple and compact design featuring high output power, power-efficient, and low-EVM is realized. In Chapter 8, the system design considerations of the proposed high-resolution, wideband all-digital I/Q RFDAC are discussed. In Chapter 9, the design procedure of a 12-bit digital power amplifier together with a class-E-based power combining network is discussed. In Chapter 10, the implemented wideband, 2×13-bit I/Q RFDAC-based all-digital modulator realized in 65-nm CMOS is presented. The 12-bit DRAC is implemented by employing the segmentation approach, which consists of 256 MSB bits and 16 LSB thermometer unit cells. The layout arrangement of the DRAC unit cell proves to be very crucial. The LO leakage and I/Q image rejection technique as well as two DPD memoryless techniques of AM-AM/AM-PM and constellation mapping are introduced, which will be employed extensively during the measurement process. In Chapter 11, the high-resolution wideband 2×13-bit all-digital I/Q transmitter will be measured. First, the chip is tested in continuous-wave mode operation. The RFDAC chip generates more than

21 dBm RF output power within a frequency range of 1.36–2.51 GHz. The peak RF output power, overall system, and drain energy efficiencies of the modulator are 22.8 dBm, 34%, and 42%, respectively. The RFDAC could be linearized using the DPD lookup table. Its linearity is examined using single-carrier 4/16/64/256/1024-QAM as well as multicarrier 256-QAM OFDM baseband signals while their related modulation bandwidth is as high as 154 MHz. Moreover, the constellation-mapping DPD is applied to the RFDAC thereby improving linearity by more than 19 dB.

In Chapter 12, the book is concluded, and sensible suggestions are presented for future developments.

CHAPTER 2

Digital polar transmitter architecture

CHAPTER OUTLINE

2.1 Introduction to Narrowband Polar Transmitters	21
2.1.1 Motivation	21
2.1.2 Contrast With Conventional Analog Approaches	24
2.2 Overview of the RFDAC-Based Polar Transmitter Architecture	25
2.2.1 Overview of the DPA	28
2.2.2 DCO Operating Frequency and CKV Clock	29
2.3 Details of Phase Modulation	30
2.4 Design Challenges for the Small-Signal Polar Transmitter	32
2.4.1 DCO Phase Noise	32
2.4.2 DCO Pulling and Pushing	32
2.4.3 VGA/DPA Nonlinearity	34
2.4.4 Amplitude-Phase Path Delay Mismatch	35
2.4.5 Amplitude-Phase Path Transfer Function Mismatch	37
2.4.6 Amplitude Path DC Offset	37
2.4.7 Amplitude Path Dynamic Range Limitations	38
2.4.8 Amplitude Path DAC Mismatches	39
2.4.9 Additional Distortions in the DPA (DAC/Mixer)	40

2.1 INTRODUCTION TO NARROWBAND POLAR TRANSMITTERS

2.1.1 MOTIVATION

Global system for mobile communication (GSM) is the most popular standard for mobile phones accounting for around 90% of the world's mobile phone market [67] with estimated 4 billion subscribers. An extension of the GSM system called enhanced data rates for GSM evolution (EDGE) provides three times higher capacity than traditional GSM system while maintaining backward compatibility with the GSM time slots and the GSM frequency planning. For many service providers, a simple software upgrade allows the capability of EDGE deployment.

With the demand for Internet connectivity increasing on handsets, the volumes for EDGE-supported phones also increase proportionally. While many of the high-end handsets are using third and fourth generation (3G and 4G) of cellular connectivity, EDGE is still popular for mid- to low-end handsets. Additionally, 3G services are not available in many places and 3G handsets revert to 2.75G (EDGE) generation in such cases. Although EDGE can be considered a relatively older technology, the GSM/EDGE spectral noise specifications at 20 MHz offset from the carrier are still considered one of the *most challenging* specifications of any radio standard. Hence, EDGE has been used as an example of the technology that can be implemented using the presented digitally intensive transmitter design approach. For any technology, the increase in the volume of sales pushes for lower cost, which can be achieved with higher level of integration of phone functionality into a single System on a Chip (SoC).

The digital CMOS process is by far the most scalable technology, following Moore's law. The digital baseband of a cellular handset is universally implemented in CMOS and cost factors push for the integration of the digital baseband with the RF radio functionality. Hence, it is very beneficial to have the RF transceiver design use the digital CMOS transistors without any process modifications. The digital circuits, when implemented using automated digital IC design flow in CMOS process, are relatively easy to migrate to the next process node. In contrast, analog/RF circuits are not as scalable and generally the design of analog/RF circuits becomes more complicated as the process nodes shrink and supply voltages are reduced. Analog intensive architectures have several other disadvantages such as difficult matching, higher leakage, susceptibility to process voltage temperature (PVT) variations, and higher current consumption. This leads to overall preference of digital circuits over analog/RF-dominated circuits. Of course, the use of analog/RF circuits cannot be entirely avoided because critical blocks such as voltage-controlled oscillator (VCO) and power amplifier (PA) are bound to be of analog nature. Hence, the best approach has been to reduce the use of analog/RF circuits as much as possible in integrated transceiver design.

2.1.1.1 Small-signal polar transmitter

There are multiple architectures that can be used to demonstrate the presented digitally intensive transmitter design. The EDGE transmitter is used as an example to illustrate the design methodology along with the specific challenges in greater detail. The small-signal polar architecture is most suitable and economic for digital implementation of relatively narrow frequency modulation such as EDGE. As mentioned in the next section, small-signal polar typically has the disadvantage of requiring several stages of mixers at IF and RF level to achieve adequate modulation quality and spectral purity at the transmitter output. In this endeavor, digitally intensive direct conversion architecture is utilized along with a digital PA (DPA)-based digital mixer. The DPA eliminates the need of analog components and hence makes the small-signal polar architecture attractive. The small-signal polar architecture simplifies the need for PA gain modulation and thus relaxes the requirements on the PA supply regulator but it also requires a linear PA, which has lower efficiency compared to the saturated or polar PA of large-signal polar architecture. To give an example, commercially available polar EDGE PA typically has efficiency of around 26% in PCS band while the linear EDGE PA offers around 24% efficiency [68, 69].

2.1.1.2 Motivation for the digitally intensive amplitude path

One of the key design challenges for the amplitude path in an EDGE transmitter is the requirement to accommodate nearly 55 dB dynamic range when including the modulation dynamic range of 17 dB with margin (Section 2.4). In a large-signal polar transmitter, most of the power control is achieved by modulating the supply of the power amplifier and hence the transmitter SoC output power fed to the PA can be nearly constant. But, for the small-signal polar transmitter, the power amplifier has constant gain and hence the transmitter SoC has to provide the entire dynamic range of the power control at its output.

Traditional small-signal polar transmitters use combinations of mixer and variable gain amplifier (VGA) gain stages to achieve the desired dynamic range. Typically the mixer and VGA gains vary over PVT conditions making the calibration and compensation of the transmitter a complex problem. Instead, in the presented transmitter, the power control is achieved by using digital scaling thus replacing the VGAs with digital multipliers. The DPA provides the 55 dB of dynamic range with the help of Sigma Delta ($\Sigma\Delta$)-based amplitude modulation (AM). The presented scheme also simplifies the data path design significantly by adaptively adjusting the resolution and dynamic range of the data path depending upon the power level, thus minimizing the word-length in the digital signal processing path. Since the majority of the design is based on either pure digital logic or digital blocks with predictable analog cores, the overall calibration complexity of the presented design is significantly reduced resulting in considerable savings in factory calibration time and cost.

To further minimize the calibration time and complexity, a new predistortion scheme is proposed as part of this endeavor. This technique eliminates the need for complex inversion calculations of DPA amplitude distortion (AM-AM) and amplitude-dependant-phase distortion (AM-PM) during the calibration phase. The predistortion scheme also adaptively changes the interpolation of the look-up table (LUT) and the predistorted data quantization resolution based on the power level. The power-based adaptive resolution featured in the transmitter significantly reduces the area and power consumption of the transceiver without compromising performance at any power level.

The digital nature of the data path allows for filtering requirements to be met without using complex analog filtering. An interpolative filter, based on a modified form of a cascaded integrator-comb (CIC) filter, allows the required filtering to be implemented without the use of any multiplier or divider.

2.1.2 CONTRAST WITH CONVENTIONAL ANALOG APPROACHES

The presented work employs the small-signal polar architecture for implementing an EDGE transmitter, which differs significantly from the traditional small-signal polar architecture. As highlighted in Section 2.1.1.1 and elaborated in detail in Chapter 2, the presented transmitter overcomes the traditional disadvantages associated with the analog centric small-signal polar transmitter by using a digital power amplifier (DPA) in combination with an all-digital phase-locked loop (ADPLL). The calibration time and complexity are significantly reduced by using a new predistortion scheme to correct for the DPA distortions and thus making it an attractive alternative to the large-signal polar transmitter. The use of the linear PA does result in small penalty in terms of efficiency but the advantages outweigh the penalty of efficiency. The small-signal polar approach eliminates the need for a complex high-bandwidth switched regulator to modulate the supply of the PA gain stages, and also significantly reduces the calibration complexity by avoiding AM-AM and AM-PM distortion arising from operating the PA in saturation mode. A possible direction for improving the efficiency in a small-signal polar transmitter is proposed in Ref. [70].

A representative small-signal analog polar transmitter was demonstrated in Ref. [71]. Following are some of the key comparative features of a competing architecture [71] contrasted with the presented architecture:

- The modulation is performed at IF first and then upconverted onto the carrier frequency in Ref. [71], which requires IF mixers and filters adding to area and current consumption requirements. The IF filters have two conflicting requirements—a narrower bandwidth helps to filter the replicas better but degrades the close-in spectrum because the phase and amplitude signals have very wide spectra. Hence, the design of the IF filter has to be very carefully optimized. Additionally, the IF spurs can penetrate at the RF output if the design/layout does not have enough isolation between modules. In the presented architecture, the information is directly upconverted to RF frequency, thus avoiding the problems listed above.
- The time alignment between the amplitude and phase paths for an EDGE transmitter has to be within 10 ns, which is very difficult to achieve due to the analog-intensive nature of the architecture in Ref. [71]. The implementation would also require complicated filtering to match the overall amplitude and phase path transfer function. In the presented architecture, the amplitude and phase paths are implemented digitally and hence the timing alignment requirements can be easily met using a binary coded flip-flop chain as described in detail in Chapter 3. Also, the transfer functions of the amplitude and phase paths can be better controlled due to the fully digital data path and hence the amplitude/phase transfer functions can be matched within the required accuracy. Additionally, the PVT variations of digital filtering are negligible compared to analog filters.
- The amplitude path in Ref. [71] has an option of using analog filtering before the amplitude is combined with the phase in the transceiver. Hence, this architecture

has a benefit of using a baseband analog filter on the amplitude path, which simplifies the design of the amplitude path and reduces constraints on the resolution of the DAC. In contrast, it is not feasible to use an analog filter in the amplitude path in the presented architecture and hence higher resolution is required from the RFDAC.

2.2 OVERVIEW OF THE RFDAC-BASED POLAR TRANSMITTER ARCHITECTURE

The motivation for the use of a digitally intensive architecture is presented in Section 2.1.1. The work covered by Chapters 2–5 focuses on innovative design solutions for the amplitude path of the small-signal polar transmitter that overcomes the challenges later listed in Section 2.4. The phase path in the presented transmitter is based on the all-digital phase-locked loop (ADPLL) disclosed in Ref. [72].

The transmitter architecture builds on the GSM transmitter proposed in Refs. [46, 73] with the addition of amplitude modulation functionality. The presented transmitter is fully compliant with the ETSI specifications for a mobile EDGE transmitter over PVT variations.

The presented transmitter is part of a fully integrated GSM/EDGE cellular phone SoC of the type shown in Fig. 2.1 [46]. The ARM7 processor and C54x DSP are part of the OMAP [74] based baseband application processor. The rest of the functionality includes various interfaces to higher layers as well as modules to communicate with the peripheral devices. An external power supply management chip is used to control the battery power management.

A functional interface diagram of the transceiver is shown in Fig. 2.2. The reference frequency (FREF) used by the ADPLL is generated by the digitally controlled crystal oscillator (DCXO) with use of an external crystal [75, 76]. The crystal frequency depends on system considerations. The presented transmitter can use an arbitrary reference frequency in the range of 8–40 MHz, but only 38.4 MHz or 26 MHz are commonly used in GSM/EDGE/WCDMA mobile phones.

The power amplifier (PA) is contained in a separate external chip which is generally part of the RF front-end module (FEM). The RF FEM also includes the switches controlling the band selection (between HB and LB) and the transmit/receive antenna path.

A more detailed transmitter system level block diagram is shown in Fig. 2.3. Section A.1 defines the EDGE 8-PSK modulation using formulae representing the symbol mapper, the $3\pi/8$ rotation, and the pulse shaping filter. These modules are interconnected in the transceiver as shown in Fig. 2.3. As mentioned in Section A.1, 8-PSK-based EDGE modulates three bits per symbol compared to one bit per symbol in the GMSK scheme of GSM. To maintain backward compatibility with GSM, the symbol rate is maintained to that of GSM −270.833 kbits/s. Hence, the EDGE symbol mapper receives a binary stream representing the raw bits at a rate of

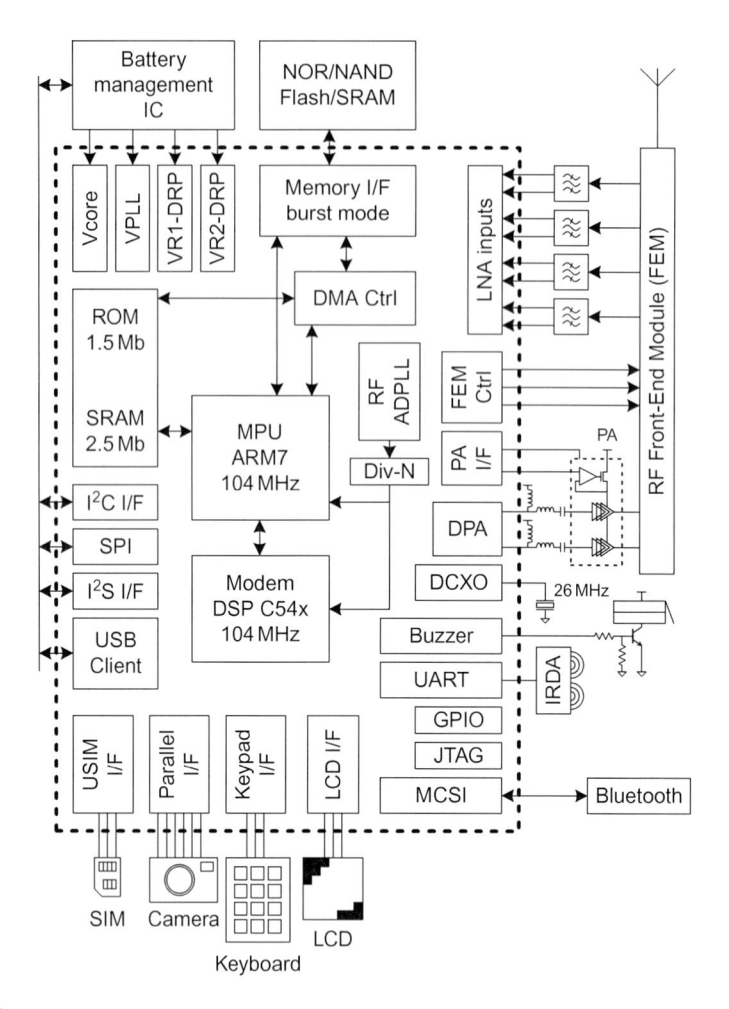

FIG. 2.1

Block diagram of SoC containing the EDGE transceiver [46].

3×270.833 kbits/s $= 812.5$ kbits/s. The output of the mapper is 270.833 ksymbols/s representing 3 bits of in-phase (I) and 3 bits of quadrature-phase (Q) information as per the constellation shown in Fig. A.1.

The output of the symbol mapper, after going through the $3\pi/8$ rotation (as shown in Fig. A.2), is represented by 4-bit wide I and 4-bit wide Q signals. After going through the pulse shaping filter, defined by Eq. (A.3), the I and Q data are converted to their equivalent polar representation (ρ, θ) using the CORDIC module [77].

One of the primary advantages of the digital architecture is the simplified power control scheme. Instead of using traditional VGA-based multistage gain control, the power control is implemented digitally using a multiplier in the data path with

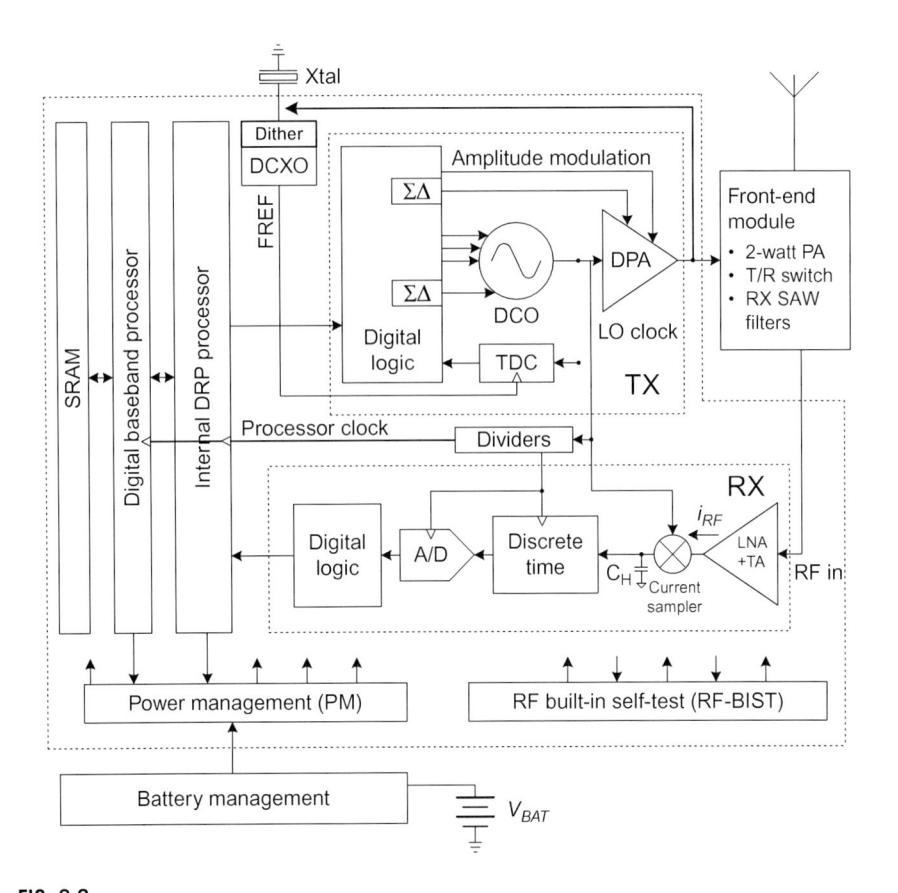

2.2 Overview of the RFDAC-based polar transmitter architecture 27

FIG. 2.2

Transmitter connectivity overview [46].

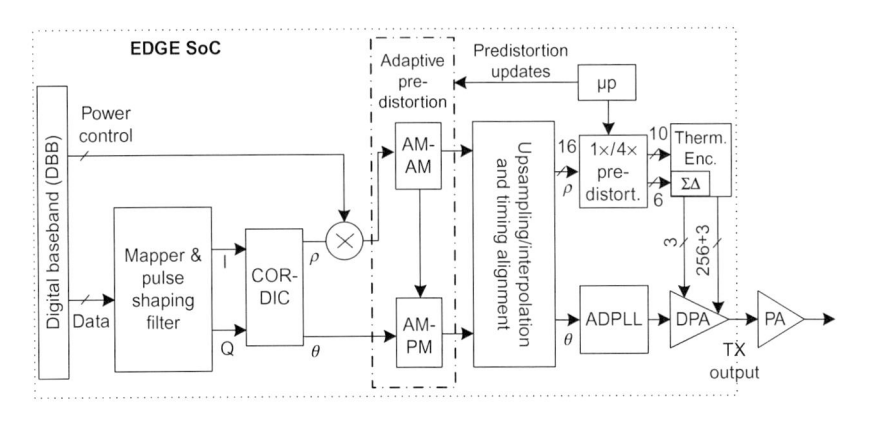

FIG. 2.3

Block diagram of EDGE transmitter.

16 bit resolution. Closed-loop power control ensures sufficient (± 0.5 dB) power accuracy over the PVT and frequency variations. Details of the power control loop implementation are considered beyond the scope of the presented work.

The digital power amplifier (DPA) acts as a combination of RFDAC and a mixer combining amplitude and phase information. The nonlinear AM-AM and AM-PM distortions of the DPA are mitigated by predistorting the amplitude and phase signals in the predistortion module [78, 79].

The RFDAC receives the amplitude signal at a sampling rate, which is equal to the carrier frequency. This allows sufficient digital filtering to be applied to suppress any replicas generated due to the interpolation. The majority of the interpolation is performed using a new interpolation filter [47], based on a form of cascaded integrator-comb (CIC) filters [80]. The interpolation filter provides upsampling by a factor of 32 with minimal replicas. The implementation of the interpolation filter is optimized to eliminate any digital multipliers/dividers as described in detail in Section 3.4.

The phase signal, after interpolation, is passed to the ADPLL [72] based phase modulator. The ADPLL fundamentally works as a frequency modulator using a digitally controlled oscillator (DCO) at its core. The phase-modulation information, represented using a frequency control word (FCW), modulates the DCO using a type-II closed-loop ADPLL. The DCO output represents a frequency-modulated RF carrier from which a digital clock is derived. The average frequency of the digital clock is equal to the RF channel carrier frequency. The instantaneous frequency is controlled by the frequency deviation represented by the FCW signal. The RF-modulated carrier clock is mixed with the amplitude signal inside the DPA. This clock is also used as a base to generate low-frequency clocks through power-of-2 dividers. An overview of the phase-modulation path is provided in Section 2.3 and more details about the ADPLL can be found in Ref. [72].

To meet the strict limits defined in the ETSI specifications for the EDGE RX band noise, at least 13 bits of amplitude resolution is necessary to eliminate any analog filtering, as covered in Section 4.3.2. To achieve this fine resolution, $\Sigma \Delta$ modulation [81] is used to add fractional resolution to the amplitude path. The integer resolution of 10 bits, combined with fractional resolution of 6 bits together provides 16 bits of resolution to drive the DPA transistor bank, as shown in Fig. 2.3 [50]. The 16 bit resolution allows for margin with respect to the 13-bit resolution requirement, to address the impairments analyzed in Chapter 4.

2.2.1 OVERVIEW OF THE DPA

The DPA comprises three integer transistors of unit ($1\times$) strength, 256 transistors of 4-times unit strength ($4\times$), and 3 unit ($1\times$) strength transistors, which are controlled by a fractional $\Sigma \Delta$ modulator. A conceptual placement layout of the DPA is shown in Fig. 2.4. The $1\times$ transistor strength is the minimum readily achievable size for a given

2.2 Overview of the RFDAC-based polar transmitter architecture

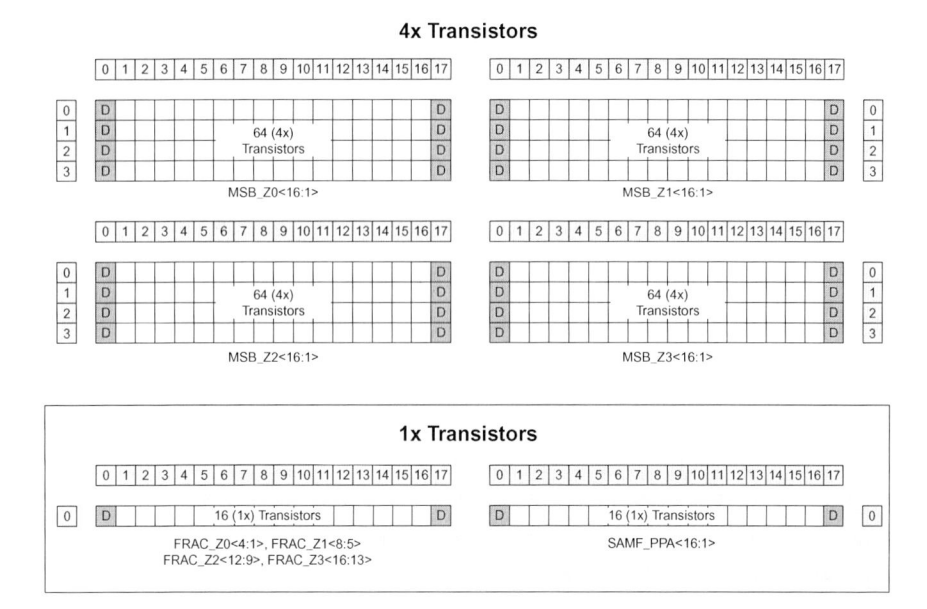

FIG. 2.4

Floorplanning diagram of DPA.

technology. The analysis and specifications of the transistor sizes and mismatches between them is one of the key aspects of the presented work.

The 4× transistors are arranged in four blocks of 64 transistors each, as shown in Fig. 2.4. Each block is controlled by row and column control signals coming from the thermometer encoder. The bottom row shows the layout of 1× transistors. SAMF_PPA represents the $\Sigma\Delta$ transistors and FRAC_Z represents the 1× integer transistors. Each 1× row contains 16 transistors to provide for dynamic element matching (DEM) functionality and additional filtering requirements.

The DPA output is further amplified in this system by an off-chip PA before driving the antenna.

2.2.2 DCO OPERATING FREQUENCY AND CKV CLOCK

The DCO operates at a frequency which is in the vicinity of 3.5 GHz. For LB, the DCO operates at 4× the carrier or channel frequency (f_{ch}), whereas for HB, the DCO operates at 2× f_{ch}, as shown in Table 2.1. The digital clock signal called CKV, derived from the DCO frequency, is an important system variable used throughout this text. CKV is generated by passing the DCO output through a buffer and divide-by-2 circuit, so the CKV frequency range spans from 1648.8 MHz to 1909.8 MHz, as shown in Table 2.1.

Table 2.1 Transmitter Frequency Bands of Operation and Corresponding DCO Operating Frequency

Band	Frequency Band	Frequency Range (MHz)	DCO Operating Frequency (MHz)	CKV Frequency (MHz)
LB	GSM 850	824.2–848.8	3296.8–3395.2	1648.4–1697.6
	GSM 900	880.2–914.8	3520.8–3659.2	1760.4–1829.6
HB	DCS 1800	1710.2–1784.8	3420.4–3569.6	1710.2–1784.8
	PCS 1900	1850.2–1909.8	3700.4–3819.6	1850.2–1909.8

2.3 DETAILS OF PHASE MODULATION

While the analysis and development of the amplitude path is the focus of this work, the details of the phase path are necessary for complete understanding of it.

A detailed block diagram of the ADPLL used in the phase path is shown in Fig. 2.5. The ADPLL frequency detector operates in the frequency domain and hence the phase signal is converted to a frequency-equivalent signal by differentiating it. The frequency signal on the reference FCW path is resampled from CKV/16 to

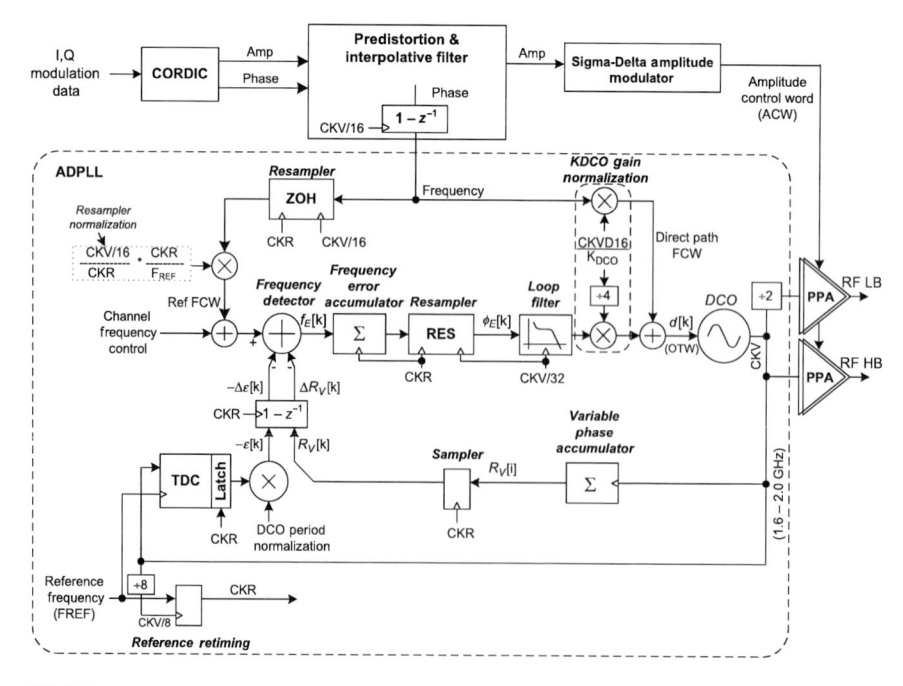

FIG. 2.5

Detailed block diagram of the ADPLL within the TX.

CKR (CKR is the FREF-based reference clock re-timed with CKV) reference clock domain. The frequency domain resampling requires gain adjustment to achieve the right frequency value in the new CKR domain. This results in a normalization factor of [(CKV/16)/CKR]. Additionally, the frequency detector operates at a normalized reference frequency F_{REF} thus adding a second normalization factor of $[CKR/F_{REF}]$. Refer to Ref. [72] for a more detailed explanation of the same.

The frequency modulation of the DCO is performed by a direct feedforward path labeled as direct path frequency control word (FCW). This removes the slower loop dynamics from the modulation path and allows wider bandwidth modulation to be used without widening the loop bandwidth of the ADPLL [72]. The error correction capabilities of the loop are still preserved though the reference path, which is filtered through the desired loop response.

The direct path FCW, after normalization by the DCO frequency conversion gain—also known as KDCO gain in Ref. [72]—is added with the loop correction signal to form the oscillator tuning word (OTW) representing the channel frequency combined with the frequency modulation information. $\Sigma\Delta$-based dithering is used to enhance the frequency resolution of the DCO output. The DCO output is represented using a digital clock running at RF carrier frequency termed as variable clock (CKV). The instantaneous frequency modulation information is effectively encoded in the variations of the periods of this clock. The phase information is extracted from this clock for the feedback path using two levels of frequency calculations—coarse and fine. The coarse calculation is performed by calculating the number of CKV clocks present in a given cycle of the resampled reference clock (CKR) effectively representing the ratio of CKV to CKR shown as $R_V[k]$.

The fine clock frequency information is extracted using a time to digital converter (TDC) which allows accurate measurements of the instantaneous phase of the CKV clock using inverter-based time estimation [72]. Both the coarse phase information, $R_V[k]$, and the fine phase information, $-\varepsilon[k]$, are differentiated to achieve a frequency equivalent representation for the feedback of the locked loop.

Since the feedback signal contains the frequency modulation, introduced by the direct path FCW, the frequency modulation has to be subtracted from it to find the error signal, that is, the extent to which the DCO output deviates from the desired value. Hence, the original frequency signal, after resampling to convert it to CKR domain, is subtracted from the feedback signal to generate the frequency error signal $f_E[k]$. The RF carrier frequency offset is added as a constant at this point, although it could be added to the original frequency signal as well.

Phase error information is derived from the frequency error signal after passing through the accumulator. The phase error is passed through the loop filter, which is constructed using proportional and integral paths—forming a proportional-integral (PI) controller. The PI controller makes the ADPLL a type-II second-order PLL loop. In addition, the loop filter contains four single stage, single-pole infinite impulse response (IIR) filters allowing further digital filtering control without sacrificing stability of the loop. Only power-of-2 coefficients are used for the loop filter to reduce the area of the implementation.

The error signal, after going through appropriate normalization for the DCO gain, is combined with the direct path FCW to achieve the final OTW signal. An additional divider is added for the LB operation to allow the DCO frequency to remain in the range of 1.6–2.0 GHz.

2.4 DESIGN CHALLENGES FOR THE SMALL-SIGNAL POLAR TRANSMITTER

This section highlights some of the key challenges in the design of a small-signal polar transmitter. The relevancy of each challenge to a specific implementation depends upon specific architectural choices. Special emphasis is given in this work to the issues affecting the design of the presented transmitter along with the implications on the transmitter specifications. The term digitally controlled oscillator (DCO) refers to the local oscillator (LO) or voltage controlled oscillator (VCO) implementation used in this transmitter. The term digital power amplifier (DPA) refers to the RFDAC used for the amplitude path, which can be thought as operating as a mixer by amplitude modulating the phase-modulated carrier.

In addition to the challenges listed below, design issues specific to the presented architecture are reported in the corresponding subsections. A major focus of this work was to achieve a solution that is predictable in terms of its performance, and is scalable across process nodes. This required detailed understanding of all the design challenges involved and the development of robust solutions for them.

2.4.1 DCO PHASE NOISE

Source: DCO phase noise originates from the time domain jitter on the sinusoidal carrier signal. The main sources of the phase noise are DCO thermal noise, shot noise, and flicker noise [72]. The DCO phase noise is represented in the frequency domain in dBc/Hz as shown in Fig. 2.6.

Degradation: The DCO phase noise is responsible for increase in the transmitter output noise floor at both close-in and far-out frequencies. Excessive phase noise at close-in frequencies, within the modulation bandwidth, can also affect the constellation quality specifications like EVM, phase error, etc.

2.4.2 DCO PULLING AND PUSHING

Source: DCO "frequency pushing" is generally caused by the sensitivity of the oscillator free-running frequency to the supply voltage variations. As the DCO uses voltage controlled capacitors (varactors), any change in their biasing point due to either V_{DD} bounce or ground bounce causing the center frequency of the DCO to move in proportion to the voltage change. Additionally, factors such as the gain of the active devices in the oscillator affect its free-running frequency resulting in nonmonotonic pushing characteristics. The final stage amplifier, in this case DPA,

FIG. 2.6

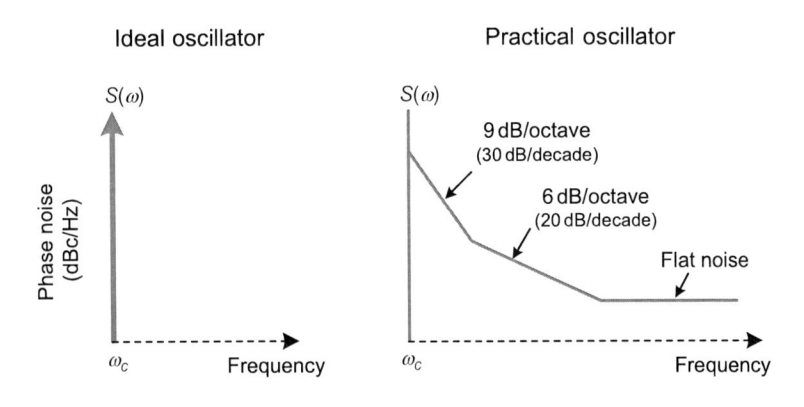

Single-sided spectrum of ideal and practical oscillators.

consumes majority of the current from the power supply. Due to the digital nature of the DPA design, the switching activity of the transistors causes higher level of supply bounce, resulting in proportional push to the DCO frequency.

DCO "frequency pulling" refers to the tendency of the DCO to injection lock to any spurious signal within its locking range [82, 83]. Since the DCO output carrier is phase modulated, this phenomenon could be relatively weaker for small-signal polar transmitter because the signal at the DCO output matches the FM component of the amplifier output in frequency domain. A second possibility is when strong amplitude-modulated RF signals find a way to couple into the core of DCO, resulting in mixing phenomenon leading to variations in frequency, which are dependent on the output power (or amplitude) and thus it is an amplitude-dependent frequency modulation (AM-FM) distortion [84].

In addition, any DCO output load impedance changes result in variations in the operating point for the oscillator feedback loop. The frequency, for which a total phase of 360° is achieved around the feedback and the oscillator gain, is changed due to the change in the loading of the oscillator. This results in frequency changes which are proportional to the number of DPA transistors turned ON, which is directly dependent on the amplitude signal. Although buffering is used to isolate the DCO output from load variations, a small amount of variations typically reaches the oscillator because the buffers do not have infinite reverse isolation. In addition, DCO pulling has multiple other possible sources, which are specific to the implementation. Fig. 2.7 shows some of the key sources of DCO pulling for the presented architecture. The term "DCO pulling" will be used as a common reference to both DCO pulling and DCO pushing in the following sections.

Degradation: DCO pulling for complex modulation, such as 8-PSK-based EDGE, causes undesired amplitude-dependent frequency variations (AM-FM). EVM and other constellation quality specifications are degraded due to the DCO pulling. The close-in spectral degradation depends on the transfer function of the coupling path

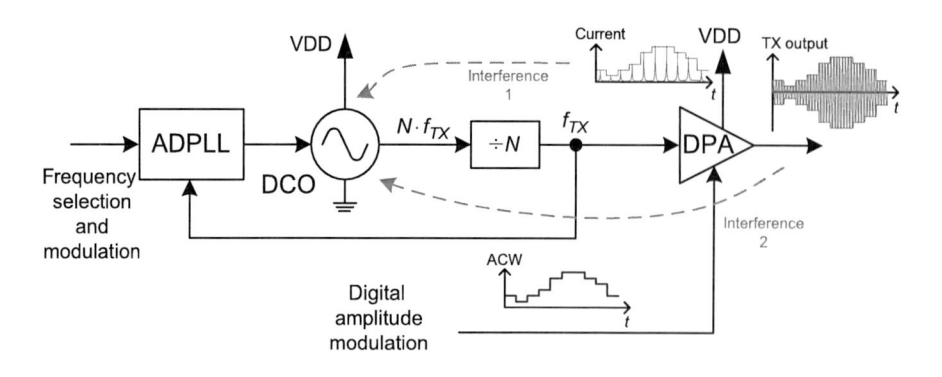

FIG. 2.7

DCO pulling through various coupling mechanisms [84].

responsible for pulling. The DCO pulling causes an amplitude-dependent frequency component, which leads to distortion in the constellation every time there is a large change in the amplitude. This distortion of the constellation results in EVM degradation that may violate the specifications. Similarly, the AM-FM caused by pulling results in the spectral performance violations. Generally, spectral degradation close to the carrier, up to about 400 kHz, is a common consequence which must be addressed [84].

2.4.3 VGA/DPA NONLINEARITY

Source: The final driver stage of the small-signal polar transmitter (DPA), shown in Fig. 2.7, delivers nearly +8 dBm peak output power and supports nearly 50 dB of dynamic range. At high-output powers, the amplifier output begins to saturate and results in a compressed nonlinear amplitude transfer function. This is referred to as AM-AM distortion. Depending upon the implementation of the amplifier, the change in amplitude can result in distortion in phase, referred to as AM-PM distortion. An increase in amplitude in the DPA is realized by adding incremental transistors, which changes the complex impedance seen by the DPA output matching network resulting in AM-PM distortion. Traditional transmitters have been using VGA instead of the DPA and the same analysis will be applicable for these implementation as well.

Fig. 2.8 shows the AM-AM and AM-PM distortions of DPA over temperature. The AM-AM response, showing the DPA output voltage versus the input codeword, clearly shows the compression of the output voltage at higher codes. The phase change in the AM-PM curve is about 30 degrees from the minimum code to maximum, which implies that an amplitude change from the minimum to maximum code, would result in 30 degrees of error in the constellation, which would severely degrade the EVM performance. The AM-PM response has a sharp phase change in the lower codes because adding incremental transistors at low powers has a much larger relative impact in cumulative phase.

FIG. 2.8

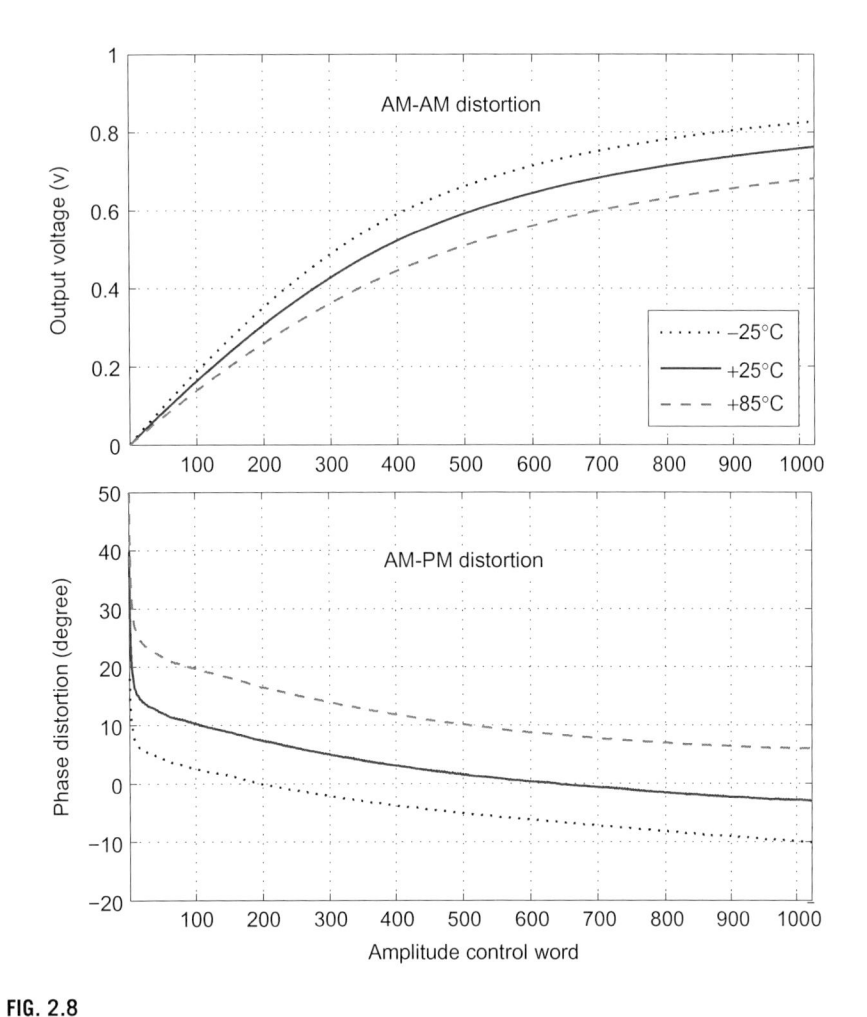

DPA AM-AM and AM-PM distortion over the digital amplitude control word.

In addition to the AM-AM and the AM-PM nonlinearities, there could be other implementation specific nonlinearities, which should be appropriately analyzed and either minimized by design or compensated through predistortion.

Degradation: The AM-AM distortion causes compression of the transmit constellation at higher power levels resulting in EVM degradation. The AM-PM distortion causes degradation in the constellation as well as the close-in spectrum.

2.4.4 AMPLITUDE-PHASE PATH DELAY MISMATCH

Source: As highlighted in Sections 1.5–1.6, both amplitude and phase have separate processing path in every polar transmitter. They are separated at CORDIC and re-combined near the output using some form of a mixer, VGA, or DPA/PA. The phase signal is used to modulate the carrier frequency using phase-locked loop while the amplitude is passed through either DC or RFDAC. For a high-fidelity reconstruction to happen, the time discrepancy between the two paths have to be kept extremely low [85]. An upper bound can be found for the delay misalignment for any given modulation either analytically or through system simulation. Such bound has been found in Section 3.4 to be 10 ns for EDGE implementation. Some of the delay elements in the data path could vary over PVT and frequency conditions which might require adaptive delay adjustment.

Degradation: The close-in spectrum at 400 kHz offset from the carrier is the most sensitive specifications to this delay mismatch between the amplitude and phase paths. System level simulation is used to quantify the degradation in the spectrum and a timing specification is derived to comply with 400 kHz offset specification with sufficient margin. Fig. 2.9 shows the simulation results indicating the need for a better than 15 ns accuracy of timing alignment. It also shows that 600 kHz performance can tolerate higher amount of delay mismatch between amplitude and phase paths.

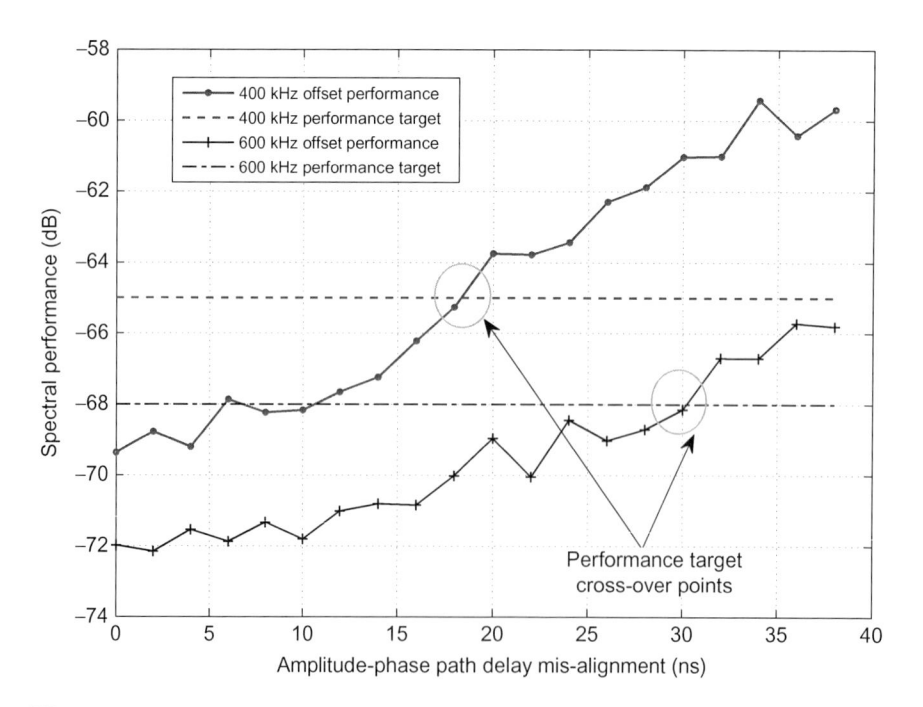

FIG. 2.9

Degradation of 400 kHz and 600 kHz offset spectrum due to amplitude-phase delay misalignment.

2.4.5 AMPLITUDE-PHASE PATH TRANSFER FUNCTION MISMATCH

Source: As mentioned in Section 2.4.4, both amplitude and phase paths undergo different signal processing which can have different delay and different transfer functions from each other. To achieve ideal combining of amplitude and phase paths in a polar transmitter, both amplitude and phase have to go through exactly same transfer function to achieve zero distortion. Any difference in the transfer functions of the signal processing paths up to the combiner will lead to distortion [85].

Degradation: The distortion observed is loosely related to the spectrum of the mismatch transfer function. If the mismatch in the transfer function happens beyond 600 kHz, the quantization noise is dominating the spectrum so the transfer function mismatch has less impact. Hence the most critical range of frequencies to match are from 0 kHz to 600 kHz where the modulation energy is fully concentrated. If the mismatch is strong at frequencies less than 200 kHz then the EVM will also get degraded. Although the I/Q signal energy is fully contained in 0–600 kHz bandwidth, the Cartesian to polar conversion leads to bandwidth expansion. Hence care should be taken to allocate at least 1 MHz bandwidth especially to the phase signal. More discussion about this will be covered in Chapter 3.

2.4.6 AMPLITUDE PATH DC OFFSET

Source: For a Cartesian transmitter, DC offset in the RF modulated complex signal refers to the presence of unmodulated carrier signal. When down-converted to baseband, the unmodulated carrier exhibits itself as a DC offset which changes the decision boundary on the receiving constellation. Most often the source of amplitude path DC offset is through the DCO carrier leaking at the transmitter output through the power supply/ground connection. In addition, there could be other paths for the carrier signal to couple to the output of the transmitter specific to an implementation.

In case of a polar transmitter, the carrier is always phase modulated and hence the presence of carrier signal leaking to the RF complex waveform is not as detrimental. But there could be a systematic DC offset on the amplitude signal itself, which gets mixed with the phase signal. When down-converted, this would result in phase-modulated carrier acting as an undesired offset being added on all the frequency components where phase-modulated carrier has presence. The source of such DC offset on amplitude could be systematic offset in digital or amplifier output offset.

Degradation: For a Cartesian transmitter, carrier leakage leads to both EVM, Modulation spectrum degradation and failure of Origin Offset specifications. But for the polar transmitter, the energy is not concentrated on the carrier frequency alone (or in other words, energy is not concentrated on the DC component at baseband) so the origin offset is not affected as much by the carrier leakage. The primary degradation in a polar transmitter is the close-in specifications around 400 kHz offset from carrier as shown by simulation in Fig. 2.10. If the amount of offset increases, it would

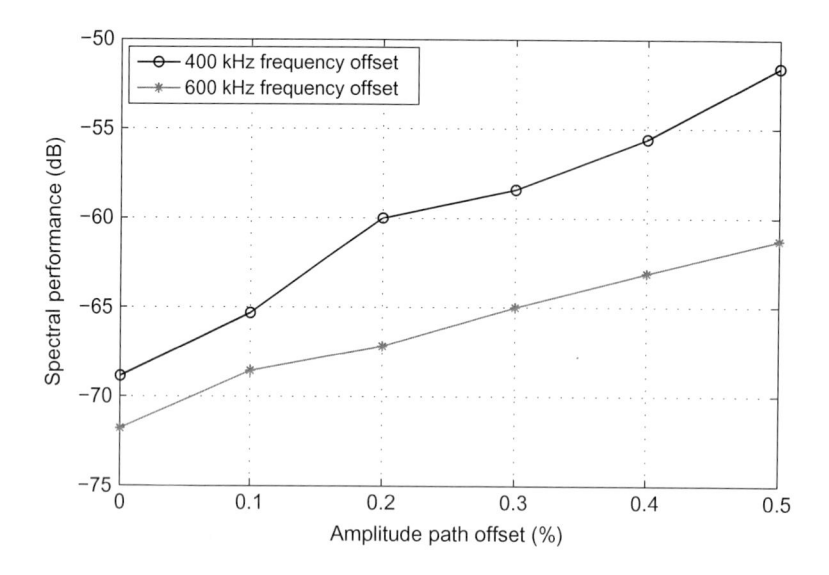

FIG. 2.10

Degradation of 400 kHz and 600 kHz offset spectrum due to amplitude path DC offset.

eventually lead to EVM degradation. If the leakage level is above −48 dBc then the transmitter will fail to meet the time mask which requires leakage to be better than −48 dBc as shown in Fig. 2.11.[1]

2.4.7 AMPLITUDE PATH DYNAMIC RANGE LIMITATIONS

Source: EDGE modulation requires about 17 dB of peak-to-minimum dynamic range for the amplitude modulation. Furthermore, the transmitter needs to support 16 power controls mentioned in Table A.3 requiring additional 30 dB of dynamic range. In total, the transmitter needs to support about 55 dB of dynamic range to maintain the margin for the PVT variations. Additionally, sufficient quantization noise floor needs to be maintained even at the low power level to meet the RF modulation spectrum requirements. The exact calculations for the quantization noise floor and bitwidth requirements will be covered in Chapter 3. In addition, the time mask for the EDGE transmission burst requires transmission to be below −48 dBm before the start of the burst as shown in Fig. 2.11. Generally it's difficult for a single stage amplifier to meet such dynamic range requirements at all power levels without use of resolution enhancement techniques such as $\Sigma\Delta$ modulation. An alternative option would be to use multilevel VGAs in a traditional analog architecture.

---

[1] In Fig. 2.11, ∗ = −48 dB. ∗∗ and ∗∗∗ are not applicable to presented work.

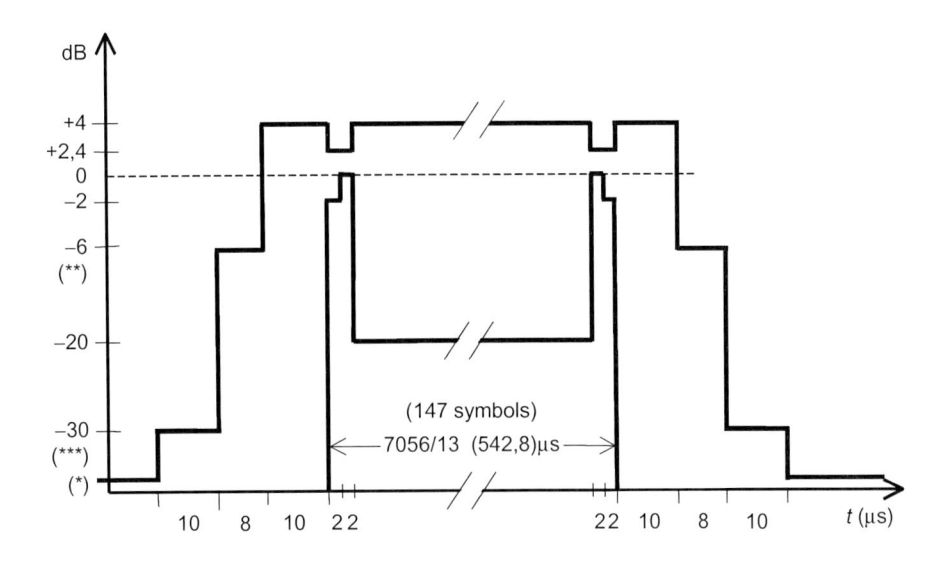

2.4 Design challenges for the small-signal polar transmitter

FIG. 2.11

Time mask specifications for a typical EDGE burst.

Degradation: If the dynamic range requirements are not met, multiple violations would prevent the transmitter from achieving standards compliance. Time mask, average burst power, 200 kHz offset spectrum, 400 kHz offset spectrum, and 600 kHz offset spectrum specifications are a few examples of the parameters that would be violated without sufficient dynamic range. For a detailed discussion of the amplifier dynamic range on amplitude path, see Chapters 3 and 4.

2.4.8 AMPLITUDE PATH DAC MISMATCHES

Source: The process mismatch between the transistors generally increases as the process feature size shrinks. Presented work is implemented in 65 nm which has significant mismatches between the transistors. In addition to the random transistor mismatches, there could be systematic mismatch among the same and different sizes of transistors. Transistor mismatch is big concern for the DPA as the mismatch directly impacts the output RF spectral performance. Significant part of this work (see Chapter 4) will focus on estimating the mismatch, defining upper bounds and measuring the mismatch in lab to further optimize it.

Degradation: Wideband noise (WBN) is primarily affected by mismatches between the transistors. 600 kHz to 6 MHz offset spectrum can have degradation due to the mismatches but generally the mismatches are not a dominant factor in this frequency range.

2.4.9 ADDITIONAL DISTORTIONS IN THE DPA (DAC/MIXER)

Source: Any analog implementation imperfections can create additional distortion over and above the above-mentioned distortions. Some of the examples of such distortions are

1. The phase-modulated clock has different propagation delay in reaching to various transistors of the DPA (or a generic DAC) resulting in phase mismatch between the outputs of individual transistors;
2. The phase of the clock is not aligned with the phase of the data representing the amplitude waveform causing severe WBN degradation;
3. If proper isolation is not achieved between various modules of a transmitter and also within various transistors of a given module, then signals from one module can couple to another creating various undesired mixing products which can be difficult to troubleshoot and fix.

Degradation: The degradation depends on the exact distortion in the DPA circuit. Analysis of some of these nonidealities and corresponding degradation will be covered in Chapters 3 and 4.

CHAPTER 3

Digital baseband of the polar transmitter*

CHAPTER OUTLINE

3.1 Overview of the TX Digital Baseband 41
    3.1.1 Pulse Shaping Filter 43
    3.1.2 Resampler 47
    3.1.3 CORDIC 49
3.2 Predistortion Module 51
    3.2.1 Overview 51
    3.2.2 Principle of Operation 55
    3.2.3 Analysis of the Quantization Effects 64
3.3 Predistortion Self-Calibration 66
    3.3.1 Effect of Temperature Variations on Predistortion 69
3.4 Interpolative Filter 69
3.5 Polar Bandwidth Expansion 75

3.1 OVERVIEW OF THE TX DIGITAL BASEBAND

The baseband part of the digital transmitter (also termed DTX) is responsible for the sub-200 MHz signal processing of amplitude and phase information. DTX generates the frequency waveform that is used for modulating the RF carrier in the ADPLL. It also generates the amplitude waveform, containing power control information, to be upconverted by the $\Sigma\Delta$ amplitude modulator (SAM) block. Connectivity of the DTX with other key transmitter modules is shown in Fig. 3.1.

The DTX is responsible for generating baseband signals, upsampled to about 200 MHz, for both GSM and EDGE modulation standards. For GSM transmission, based on GMSK modulation, the baseband data is passed through Gaussian pulse shaping filter resulting in the frequency modulation waveform. The frequency modulation is performed by ADPLL receiving the frequency control word (FCW) signal from DTX. The amplitude for GSM modulation is constant and can be controlled by a programmable digital amplitude code driving the DPA. The GSM path shares significant hardware resources with EGDE path.

---

*The authors acknowledge substantial contributions from Prof. Dr. Poras T. Balsara (UT Dallas) and Dr. Oren Eliezer (Texas Instruments).

CHAPTER 3 Digital baseband of the polar transmitter

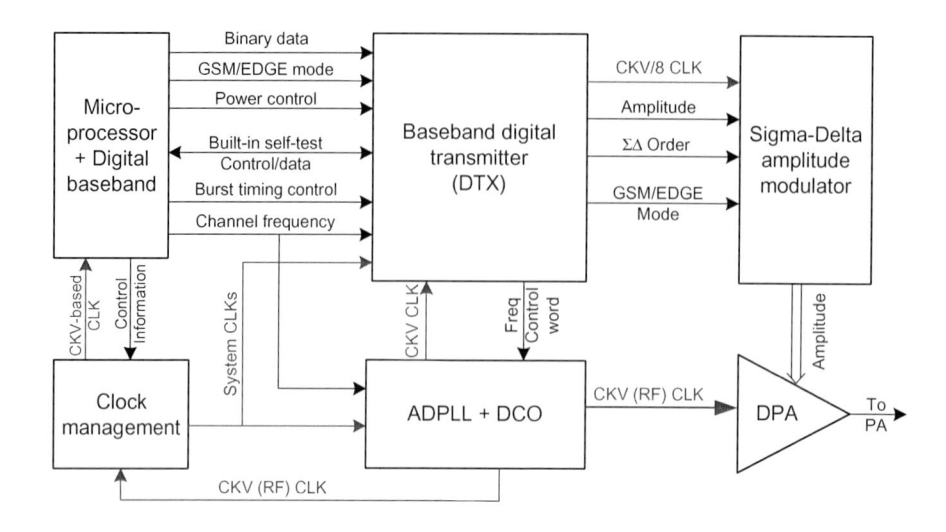

FIG. 3.1

Interface between various key modules in the transmitter system.

The EDGE modulation uses the same 200 kHz bandwidth as GSM but adds amplitude modulation to increase the data rate by a factor of 3. Representing the 8-PSK modulated complex signal in polar form requires significantly higher bandwidth than the original 200 kHz bandwidth due to the polar bandwidth expansion problem (see Section 3.5). Hence the datapath for amplitude and phase is designed to accurately reproduce both using sampling rates higher than 2 MHz. Similar to the GSM path, the frequency/phase path output is used by the ADPLL for frequency modulation. The amplitude control word (ACW) is passed through a thermometer encoder and $\Sigma\Delta$ modulator inside SAM block. The DPA is used to combine the amplitude and path information to form complex EDGE modulation centered around the RF carrier.

A detailed block diagram of the EDGE modulation in DTX datapath is shown in Fig. 3.2. The complex modulation is most conveniently generated using in-phase (I) and quadrature-phase (Q) modulating signals. The binary data to be modulated is supplied by the digital baseband (DBB) in form of a single-bit stream sampled at 812.5 kbits/s. The serial data stream is grouped into the symbols of 3 binary bits used for 8-PSK modulation. Mapper converts the 3-bit symbol to a complex constellation point using Table A.1 as a reference. The resulting constellation at the mapper output is shown in Fig. A.1. As explained in Section A.1, the zero crossing at the center of constellation is avoided by rotating the constellation with a fixed rate of incrementing by $3\pi/8$ per symbol. The rotation information is generated by a digital counter running at the symbol frequency of 270.833 kHz. The counter output is added to the original constellation generated by mapper, resulting in the constellation shown in Fig. A.2. Both I and Q signals can be represented by 4 bits to cover all the points in the constellation shown in Fig. A.2.

3.1 Overview of the TX digital baseband

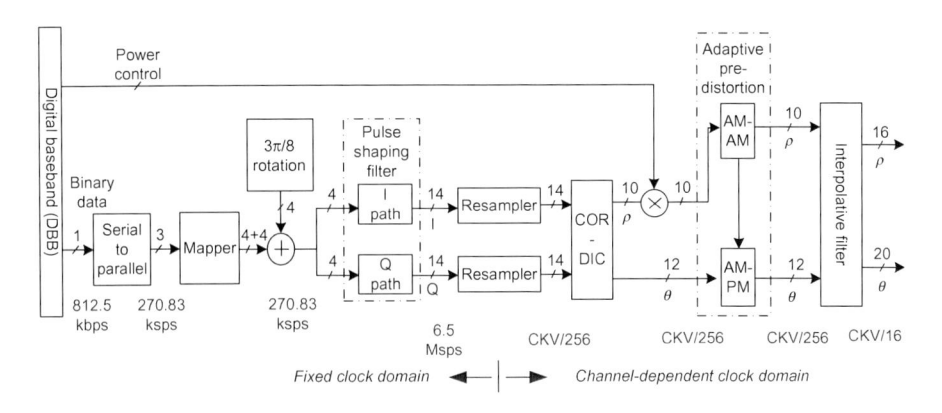

FIG. 3.2

Block diagram of the baseband digital transmitter.

3.1.1 PULSE SHAPING FILTER

The pulse shaping filter (PSF) is used to upconvert the I and Q signals, sampled at the symbol frequency of 270.833 kHz, to 6.5 MHz of fixed frequency. The PSF smoothes the transitions in going from one constellation point to another, thus restricting the complex EDGE signal spectrum to 200 kHz bandwidth. While the shape of the filter is defined by the standard (see Section A.1), the sampling frequency of the filter, length of the filter, and quantization are determined for individual design based on the margin desired over the specifications. Fixed point simulations, incorporating realistic implementation of the filter, are used to design the filter parameters as shown later in this section. The filter design directly impacts the spectral power at 400 kHz and 600 kHz offset from the carrier. Hence these two specifications are used as a guiding point for designing the filter. As a rule of thumb, 10 dB of margin over the ETSI specifications is targeted for the PSF design to ensure negligible degradation at the desired spectral offsets.

The length of the PSF determines the number of symbols that the filter impulse response covers. A longer filter provides better smoothening and thus better spectral performance. The time domain filter impulse response, defined by Eq. (A.3), is graphically shown in Fig. 3.3 for filter length $L = 5$ symbols. The impulse response of the filter nearly reaches zero at $L = 5$ and thus longer lengths do not result in any improvement. Hence $L = 3, 4$, and 5 are used for the comparison of the spectral performance in this section.

The sampling rate selection is primarily driven by the placement of replicas while oversampling to the RF frequency. For a given sampling frequency of f_{PSF}, there will be upsampling replicas at $2nf_{PSF}$ ($n = 1, 2, 3, \ldots, \infty$). If the value of f_{PSF} is chosen to be small, more number of replicas will be contained between f_{PSF} and the carrier frequency. Alternatively, higher oversampling factor will be needed to reach the RF carrier sampling frequency for a lower f_{PSF}, resulting in higher sampling replicas. On

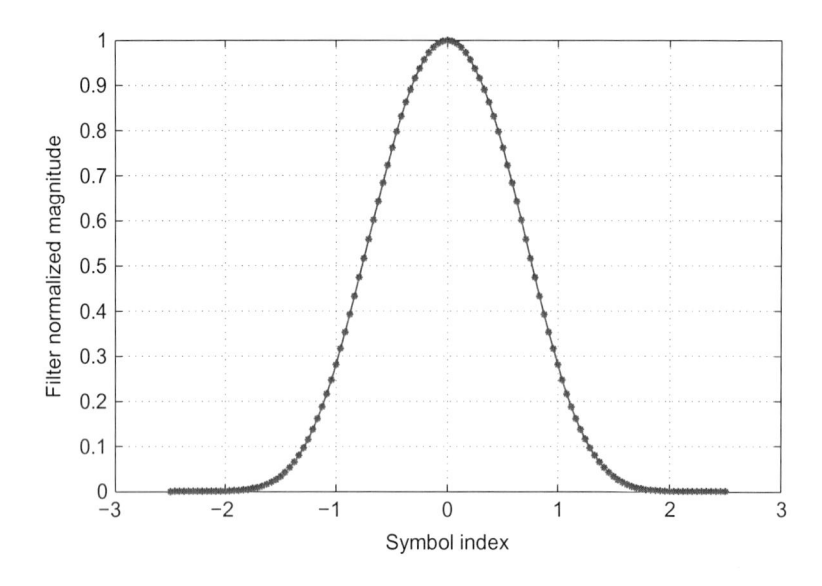

FIG. 3.3

Pulse shaping filter time domain response.

the other hand, increasing the f_{PSF} increases the oversampling ratio of the PSF thus increasing the number of filter coefficients and complexity.

In addition to the above two opposing constraints, additional restrictions are imposed by the modules following the PSF (such as the CORDIC). It is highly desirable to avoid any mismatch in the sampling rates between PSF and CORDIC to maintain the fidelity of the ρ and θ signals produced by the CORDIC module. Even though a resampler is used between these two modules, optimum resampler functionality requires the ratio of the two sampling frequencies to be as close to 1 as possible (Section 3.1.2). Additional details of constrains on the CORDIC sampling rates are covered in Section 3.1.3. Based on the implementation resource constraints of PSF and CORDIC, and the performance plots covered later in this section, the sampling rate of the PSF has been chosen as 6.5 MHz, which translates into the oversampling rate $OSR_{PSF} = 24$. This results in the number of coefficients of the filter $NC_{PSF} = (OSF_{PSF} \times L) = 120$.

Next critical design parameter for the filter design is the quantization of the filter. For the sake of simplicity in presentation, only output bits are shown as a design parameter for all the plots in this section. The internal bitwidth of the filter is optimized accordingly to ensure effective number of output bits are limited by internal resolution at any point. Fig. 3.4 shows the fixed point simulation results of 400 kHz spectral performance at the PSF output for the three filter lengths, while varying the number of bits used for the PSF output. The length of filter (L) was chosen to be five to ensure more than 99.999% of the impulse response energy is covered. The specification line shows 400 kHz specification with 10 dB of additional

3.1 Overview of the TX digital baseband

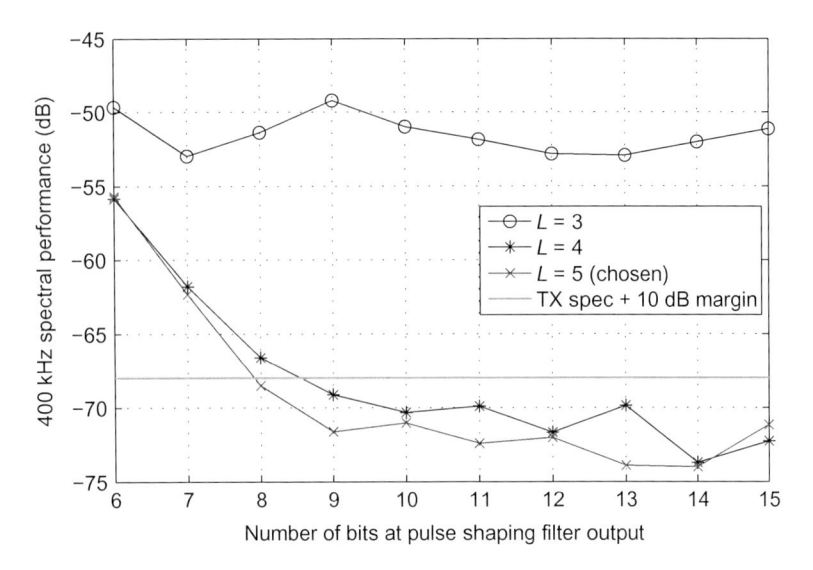

FIG. 3.4

Effect of filter length and bitwidth on 400 kHz offset spectral performance for the pulse shaping filter.

margin. Similarly Fig. 3.5 shows fixed point simulation of 600 kHz offset spectral performance at the PSF output. Based on these simulation plots, the following parameters were selected for the PSF:

- Length of the filter $L = 5$
- Number of bits for the PSF output = 14
- Oversampling rate of the filter = 24
- Filter sampling frequency = 6.5 MHz

The implementation of the PSF is optimized for $NC_{PSF} = 120$ coefficients. The total filter response is divided into five equal parts of 24 coefficients each. These parts are represented by h_C', h_B', h_A, h_B, and h_C in Fig. 3.6. Looking back at Fig. 3.3, the five parts can be mapped to coefficients as follows, where 0 is the current symbol:

- h_C' $\Rightarrow$ [−2.5 to −1.5] symbols mapping to 0 → 23 coefficients
- h_B' $\Rightarrow$ [−1.5 to −0.5] symbols mapping to 24 → 47 coefficients
- h_A $\Rightarrow$ [−0.5 to 0.5] symbols mapping to 48 → 71 coefficients
- h_B $\Rightarrow$ [0.5 to 1.5] symbols mapping to 72 → 95 coefficients
- h_C $\Rightarrow$ [1.5 to 2.5] symbols mapping to 96 → 119 coefficients

As the filter response of Fig. 3.3 completely symmetric around 0, it can be easily seen that h_B' (0:23) = h_B (23:0) and h_C' (0:23) = h_C (23:0), that is, h_B' is flipped in left-right direction version of h_B. Thus the storage requirement can be

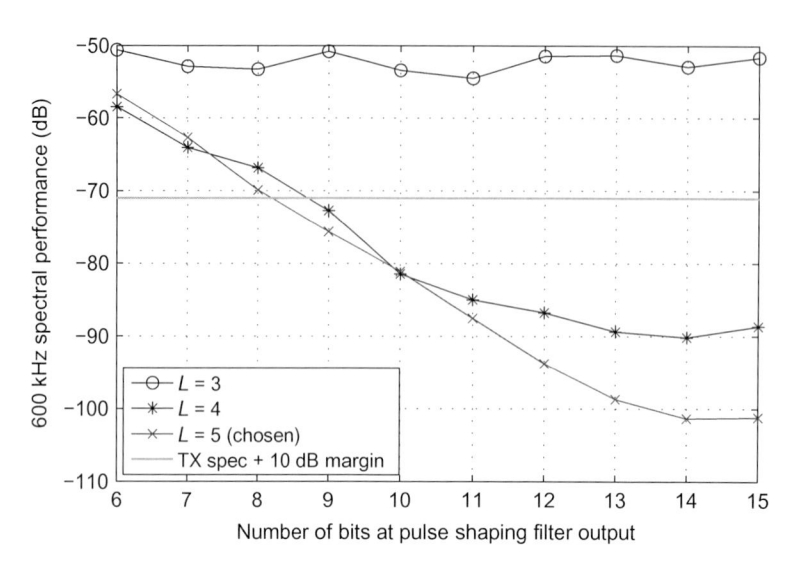

CHAPTER 3 Digital baseband of the polar transmitter

FIG. 3.5

Effect of filter length and bitwidth on 600 kHz offset spectral performance for the pulse shaping filter.

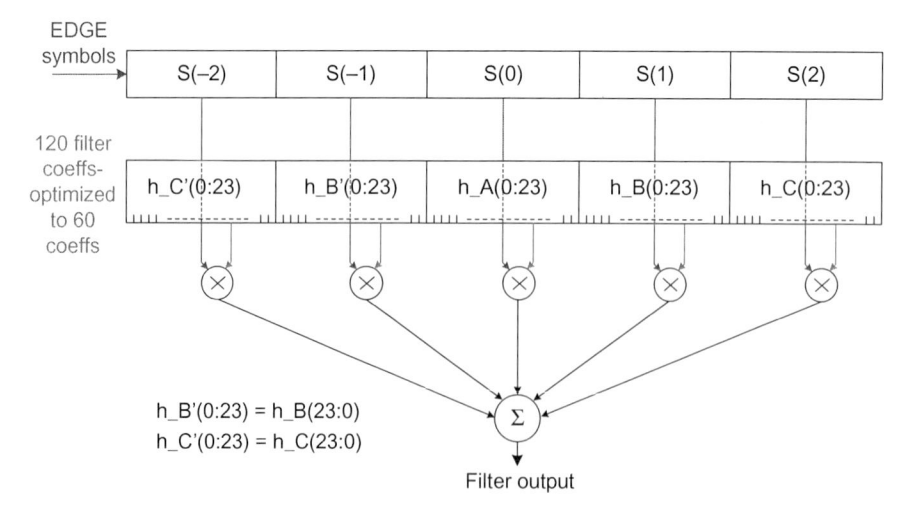

FIG. 3.6

Implementation of the pulse shaping filter.

reduced significantly by just storing h_B and h_C and obtaining h_B' and h_C' by using inverted addressing. The requirements of dedicated multipliers is reduced to five by running the PSF at 24× higher clock and utilizing a single multiplier for each part of the filter response.

The 4-bit I and Q data is stored in the shift register with a memory buffer of length $L = 5$ (same as the filter length L) as shown in Fig. 3.6. The shift register is clocked at the symbol rate of 270.833 kHz, while the coefficients are accessed at 24× faster speed of 6.5 MHz. The data and filter coefficients are multiplied and the results are added together to produces the final filter response sampled at 6.5 MHz as shown in Fig. 3.6. Naturally 24 multiply operations are performed for every input symbol due to the over sampling rate of the filter.

3.1.2 RESAMPLER
3.1.2.1 Need for a resampler
The resampler is used to convert the data, sampled at a fixed clock of 6.5 MHz, to a channel frequency-dependent CKV/256 clock [86]. The binary data stream-based symbols generated by the digital baseband are always at a fixed rate of 270.833 kHz. After going through the upconversion by a factor of 24×, the PSF output is generated at a fixed rate of 6.5 MHz. This sampling rate is independent of channel frequency as it only depends on the symbol rate specified by the EDGE standard.

The DPA, working as a digital mixer of amplitude and phase information, needs to mix the information at the channel frequency to ensure appropriate upconversion. The clock generated by the ADPLL is centered around the channel frequency and the same clock is used to sample the amplitude information. Thus the fixed clock sampled information (generated at 6.5 MHz) has to be converted to the channel frequency (CKV)-dependent clock sampling at some point in the data path before the DPA.

The resampler provides such functionality where it can transfer information, sampled with one clock, to another clock while maintaining the desired signal to noise ratio. The right placement of the resampler in the data chain is also crucial. As explained later in Section 3.5, the phase/frequency signal goes through bandwidth expansion when going from Cartesian to polar coordinates. If the signal energy is not well contained (as is the case after the polar conversion), higher order resampling will be needed to maintain the desired signal to noise ratio. Thus its better to perform resampling operation on the I and Q signal before they are converted to the polar format. Hence the logical placement for the resampler is before the CORDIC module (used to convert Cartesian to polar coordinates) and after the pulse shaping filter.

3.1.2.2 Implementation
The design is based on the resampler used in Ref. [87]. The resampler uses first-order Lagrange polynomial which is equivalent to a linear interpolation. Simplified

FIG. 3.7

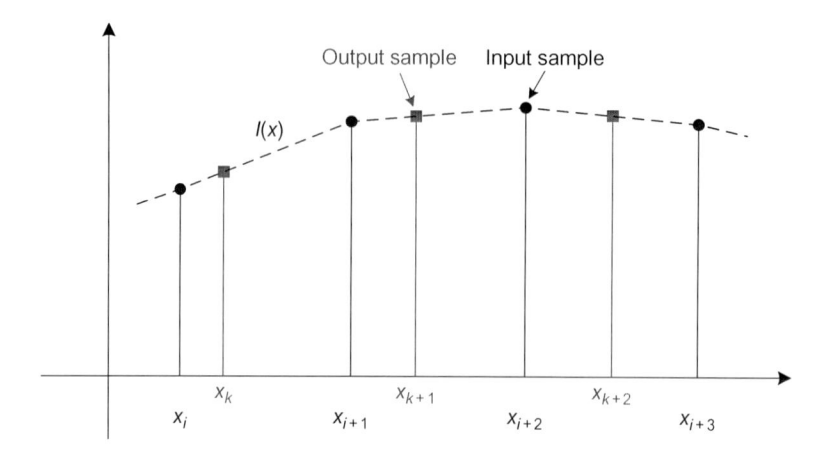

First-order resampler operation.

operation of first-order resampler is shown in Fig. 3.7 for $I(x)$ input signal. Consider the input sampling points to be represented by x_i, x_{i+1}, x_{i+2}, and x_{i+3}, sampled at 6.5 MHz in this example. The resampler performs a time domain linear interpolation between every two points of the input samples as shown by the dashed line between the samples. The output sampling frequency points (CKV/256) are represented by x_k, x_{k+1}, and x_{k+2}. The separation between x_k and x_{k+1} is equal to the period of the output sampling clock (CKV/256).

Thus the resampler operation can be equivalently represented by a linear interpolation with nearly infinite over sampling, and then decimation with the output sampling clock. Ideally the two clocks should have similar frequency to avoid any interpolation/decimation effects. If the output clock frequency is much higher than input clock frequency, the resampler will perform additional linear interpolation. If the ratio of the output to input clock frequency is greater than 2, sampling replicas will be visible with only 12 dB attenuation from carrier power due to the linear interpolation. Similarly if the output frequency is much lower than the input frequency then critical information will be lost due to the decimation. Thus it is necessary to select the CKV-based clock frequency which is the closest to 6.5 MHz input clock frequency.

To select CKV factor-of-2 derivative clock, it is necessary to analyze the full range of the CKV frequencies possible across the frequency range of operation. Section 2.2.2 calculates the variation in CKV frequency over the transmitter frequency range of operation. The variation on CKV/256 can be calculated based on Table 2.1 to be from 6.44 MHz to 7.46 MHz. This factor-of-2-derived clock from CKV is the closest frequency to 6.5 MHz and hence the resampler output rate has been chosen to be CKV/256.

3.1.3 CORDIC

The Cartesian domain I/Q signals are converted to polar domain ρ–θ signals using COordinate Rotation DIgital Computer (CORDIC) module. CORDIC algorithms perform a wide range of functions such as trigonometric, hyperbolic, linear, and logarithmic functions without using hardware multiplier. The algorithm uses shift-and-add mechanism to perform these operations using a small lookup table (LUT) [77, 88].

The conversion between the Cartesian and polar coordinates can be expressed by the following nonlinear equations where ρ denotes the amplitude and θ denotes the phase.

$$\rho = \sqrt{I^2 + Q^2}$$
$$\theta = \tan^{-1}\left(\frac{Q}{I}\right) \quad (3.1)$$

The core of CORDIC algorithm relies on iteratively calculating the formula described by Eq. (3.2). The initial value of x_0 and y_0 corresponds to I and Q inputs, respectively, while z_0 is set to 0. The desired goal of the iterations is to make y_n (after n iterations) to converge to 0.

$$x_{i+1} = x_i - (y_i \times d_i \times 2^{-i})$$
$$y_{i+1} = y_i + (x_i \times d_i \times 2^{-i})$$
$$z_{i+1} = x_i - [z_i \times \tan^{-1}(2^{-i})] \quad (3.2)$$
$$d_i = +1 \text{ if } y < 0,$$
$$= -1 \text{ otherwise}$$

where d_i is an internal variable.

After going through sufficient number of iterations (n), the value of x_n and z_n represents the desired polar coordinates with a fixed scaling as shown in Eq. (3.3).

$$x_n = A_n\sqrt{x_0^2 + y_0^2}$$
$$y_n = 0$$
$$z_n = \tan^{-1}\left(\frac{y_0}{x_0}\right) \quad (3.3)$$
$$A_n = \prod_n \sqrt{1 + 2^{-2i}}$$

Thus the value of ρ and θ can be derived from x_n and z_n using Eq. (3.4). The number of iterations (n) is set to 16 based on the desired accuracy and availability of the high-speed clock for parallel operation.

$$\rho = \left(\frac{x_n}{A_n}\right)$$
$$\theta = z_n \quad (3.4)$$

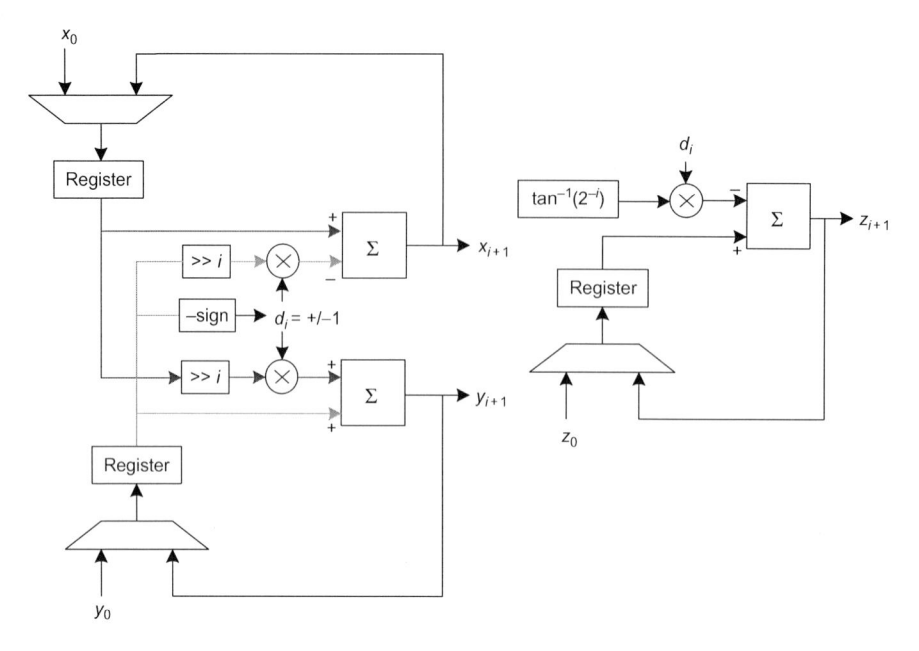

FIG. 3.8

Implementation of the CORDIC module.

Detailed implementation of the CORDIC module representing Eq. (3.2) is shown in Fig. 3.8. The architecture relies on a 16× faster clock (16× CKV/256 = CKV/16) to perform fully parallel operation using only three adders and two shifters. Initial multiplexer selects between initialization condition (when $x_i = x_0$), and iterative loop operation when x_{i+1} is the current output of x_i path. For calculation of x_{i+1}, the register provides the buffered x_i value. Similarly calculated value of y_i, after going through the appropriate shift and sign change (shown by first line of Eq. (3.2)), is used to calculate the new value of x_{i+1}. The value of d_i represents the sign of the y_i signal. The multipliers shown in Fig. 3.8 only represent a sign change based on the value of d_i. y_{i+1} is calculated using the same approach as x_{i+1}.

The calculation of z_{i+1} requires use of $\tan^{-1}$ function estimation but the input to the $\tan^{-1}$ function does not depend the real-time I and Q values. The parameter to the $\tan^{-1}$ function only depends on the iteration index (i) and can be stored in the lookup table (LUT) for range of $i \epsilon (1 \ldots n)$ where $n = 16$ for the presented architecture.

The sampling frequency of CORDIC is a critical system parameter. With any increase in sampling frequency, the need for 16× faster clock makes the power consumption of CORDIC increase rapidly. Similarly, the bitwidth at the output of CORDIC is another critical system design parameter. The 400 kHz and 600 kHz performance is directly impacted by the selection of these parameters. Fig. 3.9 shows the spectral performance at 400 kHz offset across the CORDIC sampling rate and amplitude output bitwidth. Fixed frequencies (such as 6.5 MHz) are used in this

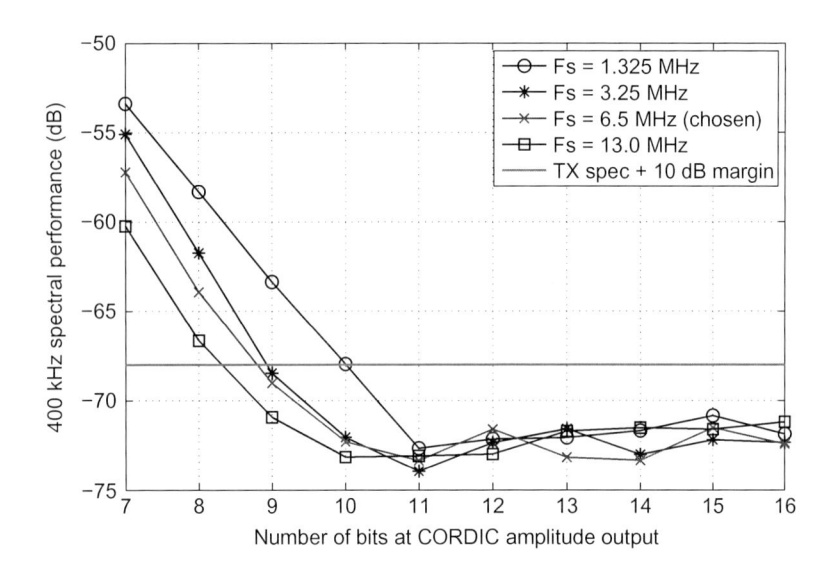

FIG. 3.9

Effect of sampling rate and bitwidth on 400 kHz offset spectral performance for the CORDIC.

plot for simplicity but as explained in Section 3.1.2, the actual frequency used by CORDIC will be the closest CKV/ratio frequency. For example, the 6.5 MHz line corresponds to CKV/256 channel-dependent clock frequency.

Fig. 3.10 shows the spectral performance at 600 kHz offset. The performance at 600 kHz offset is impacted more visibly than the 400 kHz performance while changing the amplitude path bitwidth. Based on the criteria of 10 dB margin at each spectral offset, the sampling rate of Fs = CKV/256 (around 6.5 MHz) is determined with 10 bits for amplitude output and 12 bits for phase output. The sampling frequency of CKV/256 (around 6.5 MHz) allows the pulse shaping filter and CORDIC to operate at the same rate thus simplifying the transfer between the two modules.

3.2 PREDISTORTION MODULE

3.2.1 OVERVIEW

DPA AM-AM and AM-PM distortions, highlighted in Section 2.4.3, are shown to be one of the main sources for the degradation in the EVM of the transmitter. The distortions are graphically shown in Fig. 3.11. The most common linearization method to overcome such distortion is by predistorting the signal in advance by typically using a digital predistortion module. Predistortion works by finding the

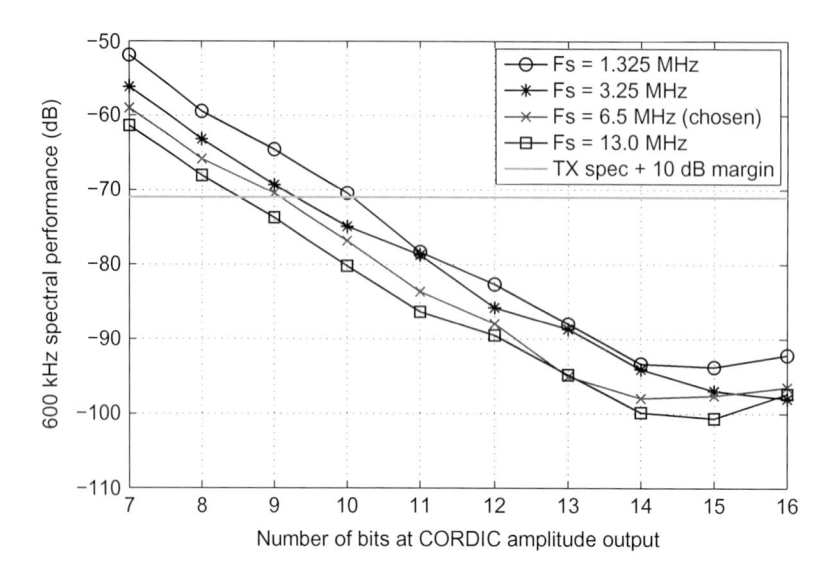

FIG. 3.10

Effect of sampling rate and bitwidth on 600 kHz offset spectral performance for the CORDIC.

inverse of the distortion transfer function and applying the inverse transfer function in the digital path before the DPA. Thus the combination of predistortion and DPA can be treated as a power amplifier with linear transfer function and without amplitude or phase distortion.

Depending upon the dynamic nature of the distortion, feedback can be used to adapt the predistortion transfer function. Predistortion designs can be classified into the following three categories based on the type of feedback.

1. *Open loop feedforward predistortion*: This type of predistortion relies on one time-preprogrammed transfer function and no updates are provided. This option can work in the cases where the distortion is completely static and can be characterized one time for each device. The AM-AM and AM-PM must remain fairly constant over temperature/voltage and frequency variations.
2. *Open loop feedforward predistortion with periodic update using closed-loop feedback (Adaptive Predistortion)*: If variations over temperature/voltage and frequency are expected then periodic updates with any change in such parameters can be used to recalculate the transfer function for the predistortion mechanism. This is the case for the presented transmitter. Since the period updates are expected to take place during the routine operation, some form of built-in self-test (BIST)-based mechanism is necessary to recalculate the transfer function [78].

FIG. 3.11

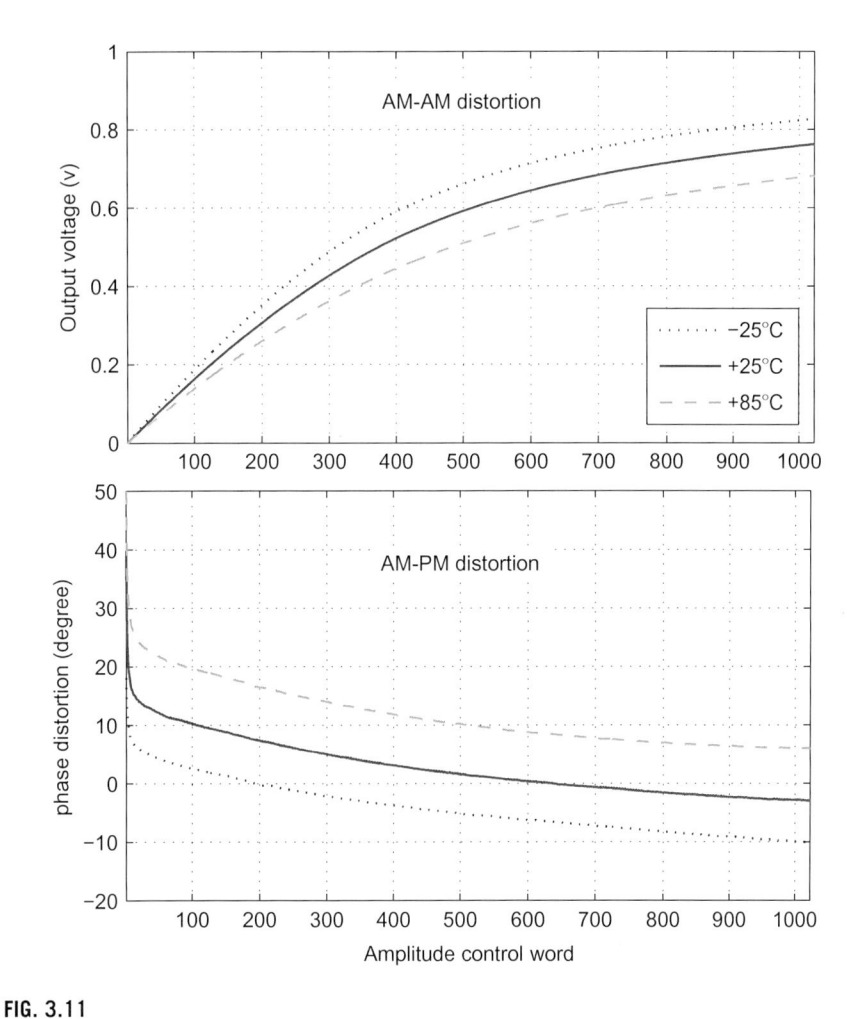

DPA AM-AM and AM-PM distortion over the digital amplitude control word.

3. *Closed-loop continuous feedback predistortion*: If the changes in the AM-AM and AM-PM transfer function can occur during the transmission, continuous closed-loop feedback can be used for performing predistortion with immediate corrections [89]. Generally the closed-loop systems are more complex to implement and the bandwidth of the predistortion is limited by the loop bandwidth. One such example of the system is shown in Ref. [90]. Since the DPA distortion is found to be stable for a given PVT and frequency condition, the use of closed-loop has been avoided for the presented design.

If the AM-AM and AM-PM distortions of the amplifier depend on the bandwidth of the signal, the amplifier is termed to have memory effects. In other words, if the output of the amplifier does not depend only on the current input then it is said to have the memory effects. For a linearization system with feedback, the memory effects are corrected by the continuously operating feedback loop within the bandwidth of the loop. For the linearization system without feedback, the distortions have to be corrected by having predistortion with memory. Instantaneous change in temperature of amplifier, proportional to change in input amplitude, can be one of the major sources of the memory effect [91]. Another source is the biasing modulation of the amplifier based on the input amplitude signal. Careful circuit simulations and test-chip-based measurements have shown the memory effect to be negligible for the DPA used in the presented design. The circuit simulations ramped up and down the DPA with multiple slopes and analyzed the variations based on the speed and direction of the ramp. This determines the dependence of the output of the history of the change of previous samples. The results have been found DPA to contain negligible memory effects. In addition, the use of linear PA in this design (refer to Section 2.1.1.1) introduces far less memory effects compared with the saturated PA [92]. Hence memory effects are neglected throughout the presented research.

The change in AM-AM/AM-PM response with change of transmitter carrier frequency is not considered part of memory effect for this work. The variations due to carrier frequency are analyzed separately and multiple LUTs are used for storing the distortions to guarantee optimum predistortion over the frequency band of interest.

The AM-AM and AM-PM distortion and predistortion information can generally be represented either as entries in look-up-tables (LUT) or in polynomial equation form by compressing the information using polynomial fitting methods. Following are some contrasting points between the two solutions [78]:

- The LUT-based implementation has the flexibility of being able to predistort nearly all types of memory-less distortions without being dependent on the specific distortion characteristics because the raw information is stored in the LUT. On the contrary, the polynomial based implementation can operate with a limited types of distortions based upon the convergence characteristics of the polynomial used [93].
- The LUT-based implementation requires accessing the LUT, typically implemented in a memory, for every predistortion calculation. On the other hand, the polynomial-based design required calculation of the polynomial which typically requires multiplier-adder-based polynomial evaluation. The required storage for the polynomial design is much lower due to the smaller number of coefficients to be stored compared to the LUTs containing AM-AM/AM-PM responses.
- If the design requires frequent updates the predistortion characteristics, polynomial fitting-based methods have to be used for every update to recalculate the polynomial coefficients. This process could be time consuming and might require dedicated hardware or complex software. On the other hand, LUT-based implementation, with the presented dynamic inversion of the LUT, requires minimal processing during the feedback-based update of the AM-AM/AM-PM distortions.

- Nonuniform sampling (where the spacing between the AM-AM/AM-PM LUT entries can be adjusted based on the nonlinearity) can be easily implemented using LUT-based design.

Although the main objective for the predistortion is to nullify the effect of AM-AM and AM-PM distortions of the DPA, the configurable nature of the predistortion allows it to be used for the external power amplifier linearization as well. The memoryless nature of the saturated PA allows the DPA and PA distortions, to be represented in a cascaded form, in a single AM-AM and a single AM-PM table. Changes have to be made in the way the self-calibration works to ensure the PA output is sampled during the self-calibration instead of the DPA output. This will be elaborated in Section 3.3.

Fig. 3.11, also shown in Section 2.4.3, illustrates the AM-AM and AM-PM distortions measured at the output of the DPA in the 1800 MHz DCS band. These were obtained for one sample transmitter by individually exercising the 1024 integer levels supported by the DPA using an amplitude control word originating from the SAM block. Section 2.4.3 highlighted the challenge of the DPA nonlinearity affecting the EVM and close-in spectrum. Section 3.3.1 highlights the strategy for overcoming the temperature variations of AM-AM and AM-PM distortions.

In summary, the distortions that are compensated for are measured at the point of interest in the transmitter, and are digitized and stored in nonvolatile look-up-tables as shown in Section 3.3. This measured data is directly used in the predistortion calculation, without necessitating a calculation of the inverse function, as is commonly done [94], thus offering a minimal-cost realization for the linearization system [78].

3.2.2 PRINCIPLE OF OPERATION

Fig. 3.12 shows the block diagram of the presented LUT-based predistortion scheme for the AM-AM and AM-PM distortions of the polar transmitter. The predistortion module comprises of AM-AM and AM-PM blocks receiving amplitude and phase inputs, respectively, from the CORDIC block sampled at CKV/256 (about 6.5 MHz) frequency. Power control is added in the path to achieve the 30 dB of the EDGE power control on the mobile unit. The power control is represented by 8 bit digital word—PCL_gain coming from the digital baseband as shown in Fig. 3.1. The truncation of the multiplier output is performed after the interpolative filter to maintain the resolution of the amplitude path as explained later in this section.

The amplitude is passed through a process that has transfer function equivalent to the inverse of the DPA AM-AM response. The predistorted amplitude output is also used for AM-PM distortion after passing through a scaling module, as explained in the following section.

Fig. 3.12 graphically illustrates the AM-AM distortion of the DPA being predistorted by the AM-AM module, resulting in a sufficiently linear relationship between the DPA output voltage and the digital amplitude word. The off-chip PA further amplifies the modulated signal to reach up to 30 dBm of peak output power at the transmitter output. The DPA module curve (Curve 4) represents the output voltage as

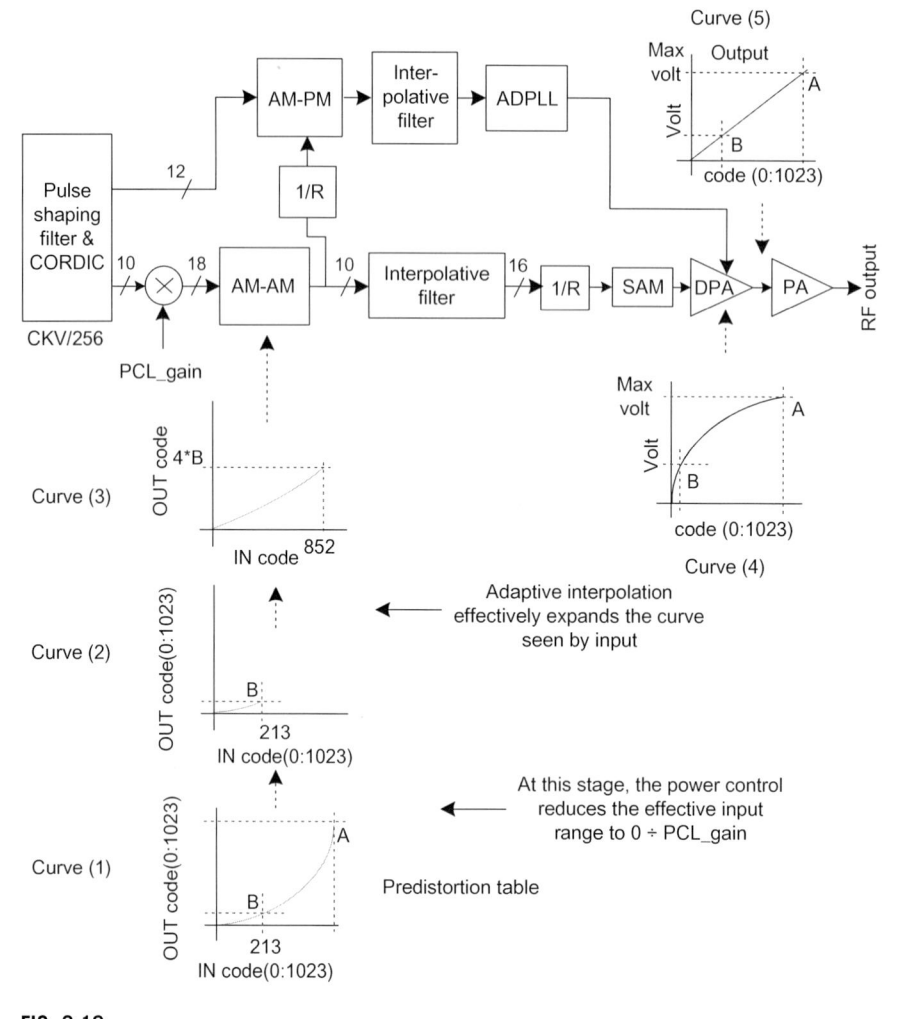

FIG. 3.12

Block diagram of the presented predistortion scheme.

a function of the input digital code. The DPA is capable of producing 1024 integer output steps as shown on the x-axis. The predistortion curves (Curves 1, 2, and 3) represent the inverse of the AM-AM distortion to be applied to the amplitude signal to achieve overall linear operation. Both axes are normalized to (0–1023) range for convenience in representation. The input to the predistortion module is given by a digital code on the x-axis, representing the desired instantaneous amplitude, and the output is the predistorted code necessary to produce that output amplitude, as shown by Curve 5 [78].

3.2 Predistortion module

The transmitter power control is realized by digitally scaling the 10-bit amplitude at the CORDIC output with the digital gain (PCL_gain). In order for the predistortion LUT to hold valid across all power levels, spanning about 30 dB of dynamic range, it is placed after the power control module. If predistortion is done before the power control, the amplitude always covers a fixed dynamic range irrespective of the power level. In that scenario, a separate LUT has to be configured for every power level which becomes extremely resource intensive. Hence the power control is applied before the predistortion module.

Since the digital power control precedes the predistortion, the number of effective bits coming at the input of the predistortion can vary based on the power level. For instance, when the output power is near maximum, the input covers more than 9 bits out of the 10 bits. But in the case of low power levels, around 30 dB lower than the maximum power level, the amplitude swing is represented by 5 fewer bits. For the predistortion to be equally effective at both the extreme power levels, it must simultaneously accommodate a wide dynamic range and fine resolution, which would be very demanding in terms of the implementation area. This is illustrated in Fig. 3.12, where two different points, A and B (shown in the curves), representing the output voltages at high and low output power levels, respectively, are considered. As can be seen in the figure, for a power level with peak amplitude corresponding to point A, the full AM-AM curve (shown in Curve 1) is usable, and hence an effective resolution of almost 10 bits is made available for the amplitude signal. On the other hand, for point B, the peak output amplitude is significantly lower, resulting in an effective resolution that is reduced to about 5–6 bits. Having 5–6 bits is not sufficient to meet the EDGE close-in spectral performance due to excessive quantization noise.

3.2.2.1 Adaptive interpolation

To solve this problem, an adaptive interpolation scheme is presented to effectively enhance the resolution of the predistortion function at low power levels by applying the appropriate amount of digital gain (represented by R) in the predistortion module. R represents the factor by which the input to the predistortion is artificially amplified, and also the factor by which the predistortion module will interpolate the amplitude response. The $1/R$ factor cancels out this amplification when appropriate resolution is present at the Interpolative Filter output, as highlighted in the following section. In the absence of such adaptive interpolation, the complete data path would have to be designed with 15 bits of resolution to maintain 10 bits of effective resolution at the lowest power levels. Hence, adaptive interpolation serves to greatly reduce the implementation area and power consumption by eliminating the need to accommodate such fine resolution.

Table 3.1 shows the adaptive interpolation-based resolution adjustment in graphical format. To increase the resolution by R, both amplitude data and AM-AM table need to be amplified by the same factor. To increase the resolution on the amplitude data signal, LSBs coming out of the digital power control multiplier are used. In a maximum output power scenario, out of the 18 bits coming out of the digital

Table 3.1 Adaptive Interpolation Operation for $R = 32$

Datapath Location	Without Adaptive Interpolation		With Adaptive Interpolation	
	# Bits Used (# Available)	Example Value	# Bits Used (# Available)	Example Value
Data before power control multiplier	10(10)	0x200	10(10)	0x200
Power Control Word	5(10)	0x010	5(10)	0x010
Power control multiplier output	15(20)	0x04000	15(20)	0x04000
Truncation at the power multiplier output	13(18)	0x01000	13(18)	0x01000
Predistortion input truncation	5(10)	0x010	10(10)	0x200
Predistortion output	5(10)	0x010	10(10)	0x200
Interpolative filter output	11(16)	0x400	16(16)	0x8000
Interpolative filter output	11(16)	0x0400	16(16)	0x8000
After $1/R$ scaling	11(16)	0x0400	11(16)	0x0400

multiplier, lower 8 bits would be discarded and the upper 10 bits will be used as an index to the 1024 element AM-AM table. But consider a very low output power scenario illustrated in Table 3.1. In this case, the power control word is effectively using 5 bits out of 10 bits available. Thus the 18-bit power control modules output contains zeros at the 5 MSBs and only uses 13 bits out of the 18. If no adaptive power control is used, predistortion input (truncated to 10 bits) will only have five effective bits and the upper 5 MSBs will be zero. To work around this issue, adaptive power control will shift the 18-bit output by a factor of 5 ($R = 2^5 = 32$) resulting in 10 effective bits at the predistortion input. After enough bits are available at the Interpolative filter output, the scaling can be adjusted back as shown in the last row of Table 3.1. The radix point for the AM-AM table indexing has to be adjusted to account for this amplification to ensure proper predistortion functionality over all the power levels.

To increase the resolution by R on the AM-AM and AM-PM table, linear interpolation is used. To illustrate this further, consider an example of point B, in Fig. 3.12, at the CORDIC output with the digital amplitude of 213 shown in Curve 1. Curve 2 is shown to correspond to the segment spanning from 0 to 213 of Curve 1 which would be used for the predistortion and power control. To represent the code of 213 (the maximum operation point B), less than 8 amplitude bits are used which are insufficient to provide the desired spectral mask performance at the transmitter output.

Based on extensive system analysis (see Section 4.3.2), a rule of thumb for the spectral compliance at 400 kHz offset from the carrier for EDGE modulation is to ensure about 9–10 bits are available for representing the amplitude information at any point in the amplitude data path. Therefore, the resolution is increased for this case through the use of interpolation by a factor of $R = 4$, resulting in Curve 3, where the maximal amplitude is shown to reach a value of 852, effectively approaching 9–10 bits of resolution. Since the AM-AM curve is smooth, the interpolation effectively adds 2 bits (because of linear interpolation by $R = 4$) on the AM path without need for significant increase in the implementation complexity. The addition of two extra bits in the presented scheme improves the adjacent channel power spectral density by as much as 12 dB, thus enabling spectral-mask compliance at 400 kHz offset without an area penalty.

The amplification has to be balanced with an attenuation at a later point in the data path to ensure the right power is produced at the output. In this example, the scaling factor of $(1/R) = 1/4$ is then applied in the signal processing path, to arrive at the correct power level at the transmitter's output. This attenuation does not result in noticeable performance-degradation since the Interpolative Filter output runs at a sampling rate that is 16 times higher and with 8 more bits of resolution. And hence losing about 5 bits in worst case will still keep the effective amplitude path resolution to more than 9 bits. Essentially, the adaptive interpolation maintains a resolution of about 10 bits in the AM path throughout the amplitude data path for all power levels, while requiring minimal hardware.

Measurements in silicon confirm the advantages of using the adaptive interpolation. Fig. 3.13 shows the measured results of comparison between the close-in spectrum with and without the use of adaptive interpolation. The carrier frequency of 1710.2 MHz and carrier power of -18 dBm is used to show the impact of interpolation at low power levels. The measurement clearly confirms more than 5 dB of improvement with the use of adaptive interpolation. The spectrum till about 300 kHz offset is not affected by the addition of adaptive interpolation because it is dominated by the pulse shaping filter response. The improvement starts to be visible at 400 kHz offset and increases significantly beyond 600 kHz because these offsets are dominated by the quantization noise as discussed further in Section 3.2.3.

3.2.2.2 Dynamic LUT inversion
To derive the inverse transfer function for AM-AM, an inverse function calculation is necessary. As an example, such a calculation is used to derive Curve 1 from Curve 4 in Fig. 3.12. The inverse calculation is extremely resource intensive and consumes significant amount of test time. Typically the inverse calculation is used for majority of LUT-based predistortion implementation [95]. Even for polynomial-based predistortion implementation, inverse polynomial has to be calculated for the AM-AM operation. Contrary to these, this work proposes a scheme that makes direct use of the distortion data that is measured during the calibration phase, thus reducing the calibration time and complexity. This is particularly important when the calibration is to be done self-sufficiently, without the use of external equipment, and cannot afford extended durations. It also helps when the distortion is changing dynamically and frequent updates are necessary to the distortion tables.

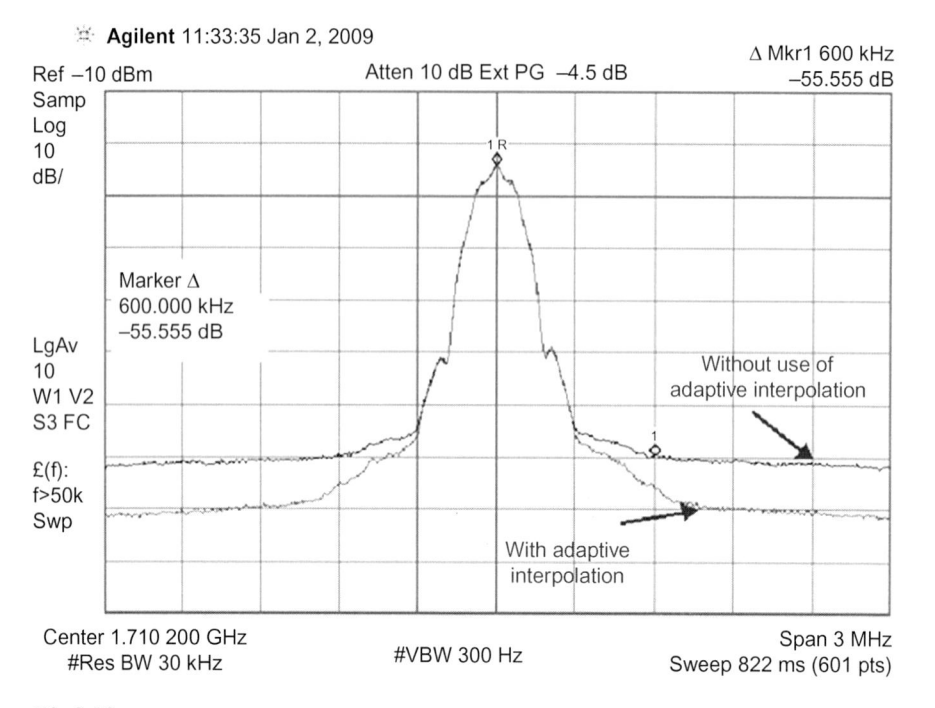

FIG. 3.13

Measurement results showing the impact of adaptive interpolation.

The predistortion values are derived from the distortion tables by using a form of content-addressed memory to perform the inversion for each predistorted sample in real-time during transmission, as shown in Fig. 3.14. First the basic operations of dynamic LUT inversion are covered using an example and later the detailed implementation of Fig. 3.14 is explained in this section.

The AM-AM distortion table contains 256 entries, each corresponding to a $4\times$ transistor in the DPA. 14 bit resolution is used to store the normalized voltage contribution of each transistor in the LUT. Similarly AM-PM table contains the phase contribution of 256 $4\times$ transistors stored with 12 bit resolution. A sample AM-AM table used by the predistortion module is shown in Fig. 3.15 corresponding to the distortions shown in Fig. 3.11. The table simply represents every fourth element in the AM-AM curve normalized to 16 bits. Similarly AM-PM table can be derived by sampling every fourth element of the AM-PM curve in Fig. 3.11.

For each input value, representing the desired instantaneous amplitude, a combination of binary and linear searches are used to determine the consecutive pair of LUT entries that are closest to the desired value. This works like a content-addressed memory operation where the AM-AM distortion values in the memory are scanned to find the value that matches closest to the input amplitude code. This is also shown graphically in Fig. 3.16.

3.2 Predistortion module

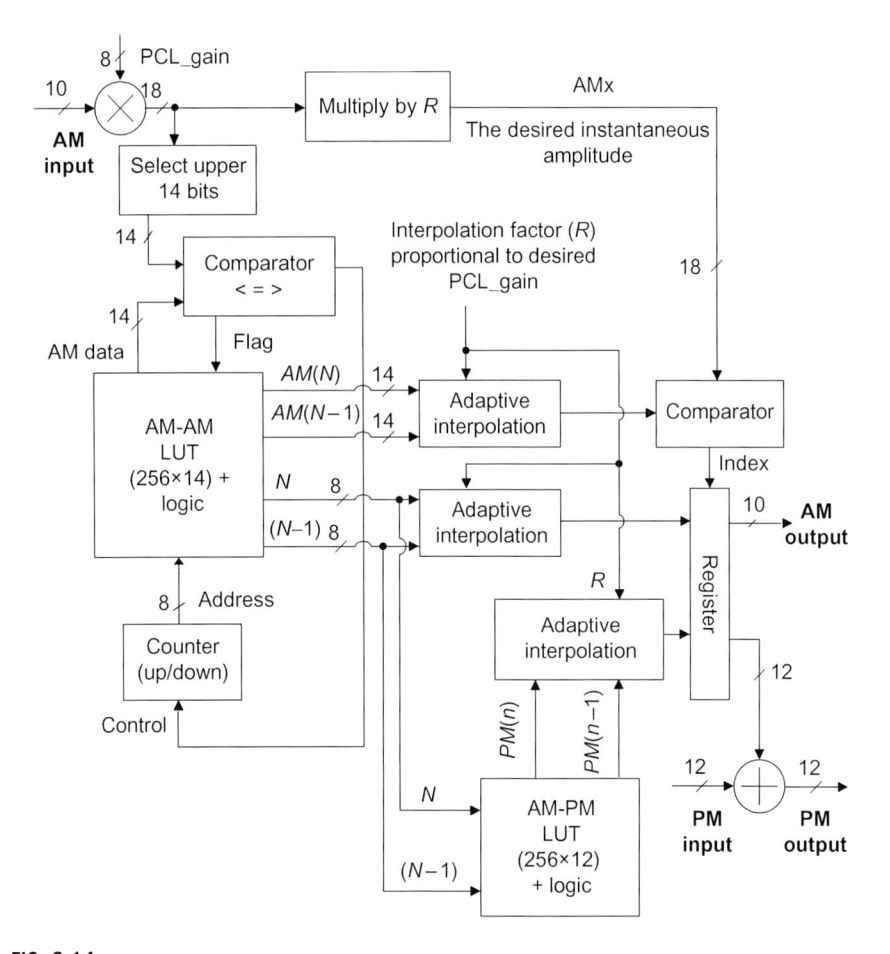

FIG. 3.14

Details of dynamic LUT inversion.

For the search mechanism to work, it is necessary to find the upper bound on the largest input amplitude change, which would determine the maximum range that the search mechanism has to cover. Based on fixed point simulations, the worst case distance between the two input points was found to be always less than 16 locations apart in the memory. Hence the search mechanism is restricted to ± 16 locations from the current pointer in the memory. The direction of the search (greater/less than current input) is determined first to limit the search to 16 locations. The 16 locations are further divided to eight locations by doing a binary search which would locate the input either in the upper or lower eight memory locations, at which point a linear search would be used to identify the two nearest points. This operation takes around 11 clock cycles after adding the memory and logic latency and hence a 16× higher

CHAPTER 3 Digital baseband of the polar transmitter

n	m	n	m	n	m	n	m	n	m	n	m	n	m	n	m
0	0	32	5806	64	10035	96	12369	128	13806	160	14795	192	15497	224	16003
1	123	33	5972	65	10131	97	12423	129	13843	161	14820	193	15515	225	16016
2	315	34	6135	66	10226	98	12477	130	13879	162	14845	194	15534	226	16029
3	507	35	6297	67	10319	99	12530	131	13915	163	14870	195	15552	227	16042
4	698	36	6457	68	10409	100	12582	132	13951	164	14895	196	15570	228	16056
5	889	37	6614	69	10499	101	12633	133	13986	165	14919	197	15588	229	16069
6	1079	38	6770	70	10586	102	12684	134	14020	166	14943	198	15606	230	16082
7	1269	39	6923	71	10671	103	12734	135	14055	167	14967	199	15623	231	16095
8	1458	40	7074	72	10755	104	12783	136	14089	168	14991	200	15641	232	16108
9	1647	41	7223	73	10838	105	12832	137	14122	169	15014	201	15658	233	16121
10	1836	42	7370	74	10918	106	12880	138	14155	170	15038	202	15675	234	16134
11	2025	43	7515	75	10998	107	12928	139	14188	171	15061	203	15692	235	16147
12	2213	44	7658	76	11075	108	12975	140	14220	172	15083	204	15708	236	16160
13	2401	45	7798	77	11151	109	13021	141	14252	173	15106	205	15724	237	16173
14	2588	46	7936	78	11226	110	13067	142	14284	174	15128	206	15741	238	16186
15	2775	47	8071	79	11299	111	13112	143	14315	175	15150	207	15757	239	16199
16	2961	48	8205	80	11371	112	13157	144	14346	176	15172	208	15772	240	16212
17	3147	49	8336	81	11442	113	13201	145	14377	177	15194	209	15788	241	16224
18	3332	50	8464	82	11511	114	13245	146	14407	178	15216	210	15803	242	16237
19	3516	51	8591	83	11580	115	13288	147	14437	179	15237	211	15819	243	16250
20	3700	52	8715	84	11647	116	13330	148	14466	180	15258	212	15834	244	16263
21	3882	53	8837	85	11712	117	13373	149	14495	181	15279	213	15849	245	16275
22	4064	54	8957	86	11777	118	13414	150	14524	182	15300	214	15863	246	16288
23	4244	55	9074	87	11841	119	13456	151	14552	183	15320	215	15878	247	16300
24	4423	56	9189	88	11903	120	13496	152	14581	184	15341	216	15892	248	16312
25	4601	57	9302	89	11965	121	13537	153	14608	185	15361	217	15907	249	16323
26	4778	58	9413	90	12025	122	13576	154	14636	186	15381	218	15921	250	16334
27	4953	59	9522	91	12085	123	13616	155	14663	187	15401	219	15935	251	16345
28	5127	60	9629	92	12143	124	13655	156	14690	188	15420	220	15948	252	16355
29	5299	61	9733	93	12201	125	13693	157	14717	189	15440	221	15962	253	16364
30	5470	62	9836	94	12258	126	13731	158	14743	190	15459	222	15976	254	16372
31	5639	63	9936	95	12314	127	13769	159	14769	191	15478	223	15989	255	16379

n	Table index
m	Table data

FIG. 3.15

A sample AM-AM table.

rate clock was used as a parallel clock to allow the search operation to be completed in one clock of CKV/256 frequency.

Fig. 3.16 explains the basic operation of dynamic LUT inversion and adaptive interpolation. Fig. 3.16A is shown to have the AM-AM distortion LUT represented in LUT format without inversion. The amplitude input to be predistorted is represented by point X on the Y-axis. As mentioned earlier, a combination of binary and linear search is used to find the two points which are nearest to the predistortion input value X. These two values, represented by $AM(N-1)$ and $AM(N)$, along with their location (index) in the table represented by $N-1$ and N are passed to the second stage. Thus

3.2 Predistortion module   63

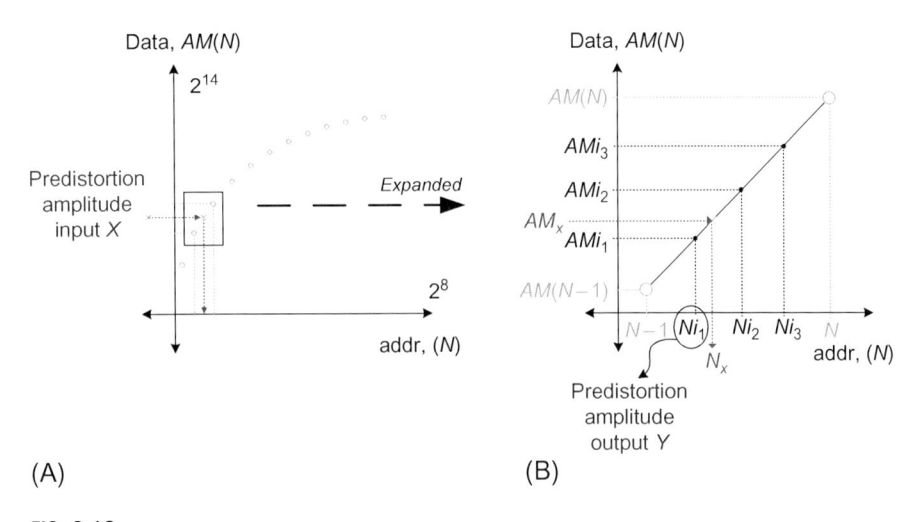

FIG. 3.16

Illustration of the AM-AM table inversion and adaptive interpolation.

the first operation represented in Fig. 3.16A is showing the dynamic LUT inversion in its very simplistic form.

The second stage is performing adaptive interpolation as shown in Fig. 3.16B. The same input amplitude X is shown as AM_x on the expanded form of the AM-AM distortion LUT in Fig. 3.16B. Linear interpolation between the values of $AM(N-1)$ and $AM(N)$ is performed to find the closest match to AM_x. The amount of interpolation performed depends on the value of R. For given R, interpolation factor is set to $2(R+2)$. R is the multiplier factor discussed earlier that is used for resolution enhancement. In addition to R, $2^2 = 4$ factor of interpolation is performed to compensate for the fact that the LUT has only 256 values instead of 1024 values. Hence the 10 bit input is indexing 256 element LUT after going through a factor of four interpolation. This will ensure the same numbers of bits are available at the output of predistortion by adding interpolative resolution. Using this linear interpolation, the value of the interpolation between $AM(N-1)$ and $AM(N)$ that matches the closest to AM_x is marked as AMi_1. While the interpolation is performed between $AM(N-1)$ and $AM(N)$, the same interpolation is performed in parallel for the two corresponding indices—$N-1$ and N as well. The interpolated value of the index that corresponds to AMi_1 is marked as Ni_1 which becomes the output of the AM-AM predistortion. This operation is equivalent of having 256 values of the LUT interpolated by a factor of $2(R+2)$ and then performing the dynamic LUT inversion of Fig. 3.16A.

As shown in Fig. 3.14, to search the 256 location AM-AM table, upper 14 bits are selected from the digital power control multiplier output and compared against the entries of the LUT. Two entries in the table which are closest to the input (both above and below the input) are selected along with their location—represented by 8 bit address (corresponding to 256 entries). This 8 bit address is the coarse predistortion amplitude value which can be applied to the DPA to produce the original amplitude desired in linear fashion. As mentioned earlier, 8 bits are not sufficient to meet the quantization requirements of the amplitude path and hence additional interpolation is used. The linear interpolation increases the resolution at the predistortion output adding two or more bits depending upon the value of R as explained earlier.

Thus for $R > 1$, the interpolation module runs at $4 \times R$ clock, producing R additional linearly interpolated values between $AM(N-1)$ and $AM(N)$. Each time, a small increment to $AM(N-1)$ is added based on R and the new value is passed to the comparator which also operates at the faster rate and selects a finer resolution point among the sequentially generated R additional values. This exact implementation is shown in Fig. 3.14. The control logic generates the address to create $AM(N-1)$ and $AM(N)$ from the AM-AM LUT from the counter module. Adaptive interpolation takes these values and provides the interpolated signal as shown in Fig. 3.16. The comparator makes the decision once the right value is found and stop the counter instrumentation. The selected values are registered and provided as the output of the block.

The AM-PM distortion is additive in nature. In other words, the phase distortion can be represented as $Distorted\_Phase = Original\_Phase + AM\_PM\ (Original\_Amplitude)$. Hence the phase distortion can be corrected by just subtracting the distortion from the phase signal. The address for the LUT is already know by the AM-AM reverse look-up. Hence the same address is used to identify the correction value for the AM-PM by using direct address-based data reading from memory. The AM-PM distortion is added out of phase with the expected DPA distortion thus canceling out the overall phase distortion. The same amount of interpolation is used on the AM-PM table as well to achieve fine cancelation of the distortion as shown in Fig. 3.14 [78].

A key feature of the presented scheme is the use of normalized AM-AM response. The predistortion does not require absolute measurement of the DPA output voltage. The predistortion uses normalized AM-AM curve always covering the full range from 0 to 1023 because the absolute distortion is not needed for the digital predistortion to work. This helps simplify the self-calibration scheme because the receiver gain does not have to be accurately known.

3.2.3 ANALYSIS OF THE QUANTIZATION EFFECTS

In order to minimize area and current consumption, the word-lengths at the various points in the predistortion module, as well as corresponding clock rates, are selected through careful analysis and optimization [96]. This section presents some of the simulation-based analysis used to determine the sizes of LUTs and the word-lengths for the presented predistortion function.

The area of the presented LUT-based predistortion implementation is dominated by the size of the RAM used to store the LUT entries. In this implementation, two

3.2 Predistortion module

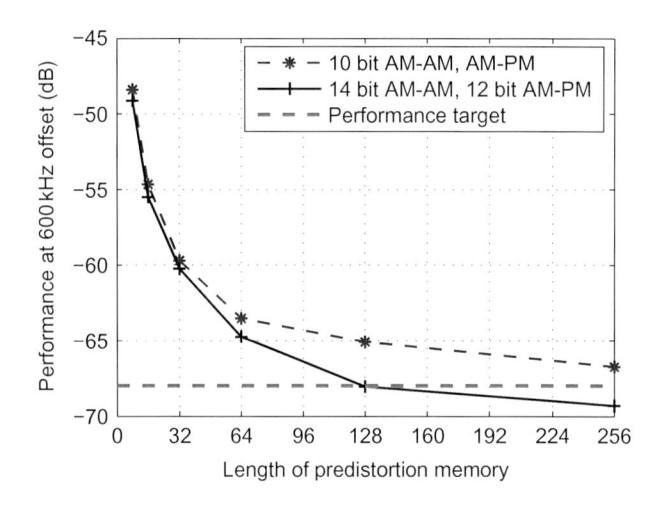

FIG. 3.17

Simulated TX spectral at 600 kHz offset for various LUT lengths.

LUTs are used for storing the AM-AM and AM-PM characteristics. As shown in Fig. 3.17, the spectral performance of the transmitter greatly depends on the size of the LUT. Fig. 3.17 shows the size of LUT without using any form of compression. Various compression schemes have been suggested to optimize the LUT storage by companding the entries [93, 97, 98]. But generally the execution of such methods add extra latency and it is difficult to use such methods along with the content-address type addressing. In addition, these techniques are mostly advantageous when the LUT sizes are much larger, such that the additional complexity associated with the compression and expansion would be worthwhile. For the calibration to be based entirely on resources internal to the SoC, it is advantageous to minimize the required processing of the self-measured distortion characteristics in the realization of predistortion. For these three reasons, use of the compression is avoided for the presented design.

To analyze the impact of various quantization and sampling parameters, 600 kHz offset spectral performance is chosen as a guiding specification. Generally the far-out spectrum is not impacted by the quality of the predistortion used unless there are gross errors in the LUT configuration. EVM is dependent on the correlation between the actual distortion and the predistortion curves but not on the quantization and sampling parameters. And 400 kHz offset is typically dominated by factors other than quantization such as AM-PM characteristics in the LUTs, amplitude path DC offset, amplitude-phase path delay alignment, etc. These factors have a lesser effect at frequency offsets above 600 kHz and hence 600 kHz performance is typically dominated by the quantization noise of the amplitude data path directly. This fact is further highlighted in the measurements results of Fig. 3.13. Hence 600 kHz offset performance has been chosen as the guiding specification in Fig. 3.17. It illustrates that the minimal size for the LUT, to ensure that the performance target is met (with 8 dB margin), is 128 entries, and hence, to allow sufficient margin, its size was set at 256 words. With 256 entry LUT, the 600 kHz performance is met with about 9 dB of margin. Other parameters such as EVM, 400 kHz offset performance, and far-out spectrum (WBN) are found to be satisfied with sufficient margin.

Fig. 3.17 also compares the performance between two cases of different word-lengths used for storing the distortion information—*(a)* word-length of 10 bits for both AM-AM and AM-PM LUTs and *(b)* 14 bits for the AM-AM LUT and 12 bits for the AM-PM LUT. Other combinations of bit-widths are also tried and these two results are presented here for the illustration purpose. Based on these simulation results, the word-lengths selected for the AM-AM and AM-PM LUTs are 14 bits and 12 bits respectively. The AM-PM distortion is additive in nature and thus having a lower resolution on the AM-PM LUT is not directly affecting the spectrum/EVM performance. On the other hand, the amplitude path output resolution is directly proportional to the number of bits used for storing the AM-AM distortion in the LUT. Coarser resolution on the amplitude path will result in lower number of effective bits at the output of the AM-PM predistortion. These word-lengths ensure adequate coverage of the required dynamic range while also meeting the resolution requirements at low power levels [78].

3.3 PREDISTORTION SELF-CALIBRATION

The DPA can be designed to minimize the AM-AM and AM-PM distortions at the expense of increased power consumption, lower efficiency and increased area. Instead, digital signal processing-based techniques are used to effectively reduce the impact of these distortions on the transmitter performance. Although digital signal processing techniques can compensate for the analog/RF shortcomings, it requires significant characterization of the analog/RF circuits to be able to characterize and compensate the distortions. The increase in the test time and complexity, along with the need for more sophisticated external equipments, has resulted in calibration cost being a significant portion of a cellular handset budget. Hence its extremely desirable to reduce the calibration time and also reduce the equipments necessary to perform the desired testing. The best way to simplify the calibration procedure is by using built-in self-test (BIST) methods for analog and RF circuit calibration involving use of on-chip modules without use of external resources. The presented transmitter is one such example of a cellular handset where BIST has been one of the design priorities at the beginning of the design cycle [79, 99].

On-chip measurements of the AM-AM and AM-PM require active use of the on-chip processor in coordination with the on-chip receiver as shown in Fig. 3.18. It is imperative that the transmitter does not produce any real on-air transmission at the PA output during the self-calibration procedure. One way to ensure this is by turning off the external PA. However, it would change the load as seen by the DPA and thus affect

3.3 Predistortion self-calibration

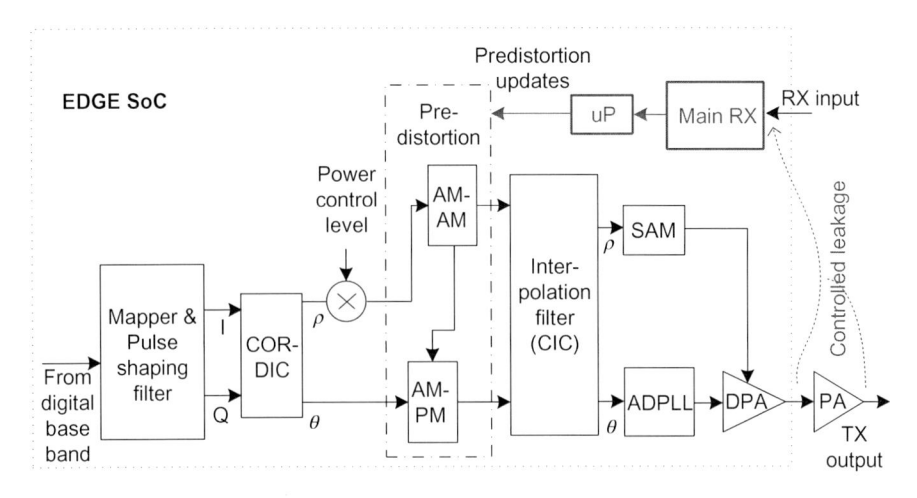

FIG. 3.18

Block diagram of the polar EDGE transmitter showing the predistortion self-calibration scheme.

the characterization outcome. Since the PA is designed with multiple gain stages, an arrangement has been made for the first stage of the PA to be turned on, while the remaining stages, including the final output stage, can be turned off. This mode of the power amplifier allows realistic loading and the impact of output impedance variations of the DPA to be accurately characterized by the on-chip receiver.

One of the key challenges in a typical on-chip receiver-based built-in self-calibration (BISC) scheme is the accurate calibration of the coupling path and the receiver path gains. The gain of the receiver changes over PVT and frequency and hence the overall complexity of the DPA calibration will exponentially increase, especially when it also requires characterization of both the receiver and coupling paths. To overcome this problem, a normalization-based AM-AM predistortion is used as explained in Section 3.2.2.2 that does not rely on the measurement of absolute values of the DPA output voltage for the linearization. The predistortion works by normalizing the peak voltage of the AM-AM response to the 2^{14} code and thus it operates on the relative scale throughout its operation. If the receiver is not able to cover the complete dynamic range of the AM-AM curve (about 60 dB), then the digital code range can be broken down into two or more separate sections with different receiver path gains by having different gain settings of low-noise amplifier (LNA). The partitioning of the code range contains overlap between the sections so the full AM-AM and AM-PM information can be obtained by "stitching" various sections together without knowing the absolute gain settings of the receiver. The I/Q mismatch and gain calibrations that are typically required for the receiver functionality would be performed anyway in advance during the receiver calibration [87] and do not need to be repeated for the transmitter DPA calibration.

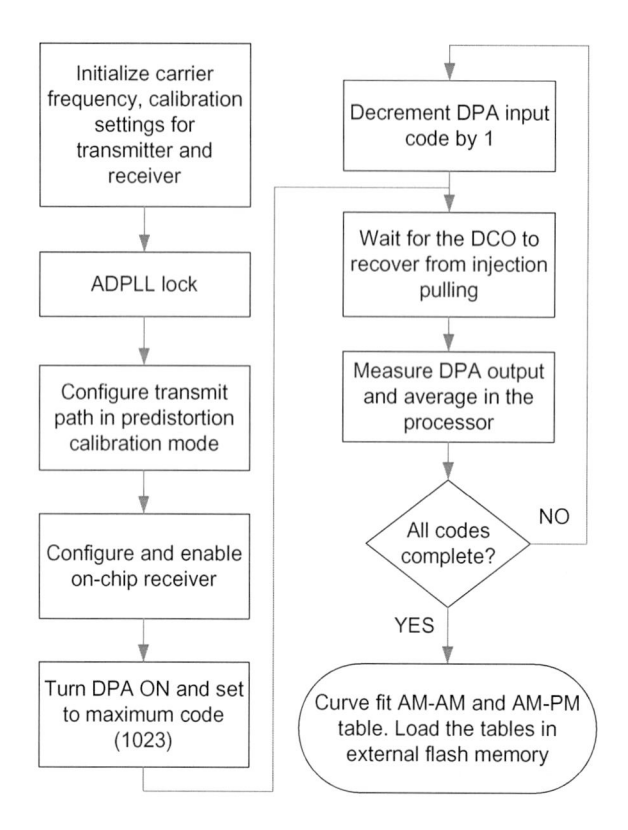

FIG. 3.19

Flowgraph of predistortion self-calibration.

Accurate estimate of the AM-PM distortion requires the ability to measure the minor changes in the phase of the DPA output synchronously [89]. Since the transmitter and the receiver are using the same DCO-based clocks, the synchronous demodulation of the phase information is achieved by design. The phase measurements have to be performed after the ADPLL has settled from the injection pulling by waiting for sufficient settling period. Setting the DPA to maximum code of 1023 before starting the measurement allows the low-dropout (LDO) linear regulators to settle and also reduce the effect of injection pulling to a small extent. This is shown graphically in Fig. 3.19. First the carrier frequency and the loop parameters for the ADPLL are initialized using the microprocessor. Once ADPLL lock is detected using the internal digital status flag, the transmit path is configured for the self-calibration mode, along with the receiver mode. As mentioned earlier, the maximum code on the DPA is set to 1023 and decremented by 1 in each measurement. The final table is compiled with curve-fit-based filtering before storing it in external flash-based memory.

The BIST-based scheme can perform long-term time averaging while taking the measurement and thus the measurements can be completed in a single iteration which would span over larger duration. The measurement of each DPA code takes about 200 μs, resulting in a total of 100 ms for the 256 step AM-AM and AM-PM characterization.

The amplitude signal, covering the range of −60 dBm to +10 dBm at the matching network output, is fully characterized using the BIST-based scheme. The AM-AM data is passed through fifth order polynomial-based curve fitting to reduce the measurement noise from the INL characteristics of the DPA. Since the AM-PM curve has sharp response at the beginning of the curve shown in Fig. 3.11, polynomial-based curve fitting does not represent the original curve accurately and hence FIR-based filtering is used to smoothen the data.

When the predistortion is to consider the distortions of the external PA as well, the coupling to the on-chip receiver can be achieved either through parasitic paths or, alternatively, an additional coupler can be added for the feedback [100], whereas in transmitters having an on-chip integrated PA, the output signal is available on-chip, such that the feedback is entirely internal [79].

3.3.1 EFFECT OF TEMPERATURE VARIATIONS ON PREDISTORTION

Characterization over temperature revealed that the change in temperature only affects the gain of the AM-AM curve but its overall shape remains unchanged. Similarly, for the AM-PM distortion, only a fixed phase shift was observed between the curves of multiple temperatures as shown in Fig. 3.11. In case of AM-AM, the presented scheme works on the normalized AM-AM response and does not take the absolute value of the output voltage into account. Likewise, the phase offset between the AM-PM curves is ignored since the modulation is insensitive to it, and the presented scheme is therefore effectively immune to the effect of temperature variations on the DPA used in this transmitter. In a general solution where there may be a significant change in the AM-AM and AM-PM characteristics over temperature, periodic re-calibration may be required.

3.4 INTERPOLATIVE FILTER

The digital nature of the amplitude modulation signal fed to the DPA eliminates the need for any continuous-time analog filtering but requires careful selection of the sampling rate and the digital filtering this signal undergoes. The interpolation filter used for the amplitude and phase path is derived from the basic cascaded integrator-comb (CIC) structure [80] with a few significant modifications. A traditional CIC filter is shown in Fig. 3.20A. CIC filters are widely used for decimation and interpolation in communication systems. One of their key advantages is the simplicity in implementing the filter as it doesn't require any multiplier or divider and uses limited memory elements. The filter operates in power of 2 operations so the

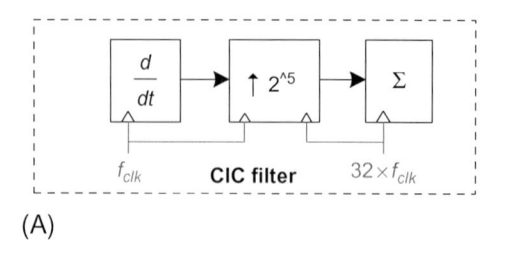

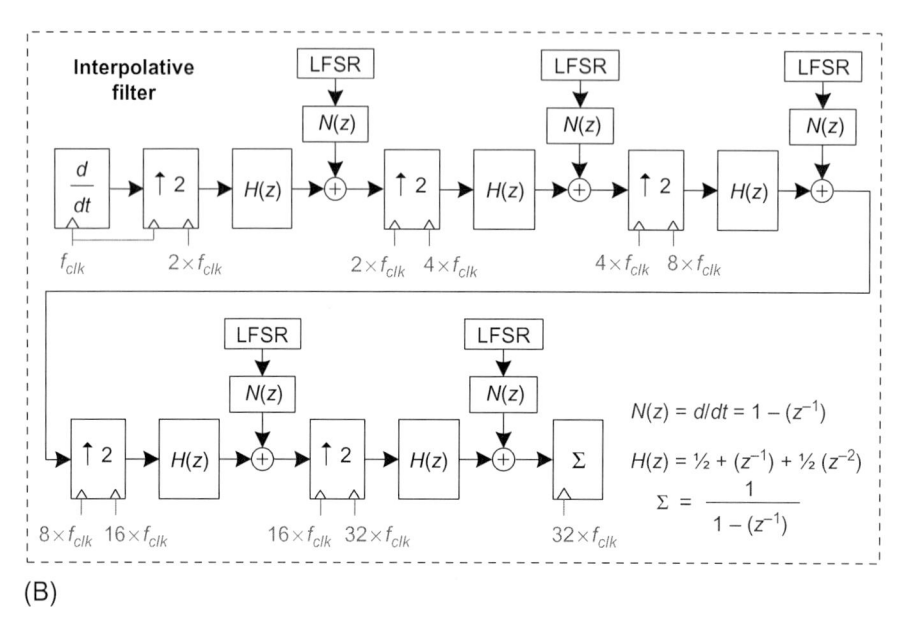

FIG. 3.20

Comparison of interpolative filter with CIC filter.

multiplications and divisions can be implemented by binary shifting. The integrator is essentially a digital accumulator while the comb is a differentiator. Thus the filter is an ideal candidate for high sampling rate filtering requirement such as the high-speed data path in the presented transmitter.

For a basic CIC filter with a factor of 2 upsampling, the z-domain response can be given by the following equation:

$$H_{CIC}(z) = \frac{1 - z^{-2}}{1 - z^{-1}}$$
$$= 1 + z^{-1} \qquad (3.5)$$

which provides first order filtering for the replica at first multiple of input sampling frequency. This amount of attenuation is not sufficient to suppress the replicas for

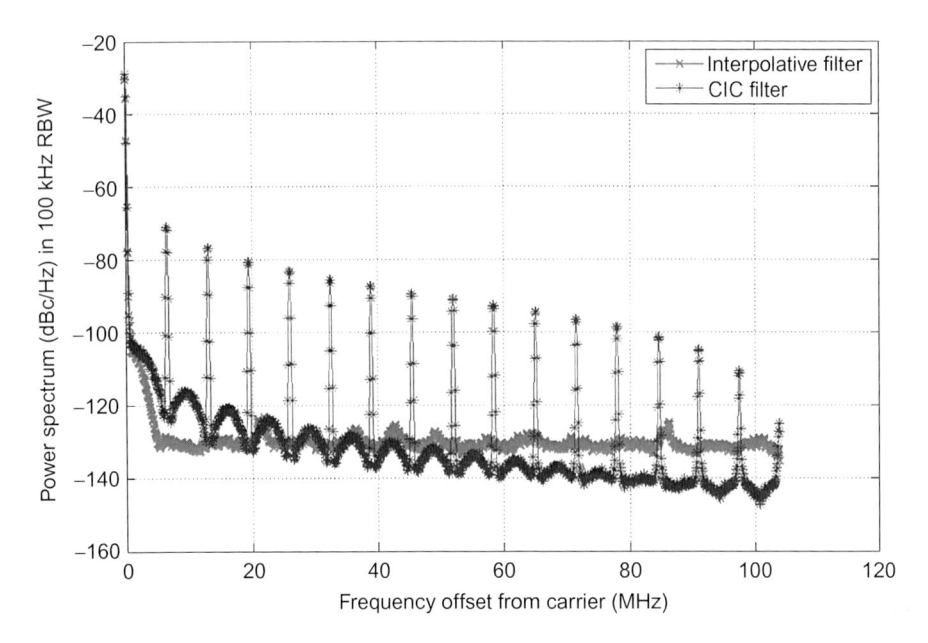

FIG. 3.21

Comparison of spectral response of Interpolative Filter and CIC filter without additive noise.

a cellular transmitter application as shown by the CIC filter response at low power (−12 dBm) in Fig. 3.21. Hence the presented interpolation filter uses two additional enhancements to the basic CIC filter to reduce the level of replicas as shown in Fig. 3.20B.

1. Linear interpolation: Both filters use differentiator as the first element and integrator as the last element but the design of the upsampler is significantly different. CIC uses zero-insertion-based interpolation where, for the example of upsampling by 32, 31 zeros are inserted after every input sample. Instead of zero-insertion, linear interpolation is used for the interpolation filter which provides two higher order of attenuation for the replicas at the frequency multiples of f_{clk}.

To illustrate the difference in z-domain, consider the same example of a filter with factor of 2 upsampling (Eq. (3.5)). The linear interpolation, replacing the zero-insertion, provides an additional $H(z)$ filtering cascaded with rest of the interpolation filter transfer function as shown in Fig. 3.20B. With the linear interpolation, the total response is given by

$$H_{IF}(z) = \frac{1-z^{-2}}{1-z^{-1}} \times \left(\frac{1}{2} + z^{-1} + \frac{1}{2}z^{-2}\right)$$
$$= \frac{1}{2} \times (1 + 3z^{-1} + 3z^{-2} + z^{-3})$$
$$= \frac{1}{2} \times (1 + z^{-1})^3 \qquad (3.6)$$

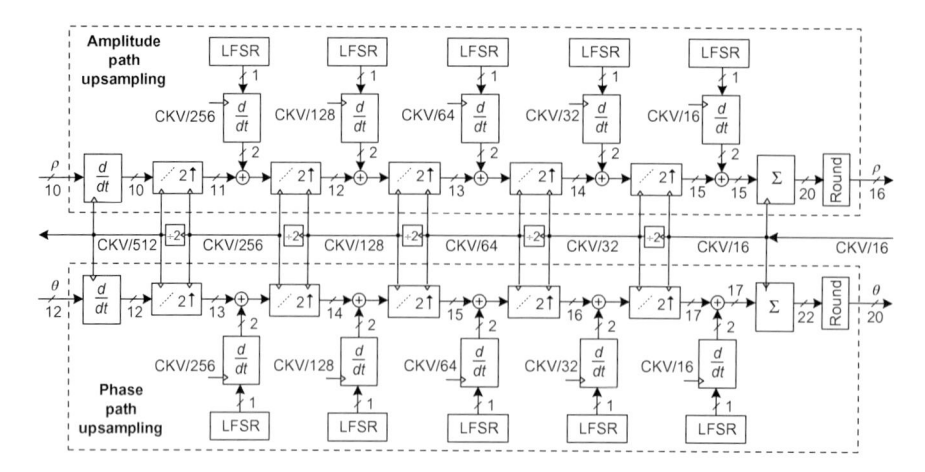

FIG. 3.22

Interpolative filter block diagram.

which gives two higher order of attenuation for the replicas compared to the basic CIC filter. The response further improves as multiple stages are cascaded as shown in Fig. 3.20B. The responses of CIC filter and the presented interpolative filter for 5 stage interpolation filter are shown in Fig. 3.21. As it can be seen, the replica suppression of the CIC filter is not sufficient and it is significantly worse than the spectral performance of the interpolative filter. The replica suppression of CIC filter is in order of ZOH response but the interpolative filter provides two order higher filtering than CIC and hence the replicas are significantly suppressed although not completely invisible. Once the finite resolution of the $\Sigma\Delta$ is introduced, it causes increase of the replicas as shown in Fig. 3.23 using the curve without highpass noise addition. More details of highpass noise addition are covered in next subsection.

The linear interpolation is broken down into five stages—each providing a factor of 2 linear interpolation. The $H(z)$ is implemented on every interpolation stage and attenuates all the multiples of f_{clk} as they are generated. The linear interpolation ($H(z)$) can be implemented using shift-and-add mechanism as well. Thus the interpolation filter does not use any multiplier or divider just like CIC filter so its extremely area efficient for the amount of attenuation it provides. The reason for implementing multiple stages of filtering is to allow highpass noise to be added later as covered in following section.

The bitwidth growth of the interpolation filter is primarily driven by the extra equivalent resolution bits added by the linear interpolation. Each linear interpolator adds 1 bit at the interpolator output and 1 bit at the accumulator output—total of 2 bits. Hence for the presented 5 stage interpolation filter, 10 bits are added on both amplitude and phase path. Truncation or rounding cannot be used on the intermediate stage outputs otherwise the errors will grow out of bound in the integration stage [80] so all the rounding is lumped at the output of the integrator.

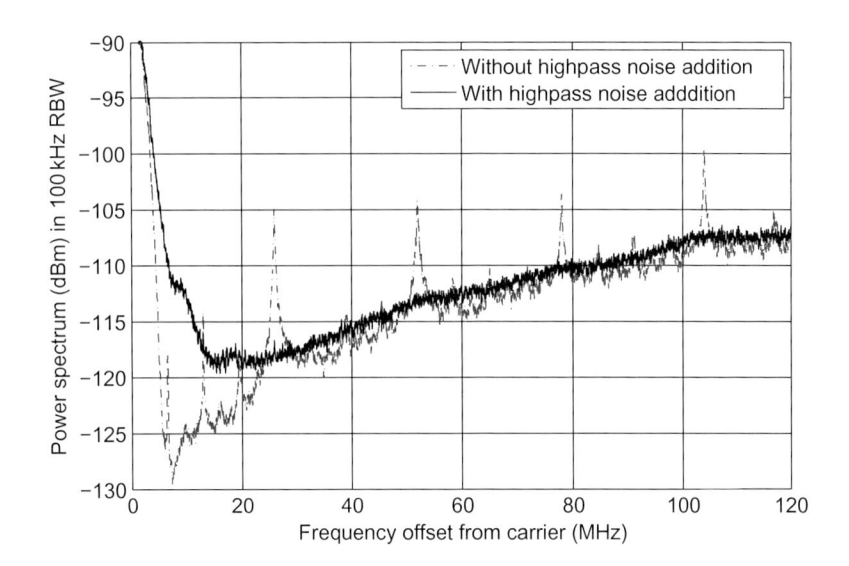

FIG. 3.23

Effect of highpass noise addition on the spectrum at low power.

Even with the addition of linear interpolation, the rounding at the output causes the replicas to re-appear as shown in "without highpass noise addition" curve of Fig. 3.23. The power control is implemented digitally by using a multiplier as shown in Fig. 3.2 so the effective number of bits available at low power decreases. This causes the replicas to become more prominent relative to the carrier power at low power levels. To eliminate this problem, the replicas are spread in frequency using dithering.

2. *Randomization of replicas*: High pass, zero mean, white noise, in the range of $(-1, 0, +1)$ is added to each stage of the linear interpolation output to randomize the LSB and thus spread the replicas introduced during the linear interpolation as shown in Figs. 3.20B and 3.22. The white noise is generated by LFSRs and it is differentiated to generate the highpass shaped noise. It is essential to perform this operation right after every linear interpolation to ensure appropriate spreading at the desired sampling frequency.

The spreading is especially effective at lower power levels where the number of effective bits are reduced as shown in Fig. 3.23. The $\Sigma\Delta$ modulator is not able to provide perfect 6 bit quantization resolution as it will be covered in Section 4.2 and hence some of the replicas that were suppressed with 16 bit resolution of Fig. 3.21 start showing up as shown in the curve labeled "Without highpass noise addition." The addition of the highpass noise effectively eliminates this replicas as well. Although there is a small increase in the noise at lower frequency offsets around 10 MHz because of addition of frequency shaped noise, this noise is typically not the dominant contributor at these frequencies. In realistic situations, the spectrum is

Table 3.2 Impact of Delay Mismatch Between Amplitude and Phase on 400 kHz Offset Spectral Performance

Delay Mismatch (ns)	400 kHz Offset Spectral Performance (dB)
0	−69.61
4.81	−69.57
9.62	−68.31
14.43	−66.73
19.24	−65.25
24.05	−63.91
28.86	−62.71
33.67	−61.63
38.48	−60.65
43.29	−59.77
48.10	−58.96

generally contaminated by other factors such as mismatch, DCO phase noise, etc. and hence the additional noise does not result in any significant degradation.

The direct cascaded power-of-two divisions of the DCO RF clock (CKV) are used to clock this function. A similar filter on the phase path provides upsampling and filtering while ensuring that the amplitude and phase signals undergo nearly identical transfer functions, to minimize distortions as they are recombined in the DPA.

One of the key requirements for a polar transmitter is the group delay that amplitude and phase path experiences from that point of separation (CORDIC) till the re-combining in the DPA. The delay has to be matched as closely as possible to ensure reliable reconstruction. In case of EDGE $3\pi/8$-shifted 8-PSK modulation at 270.833 kHz symbol rate, amplitude and phase path have to be matched very well within each other. Table 3.2 shows the simulation results for delay mismatch between amplitude and phase. The internal specification for most of the modules for 400 kHz specification is −66 dB and hence its necessary to have the delay matching to contribute less than −66 dB. As it can be seen from the table, the two signals have to be aligned within 10 ns of each other to avoid any noticeable degradation of the spectrum after re-combining at the DPA level.

While the delay through digital circuits can be estimated with reasonable accuracy, the analog circuit delays are not so easy to estimate. To provide flexibility for tuning the delay, a coarse and fine delay control circuits are used as part of the data path. The coarse delay has the resolution of CKV/512 clock in order of 6.5 MHz. The fine delay control is implemented by adding a programmable delay at each of the linear interpolation stage in the interpolation filter. The fine delay works as binary programmable delay with CKV/512 being the coarsest clock and CKV/16 being the finest clock. There is an additional delay control element after the interpolation filter the increases the resolution to CKV/8 clock period. CKV/8 clock has frequency in the order of 200 MHz which provides better than 5 ns delay alignment steps.

Fig. 3.22 also shows the details of implementation of the interpolative filter. The differentiation is implemented by using a subtractor that subtracts the previous value from the current value of the amplitude or phase signal. Interpolation by a factor of 2 is implemented by having one adder that averages the value between two adjacent samples to find the newly interpolated intermediate value. All the LFSRs are 9 bit XOR gate-based LFSRs [101] generating the randomized bits. Accumulators are traditional accumulator with an adder and a memory element contangoing the accumulated value.

3.5 POLAR BANDWIDTH EXPANSION

One of the main challenges for a polar architecture is the problem of polar bandwidth expansion. Referring to Fig. 3.24, the I and Q signal spectral content is limited to the channel bandwidth of 200 kHz and the spectral content is around −50 dB at 250 kHz offset from carrier. These are the characteristics of a well-contained signal that can be sampled with any frequency higher than 1 MHz without causing any loss of information as explained by the Nyquist sampling theorem. After going though the Cartesian to polar conversion, the resultant amplitude and phase signals are not

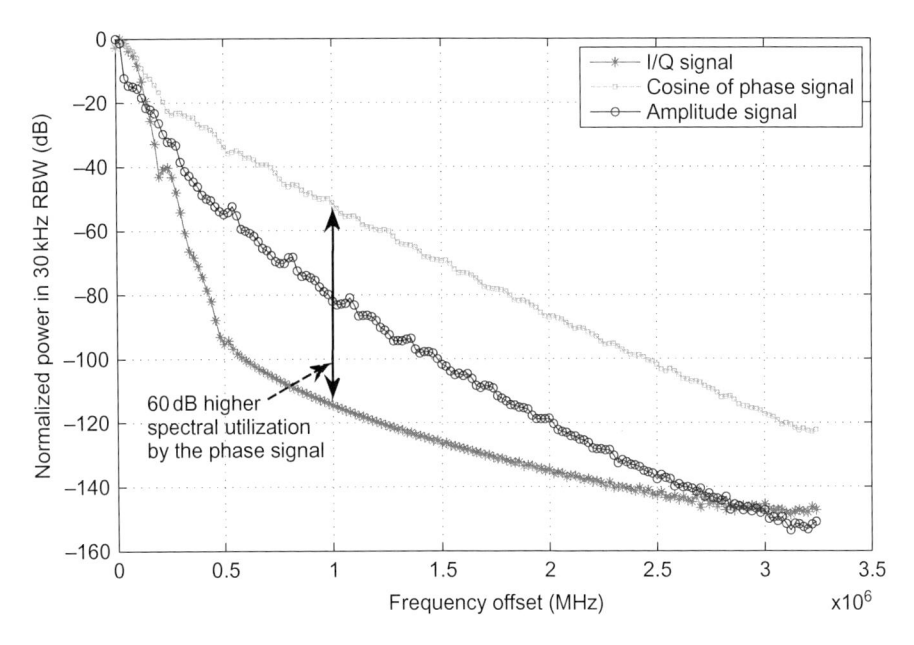

FIG. 3.24

Polar bandwidth expansion of the phase signal due to conversion from Cartesian to polar coordinates.

utilizing the same bandwidth that is utilized by the I and Q signal. As shown in Fig. 3.24, for case of EDGE, the phase signal spectrum contains spectral content of higher than −50 dB till about 1 MHz of spectrum. In contrast to the I and Q signal, this is much wider utilization of the bandwidth. This phenomenon is termed as the polar bandwidth expansion problem. For EDGE modulation, about 3.25 MHz of sampling rate is minimally necessary to ensure the fidelity of the reconstructed complex RF signal. This puts a lower limit of 3.25 MHz sampling rate for CORDIC which is used to convert the information from Cartesian to polar coordinates. Thus the sampling rate of CORDIC highly depends on the factor by which the bandwidth is expanding.

The amount of bandwidth expansion depends on the maximum frequency modulation for a given standard and the type of modulation used. Thus the polar bandwidth expansion limits the use of polar architecture for some of the newer standards (e.g., WCDMA) using wider frequency modulation. For these standards, the rate of operation for CORDIC and all the modules that operate in polar coordinates becomes very high causing increased power consumption and higher area.

CHAPTER 4

RF front-end (RFDAC) of the polar transmitter*

CHAPTER OUTLINE

4.1 Overview of the RF Front-End	78
4.2 ΣΔ Amplitude Modulation	80
4.2.1 ΣΔ Overview	80
4.2.2 Digital Design	81
4.2.3 Transfer Function and Spectrum	82
4.3 Digital Pre-PA	83
4.3.1 Overview of DPA Functionality	83
4.3.2 Analysis of DPA Quantization Noise	84
4.3.3 DPA Structural Design	91
4.4 DPA Transistor Mismatches	92
4.4.1 Amplitude Mismatch	92
4.4.2 Phase Mismatch	93
4.5 Key Categories of Mismatches and DEM	96
4.5.1 Key Categories	96
4.5.2 Simulation-Based Specifications	101
4.5.3 Dynamic Element Matching	102
4.5.4 Measurement Results	104
4.6 Clock Delay Alignment	106
4.6.1 Explanation of the Problem	106
4.6.2 Self-Calibration and Compensation Mechanism	108
4.7 Analysis of Parasitic Coupling	112
4.7.1 Possible Coupling Paths	112
4.7.2 A Novel Method of Characterizing ΣΔ Parasitic Coupling Using Idle-Tones	113
4.7.3 Relationship Between Idle Tones and ΣΔ Parasitic Coupling	114

---

*The authors acknowledge substantial contributions from Prof. Dr. Poras T. Balsara (UT Dallas) and Dr. Oren Eliezer (Texas Instruments).

CHAPTER 4 RF front-end (RFDAC) of the polar transmitter

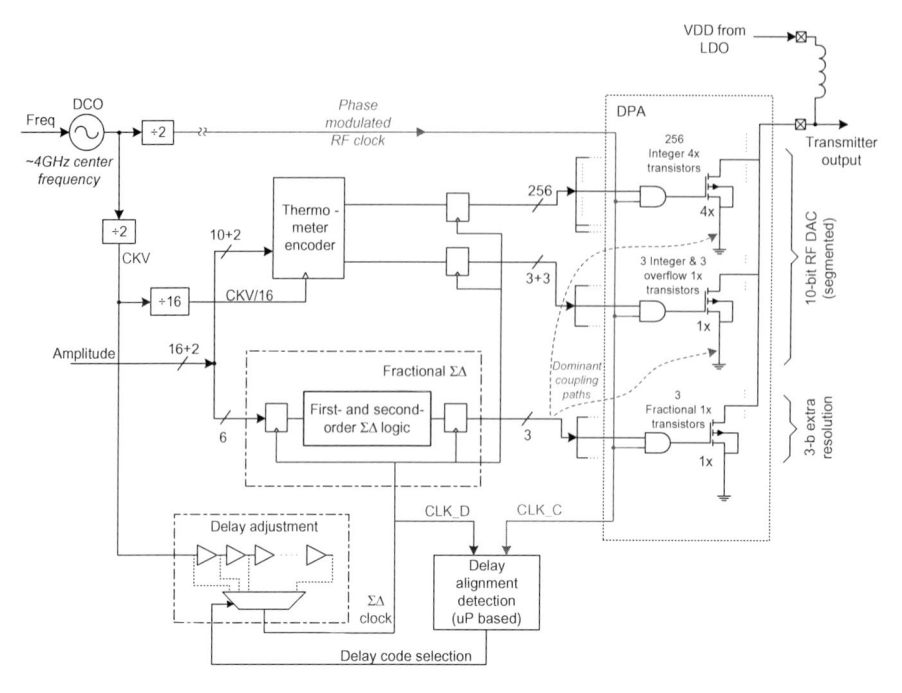

FIG. 4.1

Schematic and block diagram of the front-end.

4.1 OVERVIEW OF THE RF FRONT-END

This chapter discusses the digital RF front-end (RFE) of the presented transmitter. Since the architecture is highly digital in nature, the front-end is not similar to a traditional analog-RF-based transmitter. Majority of the modules shown in the RFE block diagram of Fig. 4.1 are completely digital modules designed using digital VLSI design flow. The $\Sigma\Delta$ module is one such block that is implemented using VLSI design flow and synthesized to run at nearly 1 GHz frequency (CKV/2).

$\Sigma\Delta$ modulation has been traditionally used to increase the effective resolution of a signal without requiring additional fractional bits for DAC and ADC designs. The $\Sigma\Delta$ modulation represents an effective trade-off between the sampling rate of the signal and quantization noise in the band of interest. Increasing the sampling rate allows the noise to be shaped further outside the band of interest and thus effectively resulting in higher signal to noise ratio (and thus higher number of effective bits) in the band of interest.

Traditional transmitters use $\Sigma\Delta$ modulation in the DAC to convert the digital transmitter information signal to the baseband analog signal. This baseband signal is later upconverted using mixer to achieve the complex RF modulation as shown in Fig. 1.5. To achieve the goal of fully digital implementation without requiring analog mixers, the DAC functionality has to be pushed as far down the datapath as possible. Since the polar transmitter combines the amplitude and phase information in the mixer as well, the optimum implementation would combine the DAC functionality, amplitude-phase combining and the RF mixer in the last stage to avoid using any analog centric components earlier. This precise goal has been achieved by the presented digital prepower amplifier (DPA).

The RFE receives digital amplitude information from the output of the DTX module, sampled at about 200 MHz (CKV/16), provides the interpolated amplitude with significantly attenuated replicas. The amplitude information is represented using 16 bit resolution, out of which 10 bits are integer (mapping to 1024 levels of the DPA) and 6 bits are fractional as shown in Fig. 4.1. In addition to the 10 integer datapath bits, 2 additional integer bits are used for additive mismatch correction as explained in Section 4.4. The integer bits are passed through a thermometer encoder block which maps the bits to appropriate control signals to be used for controlling the 1024 steps of the DPA transistors. The 1024 steps are achieve by using 3 unit weighted ($1\times$) transistors and 256 transistors with $4\times$ unit weigh strength as shown in Fig. 4.1 and further elaborated in Section 4.3.

The six fractional bits are modulated using the fractional $\Sigma\Delta$ modulation as shown in Fig. 4.1. Since fractional resolution cannot be directly achieved by using smaller size of the transistors due to process technology limitations, $\Sigma\Delta$ modulation converts the fractional bits to equivalent integer bits streams sampled at the carrier frequency.

The DPA output power is specified to be 10 dBm for LB and 8 dBm for HB. This is achieved using the above-mentioned 1024 step integer transistors with full digital control. Thus the power control and envelop amplitude modulation are achieved using the RFDAC part of the DPA functionality. Depending upon the power control word, provided by the digital baseband, the digital amplitude control word is scaled to represent the desired RMS power as shown in Fig. 3.2. For GSM path, with fixed output power, the RMS power is represented by a fixed digital code word which can be programmed to match the input power requirement of the power amplifier stage.

The DPA is also used to combined the amplitude and phase information by mixing the phase modulated RF clock with the amplitude controls as shown in Fig. 4.1. The RF clock is a digital clock, with a center frequency equal to the RF carrier frequency, along with desired phase modulation represented in variations of the pulse widths. In other words, the instantaneous frequency of the RF clock is equal to the RF carrier frequency plus the instantaneous phase modulation information. This clock is used to sample the amplitude information by using the equivalent AND gates shown in Fig. 4.1. The resulting signal is used to drive the 256 $4\times$ and 3 $1\times$ transistors of the DPA. The drain terminal of all the transistors in the DPA is connected to an internal LDO output through an external matching network. This LDO output is used as a supply voltage (V_{DD}) for the DPA module.

All the circuits in the digital RFE are operating on the clocks derived from the DCO phase modulated clock as shown in Fig. 4.1. Although the clock contains phase modulation, the highest modulation frequency is about 350 kHz for EDGE standard. For a carrier frequency of 900 MHz, this translates to only 0.04% jitter on the clock which is negligible for all the transmitter digital circuits. Additionally, the same phase modulated clock is used in the DPA so there is no loss of information or increase in noise due to this jitter. Loss of information or increase in noise can only occur if the signal sampled with this phase modulated clock is resampled with any other fixed clock. In this case, there is no re-sampling from the phase modulated clock to fixed clock and hence there is no disadvantage of using the phase modulated clock. Using the phase modulated clocks allows synchronous operation between multiple RF modules eliminating the problems associated with multiple clock domains such as the spurs arising from the mixing of the two clocks. Additionally, it eliminated the need for any resampling in last stages because the DPA has to operate at the phase modulated clock. At frequencies in order of 2 GHz, the propagation delay through the chip can affect the performance of the RF spectrum and hence a new clock delay alignment scheme (Section 4.6) is introduced to ensure optimum performance.

The rest of the chapter highlights the innovations of various modules used in the RFE, explain some of the challenges faced in the fully digital design and propose their solutions. Various mismatches and other nonidealities of the DPA are analyzed. Since the concept of RFDAC is new, no such other analysis is known to the authors. Through a system-level analysis, specifications are provided to meet the targeted spectral compliance requirements. Following the specifications, calibration and compensation schemes are introduced to overcome the analog design and layout limitations.

4.2 $\Sigma\Delta$ AMPLITUDE MODULATION
4.2.1 $\Sigma\Delta$ OVERVIEW

The presented transmitter uses $\Sigma\Delta$ modulation on fractional bits to increase the resolution of the amplitude and phase path by adding subunit transistor amplitude contribution for AM path and subunit varactor frequency deviation for the phase path. This section describes the amplitude path $\Sigma\Delta$ modulation based on Ref. [46]. Phase path $\Sigma\Delta$ is covered in Ref. [14].

A traditional $\Sigma\Delta$-based ADC is shown in Fig. 4.2. The $\Sigma\Delta$ noise shaping is proportional to the oversampling ratio—defined as the ratio of the $\Sigma\Delta$ oversampling clock to the effective signal bandwidth. In the case of ADC, 1-bit ADC is used to quantize the analog signal which becomes the single bit-stream digital output. The same output is used as negative feedback input by passing it thorough 1-bit DAC (typically implemented by a level shifter). The DAC output is compared with the original analog signal to find the instantaneous error. The error is integrated by an analog integrator before being quantized by the 1-bit ADC. A digital filter typically follows the $\Sigma\Delta$ ADC to filter the out-of-band noise generated by the $\Sigma\Delta$ noise shaping.

This concept is modified to create a digital $\Sigma\Delta$ modulator as shown in Fig. 4.3. The analog input is replaced by a quantized digital input. The number of bits used to represent this digital input provided the upper limit on the resolution enhancement provided by the $\Sigma\Delta$ modulator in the band of interest. The actual resolution

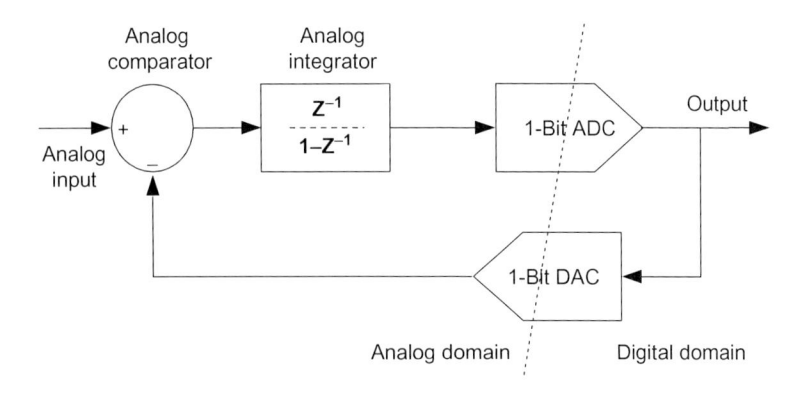

4.2 ΣΔ amplitude modulation

FIG. 4.2
Block diagram of first-order ΣΔ ADC.

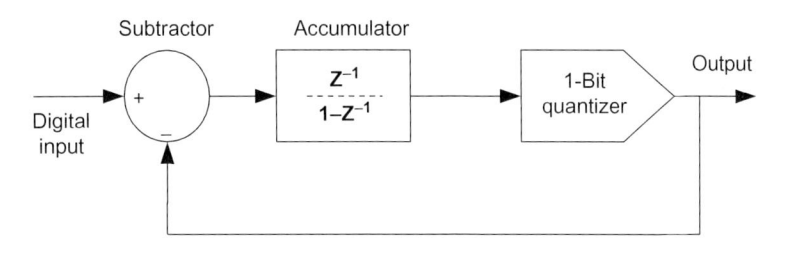

FIG. 4.3
Block diagram of first-order digital ΣΔ modulator.

enhancement depends on the oversampling ratio of the clock as was the case with the analog ΣΔ. The analog integrator is replaced by a fixed point digital accumulator which provided a continuous running digital output. The accumulator output is passed through a digital comparator with 1 bit output to provide the quantized output. It can be shown that the analog and digital ΣΔ modulators behave in identical fashion as far as the noise shaping is concerned. If a DC input is applied at the input of both ΣΔ modulators, the single bit-stream contains repeated pattern with the time average corresponding to the desired input DC value. Thus the ΣΔ effectively generates a pulse code modulated bit-stream with instantaneous time window average corresponding to the instantaneous input value.

4.2.2 DIGITAL DESIGN

The same concept is used to implement the first and second-order ΣΔ modulators for the presented transmitter as shown in Fig. 4.4. The second-order ΣΔ provided higher-order noise shaping while reducing the repetitiveness in the signal. This will be elaborated further at the end of this section.

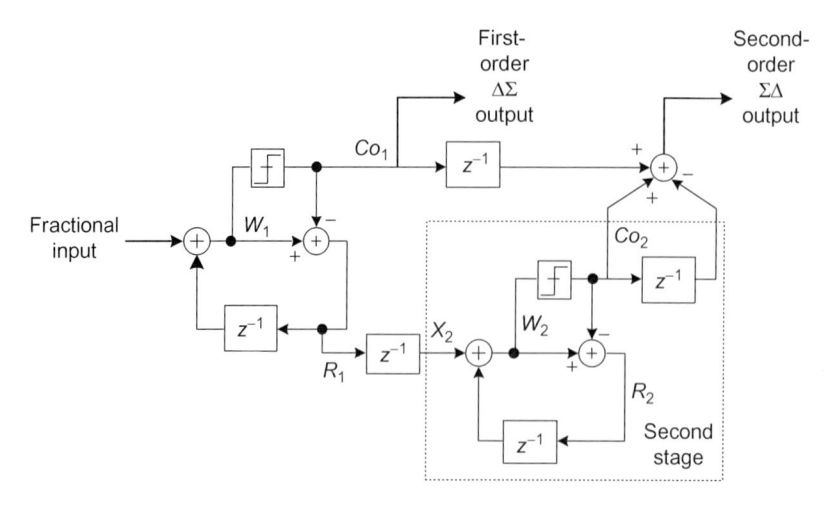

FIG. 4.4

Block diagram of first- and second-order $\Sigma\Delta$.

The digital $\Sigma\Delta$ of Fig. 4.3 can be represented by a digital accumulator with carry bit used as the output of the $\Sigma\Delta$. This concept is shown in Fig. 4.4 in MASH $\Sigma\Delta$ architecture, where the first-order $\Sigma\Delta$ output is directly taken as the carry of the accumulator (Co_1) (shown as the output of the accumulator after passing it through a threshold detection mechanism). The remainder (R_1) is used as an input to the second-order $\Sigma\Delta$ along with the carry signal (Co_1). Second-order $\Sigma\Delta$ digital implementation is identical in structure to the first-order $\Sigma\Delta$ as shown in Fig. 4.4. The output of the second-order $\Sigma\Delta$ is generated by combining the first-order $\Sigma\Delta$ output with the carry of the second-order $\Sigma\Delta$ accumulator as shown in Fig. 4.4. Since the second-order $\Sigma\Delta$ combines three separate 1-bit streams, the output contains three digital levels which are represented by 2-bit integer output. The 2-bit binary output is connected to three adjacent 1× transistors in the DPA fractional transistor bank after going through thermometer encoding as shown in Fig. 4.1.

4.2.3 TRANSFER FUNCTION AND SPECTRUM

A well-known first-order $\Sigma\Delta$ transfer function [102] is given by

$$Y(Z) = Z^{-1} \cdot X(Z) + (1 - Z^{-1}) \cdot E(Z) \tag{4.1}$$

where $Y(Z)$ represents the total transfer function, $X(Z)$ is the input signal, and $E(Z)$ is the quantization noise—approximated to be independent of the input data $X(Z)$. From Eq. (4.1), the quantization noise undergoes highpass transfer function $(1 - Z^{-1})$ thus shaping the noise out of the band of interest. The noise shaping of a first-order $\Sigma\Delta$ is graphically shown in Fig. 4.5. For the band of interest (around 20 MHz+ offset),

FIG. 4.5

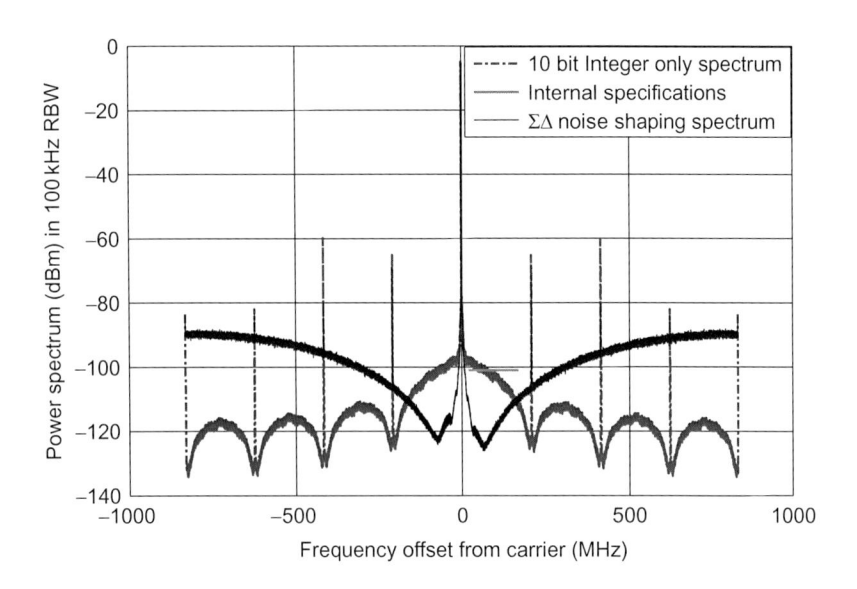

First-order $\Sigma\Delta$ spectrum.

there is about 10 dB of improvement. The transmitter would not be able to meet the specifications without using $\Sigma\Delta$ modulation as shown by the integer-only spectrum.

To achieve high oversampling ratio (OSR), the digital $\Sigma\Delta$ is clocked at the frequency of nearly 1 GHz. The absolute frequency is carrier frequency dependant. For case of lowband (GSM band: 824.2–848.8 MHz, EGSM band: 880.2–914.8 MHz), the clock frequency is the same as the carrier frequency. In the case of highband (DCS band: 1710.2–1784.4 MHz, PCS band: 1850.2–1909.8 MHz), the carrier frequency is passed through a divide-by-2 divider before being used for the $\Sigma\Delta$ modulation. Since the interpolation filter data also runs on carrier frequency dependant clock, the OSR is fixed to 8 by design irrespective of the carrier frequency.

Higher-order digital $\Sigma\Delta$ implementation can provide better noise shaping performance but the number of bits at the output of higher-order $\Sigma\Delta$ increases and the matching between such bits becomes impractical. In addition, the output spectrum is a combination of amplitude and phase spectrums combined in polar fashion. Hence even if the amplitude path represents quantization noise of better than 16 bits by using higher-order $\Sigma\Delta$, the DCO phase noise will dominate for such frequencies.

4.3 DIGITAL PRE-PA
4.3.1 OVERVIEW OF DPA FUNCTIONALITY

The functionality of RFDAC is achieved by virtue of multiple banks of transistor switches, whose RF resistance and current strength is controlled by integer and fractional (i.e., $\Sigma\Delta$ modulated) input codewords [47]. Fig. 4.6 explains the basic operation of the DPA. Drain terminal of *Transistor 1* is connected to the external V_{DD} supply through external load in form of a matching network. The current flowing through the load determined the output AC voltage at the transmitter output. Similar to *Transistor 1*, other transistors of two different sizes—$1\times$ and $4\times$ are connected at their drain terminal and connected to the same output load. *Transistor 2* is shown as an example of an additional transistor connected at drain terminal to simplify the explanation of the DPA operation.

As shown in Fig. 4.6 the phase modulated clock (C) is connected to the gate (G) of the *Transistor 1* through a pass gate. The pass gate is enabled through the data control (D) signal which is driven by the $\Sigma\Delta$ and thermometer encoder output as shown in Fig. 4.1. An additional pull-down NMOS transistor is connected to the gate of *Transistor 1* NMOS to pull the gate to ground when the data control (D) signal is zero. The phase modulated clock (C) signal is identical for all the DPA transistors. But the data control (D) signals are individual for all the transistors enabling or disabling their contribution to the output load.

A sample phase modulated clock (C) waveform is shown at the bottom half of Fig. 4.6. Similarly a data control (D) waveform is shown in the same figure. Notice that the D signal lasts for a whole pulse of the clock C. The data control (D) signal is sampled using the same phase modulated clock as explained in Section 4.6. The gate (G) terminal signal resulting from the pass gate logic is shown to contain phase modulated clock (C) pulses for the duration when the data control (D) is active. This circuit with the pull-down NMOS transistor effective works as a logical AND gate with C and D as inputs and G as the output. A clock pulse at the input of the gate (G) terminal causes the NMOS transistor to turn ON and resulting in current flowing from external V_{DD}, through the load and through the *Transistor 1*, to ground. The current waveform is shown with the label "Load current Transistor 1." Note that even though the drain output (D) is out of phase with the gate (G) signal, the current pulse will be in phase with the gate (G) terminal. The current pulse flowing through the load results in voltage cross the load which is used as the output of the transmitter.

A similar *Transistor 2* current contribution is shown with the label "Load current Transistor 2." Since both the transistors are connected to the same load through the drain terminal, their current is effectively added. This is shown by the last waveform in the figure. Notice the combined current acts as a cumulative contribution of both transistors modulating the amplitude. Extending this to 1024 level DPA transistors, the final output can be generated by combining their amplitude modulated contribution. Thus the operation of the DPA is very similar to the thermometer-coded DAC operation where multiple resistors or current-sources are combined at the output to produce the desired analog signal.

4.3.2 ANALYSIS OF DPA QUANTIZATION NOISE

System-level simulations show the need for 10 integer and 6 fractional bits to cover the full dynamic range of EDGE modulation along with the required 30 dB

FIG. 4.6

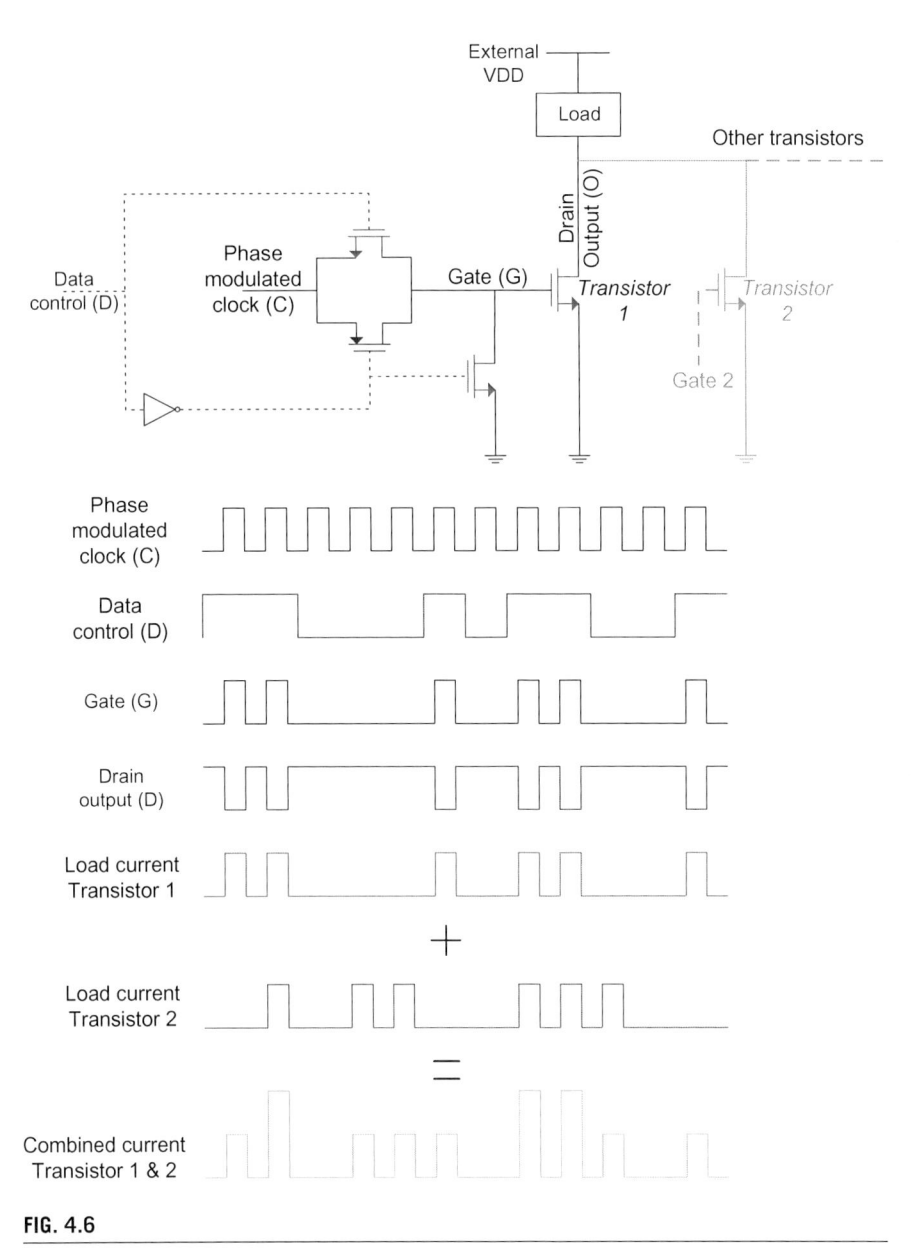

Basic DPA operation.

of power control as shown. This section highlights the selection procedure for various quantization parameters (such as number of integer/fractional transistors, bit-widths, etc.).

The selection of 10 bits for integer portion is driven by the availability of minimum size transistor in the CMOS manufacturing process used for the fabrication of the presented transmitter. The smallest size that can be reliably manufactured while meeting the mismatch specifications (Section 4.5) is referenced as the integer LSB. Adding 1024 such transistor contributions (with compression) (Section 2.4.3) is found to be enough to meet the total output power specifications (Section A.2) of the transmitter. Thus 10-bit binary coded integer amplitude word can control the 1024 transistors and hence 10 bits are used for the integer path. The fractional path resolution is driven by the ability to add fractional resolution using $\Sigma\Delta$ modulator. The 10 bits are mapped to 1024 level using multiple thermometer encoders as covered in Section 4.3.3.

Fig. 4.7 shows the cumulative power consumptions of the 1024 steps in the DPA design. The partition of the 1× and 4× transistors is covered in Section 4.3.3. Fig. 4.7A shows the contribution of 1024 transistor steps in cumulative fashion. Notice the first transistor adds about −64 dBm output power once turned ON. The peak power reached with the 1024 steps is +5.7 dBm. The same graph is shown on logarithmic x-axis in Fig. 4.7B. As it can be seen in this figure, the voltage contribution for initial codes is fairly linear and hence the power contribution looks linear when x-axis is in logarithmic steps. Toward the end of the codes, compression becomes very clear.

First 64 transistor steps are shown in detail in Fig. 4.7C with the logarithmic x-axis of the 64 transistor steps in Fig. 4.7D. As explained in Section 4.1, The 1024 integer steps of the DPA are covered using 256 4× transistors and 3 1× transistors. The use of these transistors is shown in Fig. 4.7C, first 3 codes of the DPA 1024 steps are covered with the 1× transistors. At fourth code, the three 1× turn OFF and one 4× transistor turns ON. For fifth, sixth and seventh step, there additional 1× transistors are turned on while the first 4× transistor stays ON. For step 8, all the 1× transistors are turned OFF and a second 4× is turned ON. This process is continued for all of the 1024 steps. The mismatch between the contribution of these transistors will be systematically analyzed in Section 4.4.

The individual transistor compression is highlighted in Fig. 4.8. The x-axis shows the number of 4× transistors turned ON in the DPA. As the number of 4× transistors increase, the output power increase as shown in Fig. 4.7 and hence the DPA output starts compressing because the individual transistors are compressing. The three 1× transistors start with about 2.2 mV contribution at the beginning of the AM-AM curve with zero 4× transistors turned ON. The same transistors only contribute 0.13 mV which shows compression by more than a factor of 16. Similarly 4× transistor contribution reduced from 8.8 mV to 0.82 mV which is a factor of 11. This represents mismatch between the compression behavior of the two type of transistors and it will be analyzed in Section 4.4.

To better understand the system-level requirements for the DPA, fixed point simulations were used to analyze the effect of realistic quantization noise of the DPA

FIG. 4.7

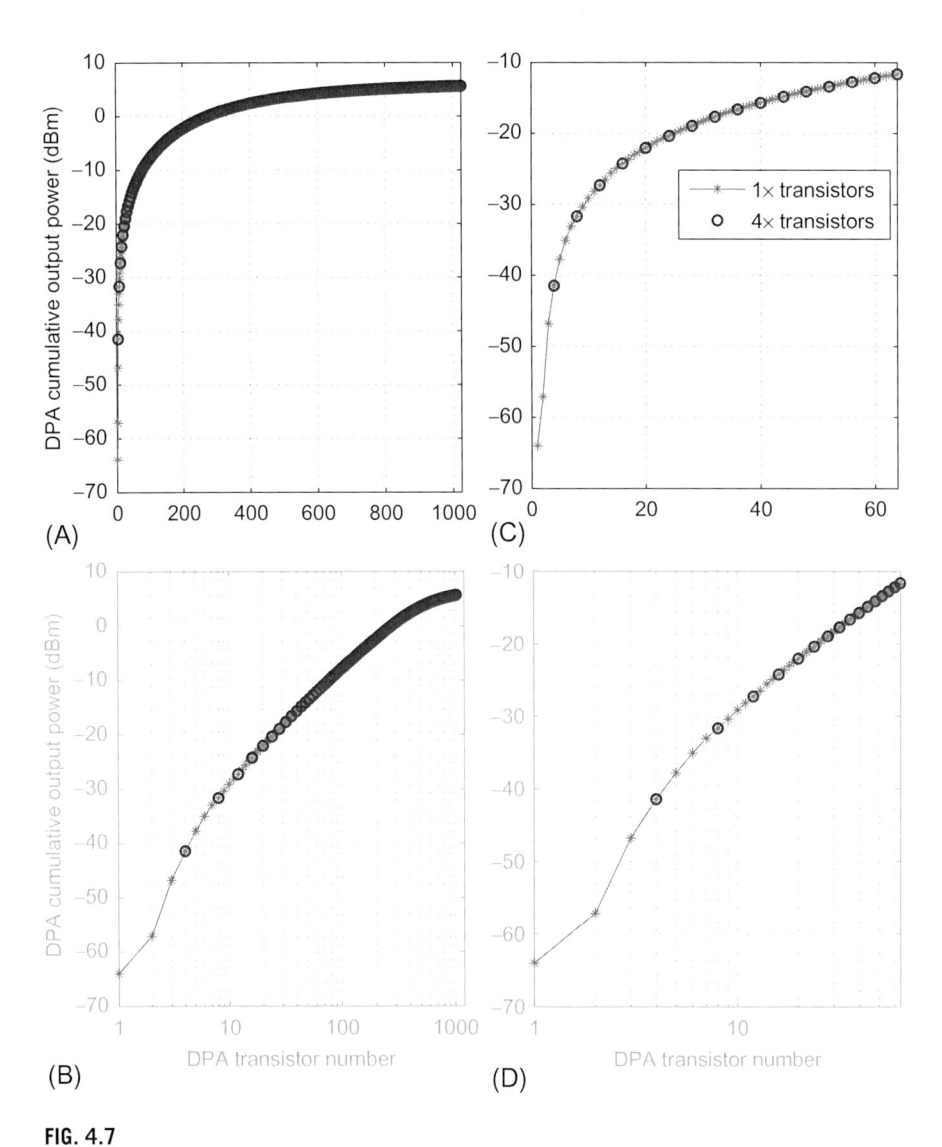

DPA transistor cumulative output power in highband.

on the final RF spectrum. Fig. 4.9 highlights an ideal simulation where all the integer bits are used to show the lower limit on the RF spectrum. The total power delivered is kept constant during this simulation. In other words, the resolution enhancement in going from 11 bits to 13 bits in this figure is completely achieved by adding smaller LSBs with ideal matching. It is worth pointing out that such resolution enhancement is not possible beyond 10 bits in practical situations.

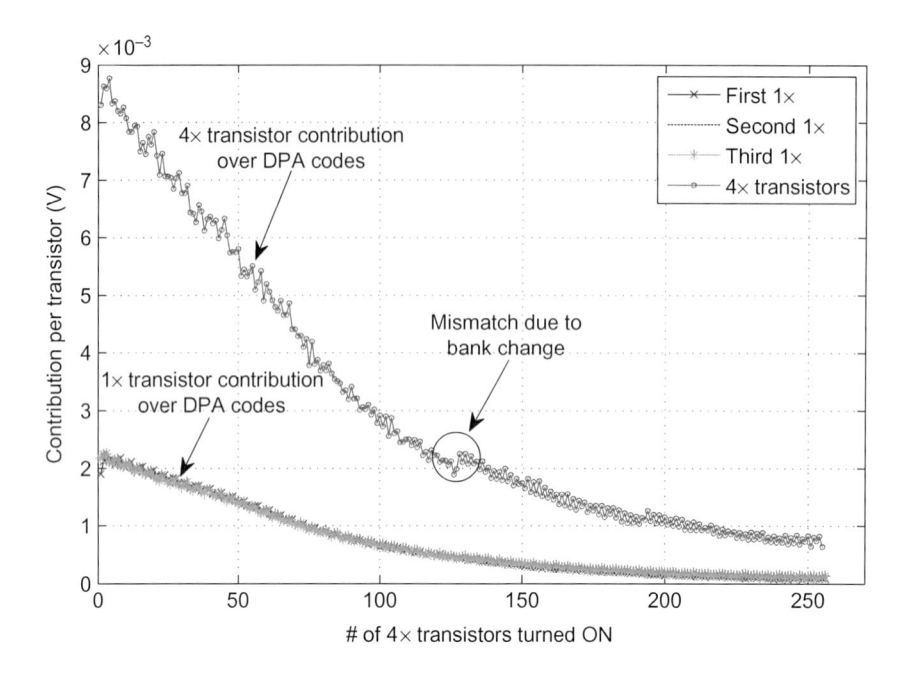

FIG. 4.8

DPA individual transistor contribution over DPA 4× codes.

Based on Fig. 4.9, at least 13 bits are necessary for the DPA to meet the wideband noise (WBN) specifications at 20 MHz offset. With 13 bits, more than 9 dB of margin is achieved at 20 MHz offset over ETSI specifications. Adding further resolution (15 and 17 bits) increases the margin at frequencies beyond 20 MHz. This puts a lower bound on the number of effective bits that need to be achieved by the DPA. As mentioned earlier, only 10 bits can be provided by the realistic DPA in the CMOS process used. Hence additional resolution of 3 bits is added by using $\Sigma\Delta$ modulation. Even though 3 bits are needed from the $\Sigma\Delta$ modulator, 6 bits are generated from the baseband digital transmitter to ensure digital circuits are not a bottleneck in the system performance.

Fig. 4.10 analyzes the effect of adding fractional resolution using $\Sigma\Delta$ modulator. As mentioned in Section 4.2, the resolution enhancement achieved by using $\Sigma\Delta$ is limited by the amount of oversampling used. For the case of oversampling by 8, as is the case in current design, Fig. 4.10 quantifies the effective resolution enhancement. In this figure, 10 integer bits are assumed to be connected to 1024 ideal DPA transistors while the fractional bits are connected to $\Sigma\Delta$ transistors after going through $\Sigma\Delta$ modulator. As it can be seen, the resolution enhancement provided by the $\Sigma\Delta$ modulator diminishes beyond 5 bits. The effective number of bits added

FIG. 4.9

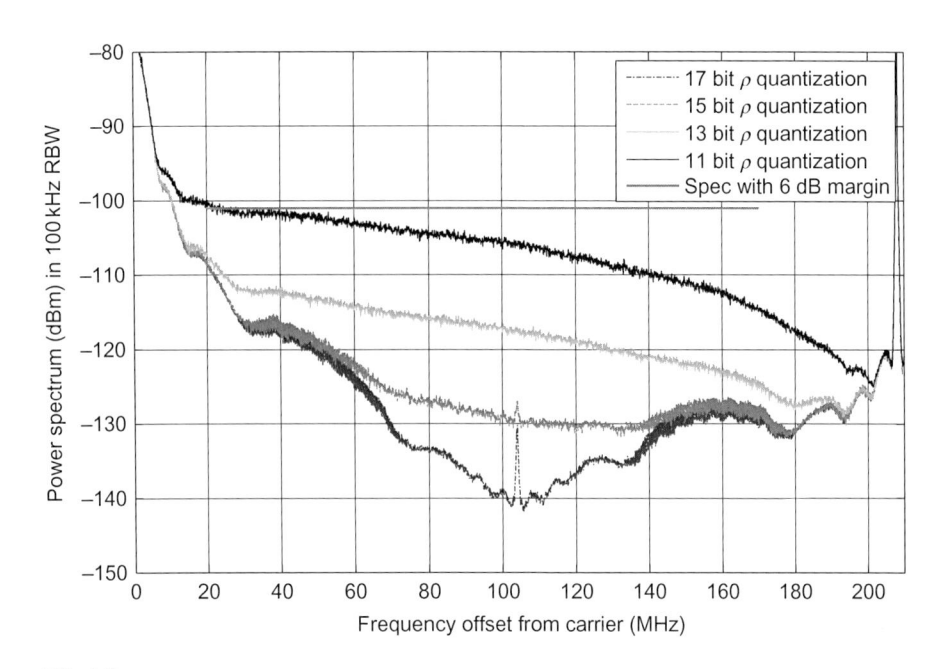

Amplitude quantization spectrum without the use of $\Sigma\Delta$.

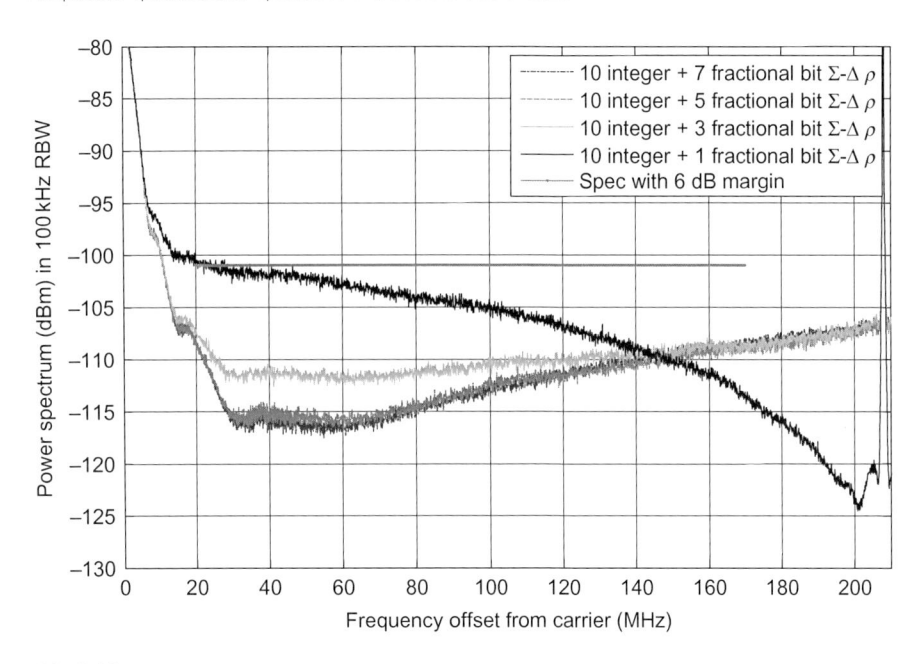

FIG. 4.10

Amplitude quantization spectrum with $\Sigma\Delta$ for fractional bits.

by $\Sigma\Delta$ is between 3 and 4 bits which is sufficient to meet the specifications of 3 additional fractional bits.

The use of $\Sigma\Delta$ causes the high-frequency noise to increase. Comparing Fig. 4.9 and Fig. 4.10, the noise around 170 MHz offset has increased by more than 15 dB for the 13 bit resolution curves. This highlights one of the major disadvantage of the $\Sigma\Delta$ modulation in the presented architecture. Although the margin at 170 MHz is more than the margin at 20 MHz, adding $\Sigma\Delta$ modulation causes concerns at both ends of the spectrum and it can be tricky to meet any additional specifications beyond RX band noise (such as WLAN, Bluetooth, etc. specifications for co-existence).

Fig. 4.11 summarizes these results. The spectrum with 10 integer bits shows that integer-only architecture cannot meet the WBN specifications without using external SAW filter. If we had the freedom to use 17 ideal bits, the spectrum will still have the same margin as we would be getting by using 10 integer and 6 $\Sigma\Delta$-based fractional bits—as it is the case with presented design. Note that the simulation does not include the DCO phase noise which would limit the transmitter spectrum to around −110 dBm. Excluding DCO phase noise allows better understanding of the amplitude path quantization effects. Also it is worthwhile to note that the interpolative filter has effectively eliminated all the replicas present up to CKV/8 frequency and a CKV/16 replica is barely visible at around 105 MHz only while simulating with 17 ideal bits. The replica at CKV/8 is present and it is only attenuated by the zero-order-hold (first-order filtering) of the integer path. Since CKV/8 frequency is around 200 MHz, the RX band is not contaminated by it.

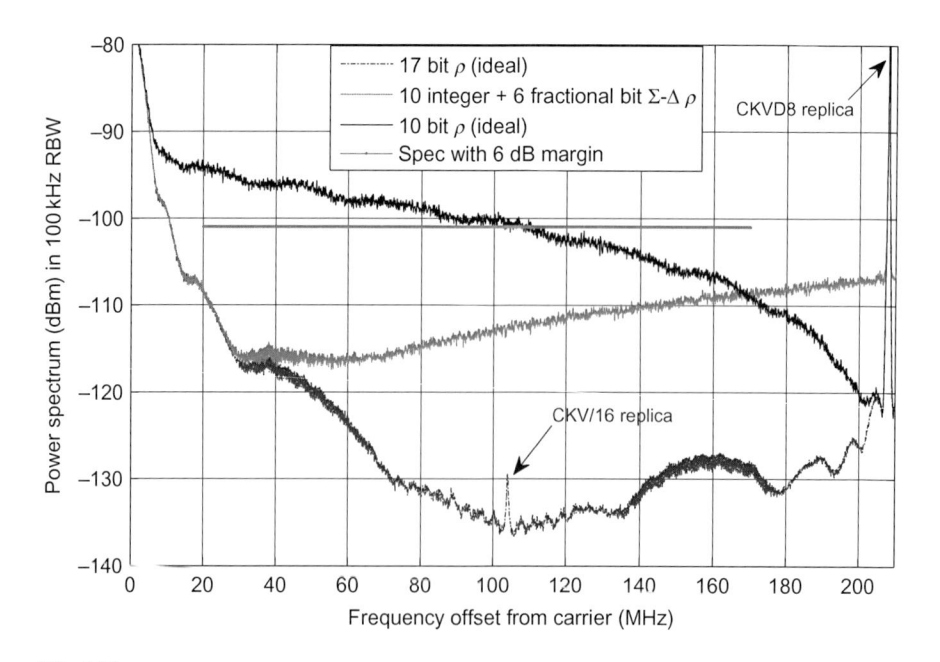

FIG. 4.11

Amplitude quantization spectrum comparison with and without $\Sigma\Delta$.

4.3.3 DPA STRUCTURAL DESIGN

Based on above analysis, 1024 integer steps are determined to be optimum to meet both output power and quantization specifications. As highlighted in Section 2.2, selection of 3 unit weighted 1× transistors are used along with 256 transistors having 4 times (4×) the strength of the unit transistors. This allows optimum trade-off between the implementation area and mismatch of the devices with different size. The 1× transistor strength corresponds to the minimum readily achievable size for a given technology. To achieve better matching, it would be desirable to keep all the transistors to 1× strength and hence have the total of $2^{10} = 1024$ unit transistors. However, this would be prohibitive due to the problems related with large area and clock routing. At the other extreme, having a completely binary-coded system (i.e., 1×, 2×, 4×, 8×, etc.) would make the task of matching these transistors of different sizes extremely challenging. Especially during the saturation of the DPA output power (high output drain voltage), the binary-coded transistor behavior would be different from that during the linear operation (i.e., at low output drain voltage). This makes the binary-coded transistor scheme prohibitive for RFDACs. Additionally, switching noise created by potentially high-strength transistors, such as 16×, could easily couple to and distort lower-strength transistors.

Consequently, based on the desired output power and desired quantization steps, the compromise of 3 (1× or LSB) and 256 (4× or MSB) transistors is chosen. The thermometer encoder converts the 10-bit binary data to thermometer coded lines that can individually drive the integer transistors. The thermometer encoder converts the upper 8 bits to 256 thermometer lines that drive the 256 4× transistors. Similarly it converts the lower 2 integer bits to 3 thermometer lines used to drive the 3 1× transistors. A total of 256 (4×)+3 integer 1×+3 fractional 1× (for $\Sigma\Delta$)=262 transistors are the core transistors used for the core modulation.

A conceptual layout of the DPA transistors is shown in Fig. 4.12. The 4× transistors are arranged in four blocks of 64 transistors each, as shown in Fig. 4.12. Each block is controlled by the thermometer encoded signals. Dummy cells are placed on the boundary of each row (marked as D) to ensure good matching of the corner cells of the rows. Each block contains 4 rows of 16 4× transistors each. The sequential access of these transistors is in the order of the Bank number—starting with Bank 0 and going to Bank 3 in counter-clockwise fashion to avoid large geometrical jumps while switching between the banks.

The bottom row shows the layout of the 1× transistors. The fractional 1× bank represents the $\Sigma\Delta$ modulated transistors [103] and integer 1× bank represents the 1× integer transistors. In addition to the need for 3 1× integer transistors, along with 3 1× integer transistors for the 1×-4× predistortion (see Section 4.5.1), 10 additional transistors are used for the dynamic element matching (DEM) (see Section 4.5.3) making a total of 16 1× integer transistors on the right half of the figure. Similarly 3 1× fractional bits are used for the $\Sigma\Delta$ modulator output while the rest of the 13 bits provide support for DEM. The delay alignment of the clock is discussed in Section 4.6.

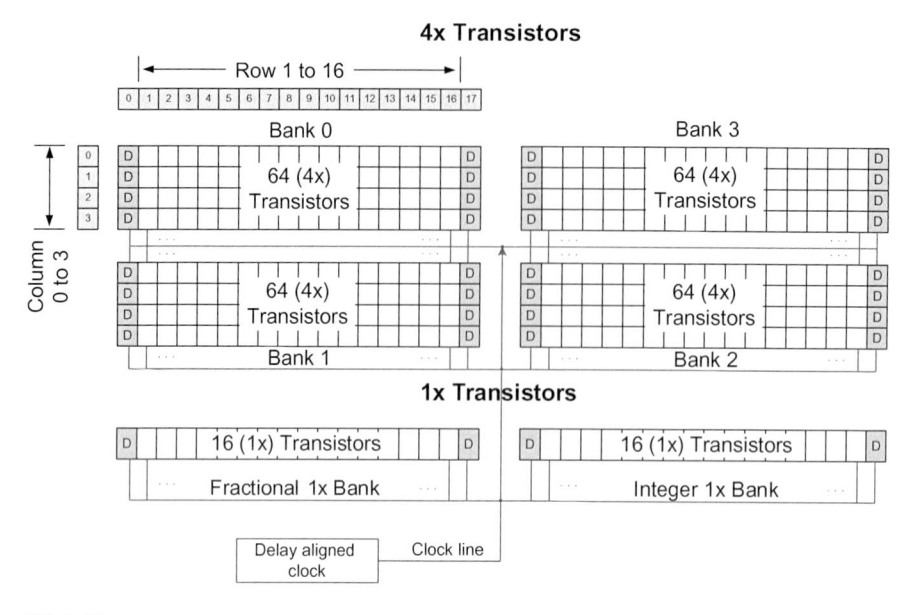

FIG. 4.12

Conceptual layout of the DPA transistors.

4.4 DPA TRANSISTOR MISMATCHES

As with any conventional baseband DAC, the RFDAC also experiences various forms of mismatches and nonidealities that can degrade the overall spectrum of the transmitted output, along with possible degradation of the modulated constellation [104]. In addition to the typical challenges faced by a DAC design, the RFDAC exhibits specific issues related to the high-frequency nature of the circuits operating at RF passband frequency.

The transistor threshold voltage (V_T) mismatch has become one of the dominant design concerns for a CMOS technology at 65 nm and beyond due to quantum mechanical limitations of achieving parametric precision [105]. Various process, device and circuit level methods have been used to minimize the impact of the mismatches but with a limited success.

Typically, most DAC designs focus on the V_T mismatch resulting in the output current variations [106], termed here as amplitude mismatch. In addition to the traditional amplitude mismatch, this work introduces the concept of phase mismatch, which is elaborated in the following section.

4.4.1 AMPLITUDE MISMATCH

Amplitude mismatch or mismatch of the output current of various transistors is one of the most commonly analyzed mismatches. It originates from the process-related variations due to fluctuations in substrate doping, implantations, fixed oxide charge, mobility, lithography, and gate oxide [107]. These fluctuations affect the transistor operation by influencing the threshold voltage (V_T) and, in-turn, the drain current. This results in variations in the output voltage/power delivered to the drain connected load as shown in Fig. 4.1.

The mismatch in these parameters is governed by the Pelgrom's law [107]:

$$\sigma_{\Delta P} \propto \frac{1}{\sqrt{WL}}$$

where $\sigma_{\Delta P}$ represents the standard deviation of the fluctuations in the various parameters, W is the width of the transistor, and L is the length. Hence, it is possible to reduce the amount of mismatch by increasing W or L. However, increasing W results in higher current and hence higher output voltage/power delivered to the load per unit weighted transistor. Thus increasing W will effectively reduce the size of the smallest transistor from the resolution point of view and result in worse quantization noise. Hence increase in W is not a feasible option. Increasing L reduces the current and channel resistance but also leads to speed reduction through the increase in parasitic capacitances. Since the designed DPA runs at nearly 2 GHz frequency, the slower transistor design will lead to rise/fall time problems causing distortions in the mixer functionality of the DPA. Thus, sizing of the transistors is not an easy option to pursue to reduce the mismatches.

Since the device matching limitations have been reached while designing this block, it is critical to systematically analyze the impact of this mismatch over multiple conditions and identify a specification that can be tolerated by the DPA. Additional techniques such as dynamic element matching (DEM) are also considered to work out a system-level solution.

Since the DPA works as a digital-code-to-current converter, the variations in the drain current lead to the effective voltage variations across the load at the DPA output. In other words, if the digital code change command requires $4\times$ transistor equivalent incremental voltage to be delivered to the load, the transistor might only deliver $3.92\times$ equivalent voltage in the case of 2% mismatch. This mismatch manifests itself as a dynamic nonlinearity (DNL) in the AM-AM curve and hence it is referred to as amplitude mismatch.

A systematic analysis of this mismatch along with creation of the specifications for matching are covered in Section 4.5.2.

4.4.2 PHASE MISMATCH

Phase mismatch is unique to an RFDAC. It arises from the timing path differences between the various transistors being engaged. For example, if two transistors outputs are combined to create the DPA output as shown in Fig. 4.6 and if output of *Transistor 2* is delayed by a fixed time (or fixed phase) with respect to the output of *Transistor 1*, then the two outputs will not combine in completely constructive fashion. This mismatch will result in a DNL distortion on the AM-PM curve response; hence it is termed phase mismatch. One source of phase mismatch is due to the process variations resulting in different output impedance between the engaged transistors. Referring to Fig. 4.1, any variations in the transistor output impedance Z_{out} translate into variations in the phase of the signal coming out of that particular transistor, thus resulting in the phase variations. Another major source of phase mismatch is the RF clock propagation delay to the various transistors. The RF clock running at multi-GHz frequency invariably experiences some skew in reaching the individual transistors. Thus if there is a delay mismatch between the clock inputs of two transistors, the same mismatch will show up in the output current waveform on the load. This time delay on the output contribution can be converted to equivalent difference in the phase at the output.

Phase mismatch is shown in a vector diagram showing the DPA complex output in Fig. 4.13. The center curve shows the ideal contribution of three transistors assuming zero phase mismatch between them. Although there is general AM-PM distortion as represented by the slope $\Theta_{am\text{-}pm}$, this slope is constant and will be removed by the AM-PM predistortion. But the variation from $\Theta_{am\text{-}pm}$ by individual transistors represents the phase mismatch. For example, transistor 1 amplitude contribution is

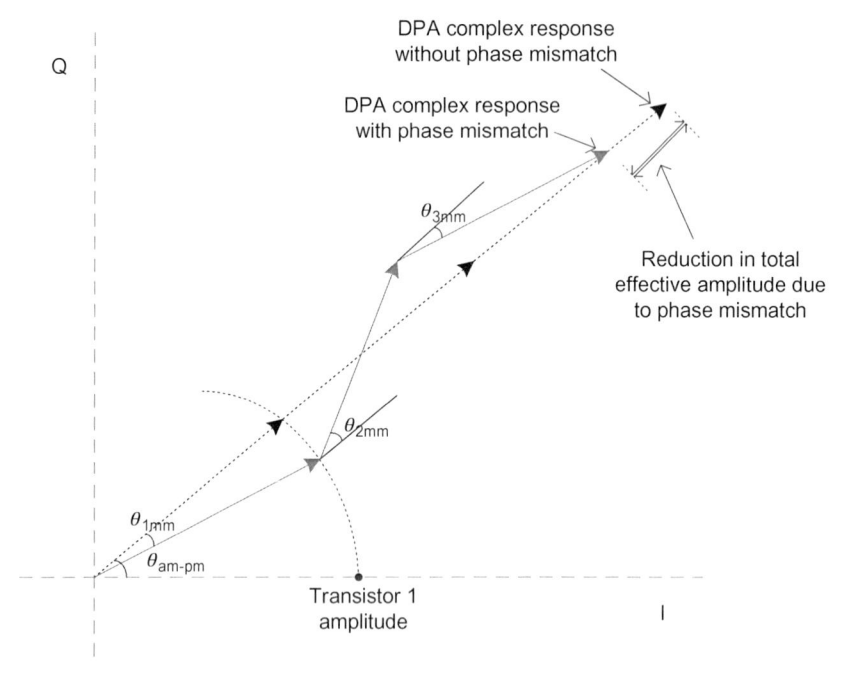

FIG. 4.13

Vector diagram showing the effect of phase mismatch.

shown by the half-circular amplitude vector. If this transistor has mismatch of Θ°_{1mm}, it would deviate from the main complex waveform by the same angle. Similarly Θ_{2mm} and Θ_{3mm} represent the phase mismatch from transistors 2 and 3, respectively. The effective contribution of these three transistors will only be the contribution that is projected on the main $\Theta_{am\text{-}pm}$ angled curve. In other words, because of the phase mismatch, the output amplitude will also be reduced as some of the output will be orthogonal to the carrier. This is graphically shown in Fig. 4.13 where the combined contribution of the three transistors falls short of the case where there is no phase mismatch. Thus phase mismatch translates into amplitude and phase distortions on the output waveform.

To gain perspective into the phase mismatch, Eq. (4.2) converts the phase mismatch to an equivalent time-domain skew/jitter observed on the clock of frequency f_c and period $T_c = 1/f_c$. For a given phase mismatch θ_{mm} in degrees, the mismatch in time is given by:

$$t_{mm} = \frac{\theta_{mm}}{360} \cdot T_c \qquad (4.2)$$

For example, for $f_c = 1784.8$ MHz and $\theta_{mm} = 3°$, the timing mismatch is $t_{mm} = 4.7$ ps, which represents a 0.8% skew on a $T_c = 560.4$ ps clock. This level of timing mismatch might be difficult to achieve. The mismatch allocated to the clock skew alone must be substantially smaller in order to ensure that this is not the dominant source. Thus, if a 2° degree mismatch is allocated to the clock skew, the layout of the clock traces to the individual transistors must have less skew than 3.1 ps. It is critical to note that this mismatch has to be ensured among the adjacent transistors in the digital access pattern. Hence, the clock routing should be proportional to the order of the DPA access sequence, thus ensuring the adjacent transistors receive the clock at nearly the same time.

In addition, there exists a systematic mismatch between the various sized transistors due to challenges involved in matching the transistor current and output impedance simultaneously. These are elaborated in Section 4.5.

Fig. 4.14 shows the impact of the phase mismatch on the emitted RF spectrum. A sine wave is used to amplitude modulate the RF carrier amplitude without any intentional phase modulation. Phase mismatch is assumed to be around 4° RMS among the DPA transistors. The amplitude modulation would cause the transistors to toggle in a fixed pattern depending upon the frequency of the sine wave. The phase mismatch would result in unintended phase modulation which distorts the RF spectrum. Using this scenario, as much as 20 dB of the resulting distortion can be observed. The distortion is especially noticeable at lower frequency offsets around 40 MHz, which is the critical frequency range where the receiver desensitization must be avoided. The phase distortion significantly increases the spectral content in the regions where the $\Sigma\Delta$ noise shaping provides the most resolution enhancement.

FIG. 4.14

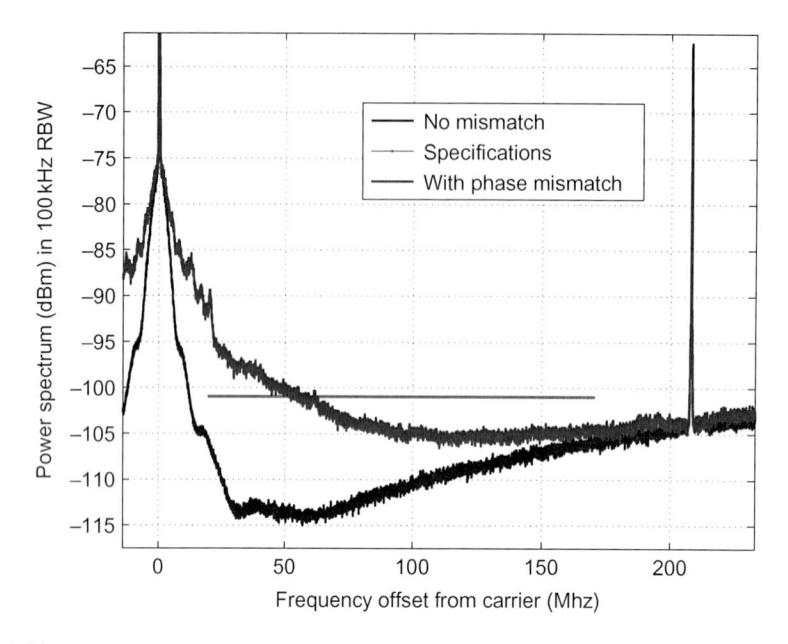

Impact of phase mismatch on the emitted RF spectrum at $f_c = 1784.8$ MHz and -2 dBm output power.

4.5 KEY CATEGORIES OF MISMATCHES AND DEM

The DPA uses combination of $1\times$ and $4\times$ transistors to achieve 1024-integer and 6-bit fractional resolution. There could be various types of mismatches (both amplitude and phase) between these transistors. This section categorizes these mismatches and highlights the most critical categories that dominate the DPA spectral performance. System-level analysis with simulations is used to quantify the impact of the mismatches on the spectrum and tolerable limits are derived for each of these categories.

4.5.1 KEY CATEGORIES
4.5.1.1 MSB ($4\times$) transistor mismatch

The 256 MSB transistors of $4\times$ strength can have various types of mismatches across them. Process variations will introduce mismatches between the transistors as highlighted in Section 4.4 for the V_T mismatch. In addition to the random variations, there can be systematic variations due to the location of the various transistors. These variations are termed as intra-die variation happening due to the spatial separation of the various transistors from each other. This can arise from a number of various manufacturing process sources [108]. For example, many deposition processes suffer

4.5 Key categories of mismatches and DEM

from a gentle "bowl" or concentric ring patterns in thickness from the center of the wafer outward [108]. Such intra-die variations also introduce mismatches in the DPA based on the size of the layout [108].

The sequence of accessing the $4\times$ transistors is always known. The access is always starting from transistor 1 and going to transistor 256 or vice-versa. In other words, if additional transistor needs to be added to the current DPA code, it would always be the next digital code numbered transistor that will be added (i.e., if the last transistor added was 100, the next transistor will always be 101 if the code is being incremented. Otherwise it would be transistor 99 if the code is being decremented). Thus, to overcome the intra-die variations, we need to ensure the transistors which will be accessed consecutively (for example transistor number 100 and transistor number 101) are always placed next to each other in the layout. Thus the solution is introduced by controlling the transistor access sequence to align with the spatial placement of the transistors.

If the access pattern is not aligned as proposed, the systematic gradient-based variations will result in systematic noise at the output (in other words spurs) along with increase in the noise floor. If the gradients are aligned with the transistor access pattern then the mismatch impact is nearly nullified. It will not result in any noticeable degradation of the spectrum because the adjacent transistor are typically immune from the intra-die variations. The systematic gradients are not removed and hence the overall shape of the intra-die variation (e.g., "bowl") will exhibit itself over the course of 1–256 transistor access pattern at much lower frequency. This response will be visible as part of the AM-AM and AM-PM response and hence they can be corrected by the predistortion module. Even though the access pattern was carefully aligned for the presented design, a intra-die variation-based mismatch can be seen in Fig. 4.8 on the $4\times$ transistor contribution line.

On the other had, the statistically random mismatches between adjacent transistors in a row will cause increase in the noise floor. All the mismatches including MSB transistor mismatch are analyzed in system simulations and appropriate specification limits are derived in Section 4.5.2. The random mismatch measured on the DPA AM-AM response can be seen in Fig. 4.8.

4.5.1.2 Unit (1×) transistor mismatch

LSB size unit transistor mismatch ($1\times$-$1\times$) refers to the mismatch between the $1\times$ transistors. As explained in Section 4.3.3, three individual $1\times$ integer transistors and at least one $1\times$ fractional transistor are used in conjunction with 256 $4\times$ transistors to achieve the desired 10-bit integer + 6-bit fractional resolution. Since these transistors are exercised in a systematic fashion, that is, the same three integer transistors will be cycled through while increasing or decreasing the codes and thus producing a repetitive error pattern, it can result in spurs in addition to the increased noise floor. The use of DEM by utilizing the rest of transistors on 16 $1\times$ transistor row reduces the spur level by minimizing the repetitiveness but it increases the overall noise floor thus representing a difficult trade-off as explained in Section 4.5.3.

4.5.1.3 Systematic mismatch between 1× and 4× transistors

The systematic amplitude mismatch between the unit size and MSB transistors (termed as 1×-4×) represents the difference between the incremental 1× transistor output and $\frac{1}{4}$th of the incremental 4× transistor output at a given point on the AM-AM curve. The mismatch could vary depending upon the amount of compression at the output which in-turn depends on the number of 4× transistors turned ON in the AM-AM curve. Similarly to the amplitude, the 1×-4× phase mismatch represents the phase difference between the 1× transistor output and the 4× transistor output at given point in AM-PM curve as shown in Fig. 4.15. The main figure shows the DPA complex response with 1×-4× mismatch (*solid line*), overlapped with ideal line (without 1×-4× mismatch) (*dashed line*). If there were no 1×-4× mismatch, the 1×-4× mismatched line (*solid line*) would follow the ideal line (*dashed line*) without creating the triangle shape of an error as shown in the insert at the top-right corner of the figure. The figure shows the error increasing while going from first to second to third transistor. After third 1× transistor, all 3 1× are turned OFF and a 4× transistors is engaged which brings the mismatched curve back on the ideal line

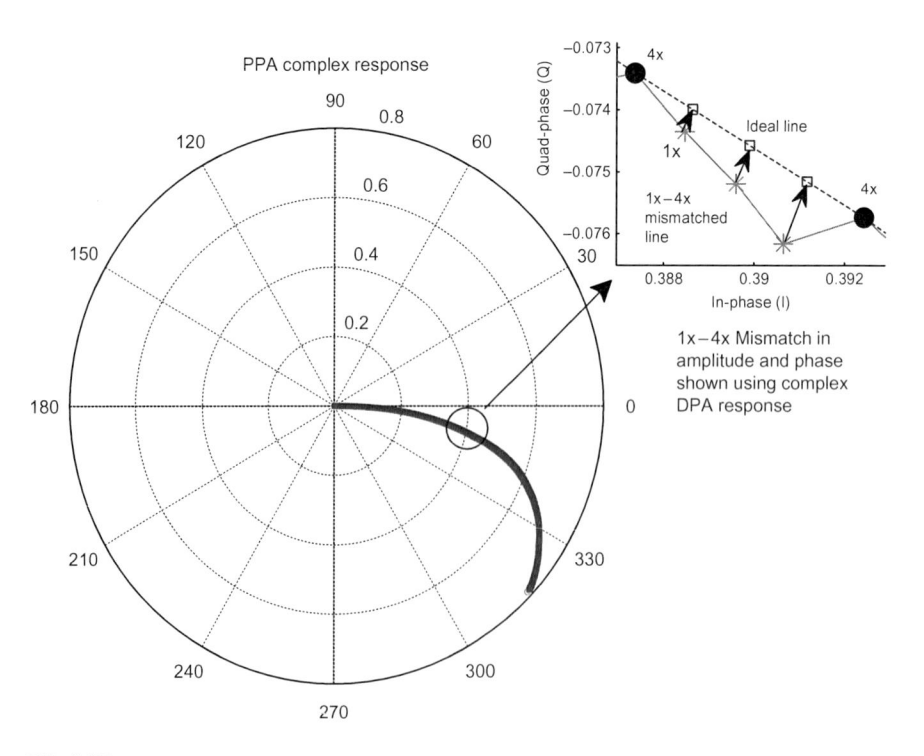

FIG. 4.15

Complex response of the DPA showing 1×-4× mismatch.

as shown in Fig. 4.15 (black dots). This step creates a big jump in both AM-AM and AM-PM response which leads to spectral distortion.

In addition, as explained in Section 4.3.2 and shown in Fig. 4.8, the $4\times$ and $1\times$ transistor behave differently to the compression of the DPA output. The $1\times$ compresses by a factor of 16 while the $4\times$ only compresses by a factor of 11. This creates additional mismatch between the two transistor types as the DPA output undergoes compression. The solution to this problem is shown in the remainder of this section.

The presented design is extremely vulnerable to the $1\times$-$4\times$ mismatch. While the amplitude difference between the output of $1\times$ and $4\times$ transistors can be problematic, the phase difference can lead to severe spectral degradation in the far-out frequencies. It is difficult to align the phase of the $1\times$ and $4\times$ transistor contributions precisely due to the difference in the impedance of the transistors (due to different sizes) (The $1\times$ and $4\times$ contributions might look like the ones shown in Fig. 4.13). Second reason for the phase difference is the delay difference in the propagation of the clock leading to the phase difference at the output, as explained in Section 4.4.2. Special sizing of the transistors with careful manual layout was used to reduce this mismatch.

Circuit-level SPICE simulations combined with system-level simulations shown in Fig. 4.17 indicate that the $1\times$-$4\times$ mismatch is one of the most dominant and one of *the hardest* mismatches to satisfy. Hence, a digital predistortion scheme is presented to correct for this mismatch. The mismatch is first characterized with an assistance of an on-chip receiver using the same mechanism as described in Ref. [78]. The mismatch is stored as the ratio of the expected $1\times$ contribution to the actual $1\times$ contribution throughout the AM-AM curve. Referring to Fig. 4.15, this number will be the ratio between the ideal line points (*dashed line*) and the $1\times$-$4\times$ mismatched line (*solid line*) points in the insert. The ratio correction will effectively move the amplitude of the $1\times$-$4\times$ mismatched line to match with the ideal line. The phase mismatch will not be corrected so there will still be distortion in the AM-PM response.

The value of the ratio changes significantly over the AM-AM curve due to the change in loading, phase alignment of transistors, etc., and hence 256 correction values are stored corresponding to the 256 $4\times$ points on the AM-AM curve. At the time of compensation, the real-time value of the amplitude code is observed and the correction factor is determined based on the location on the AM-AM curve and the number of $1\times$ transistors used, as shown in Fig. 4.16. As an example, consider a case where the number of $4\times$ transistors used for current code are 100, and at this code, the $1\times$ transistors are contributing 40% less than the $4\times$ transistors. Using this example, the correction factor will be 1.4 that need to be multiplied to the $1\times$ transistors. If the number of $1\times$ turned ON for the present code is 3 then the final correction will be $1.4 \times 3 = 4.2$. Since only three integer transistors can be supported using the 2 bit binary $1\times$ integer datapath, additional overflow transistors are used to contain the overflow. Hence in this case 1 overflow transistor will be turned ON. The remaining 0.2 transistor contribution will be added to the fractional contribution. The same correction factor is also multiplied to the 6 fractional bits because they

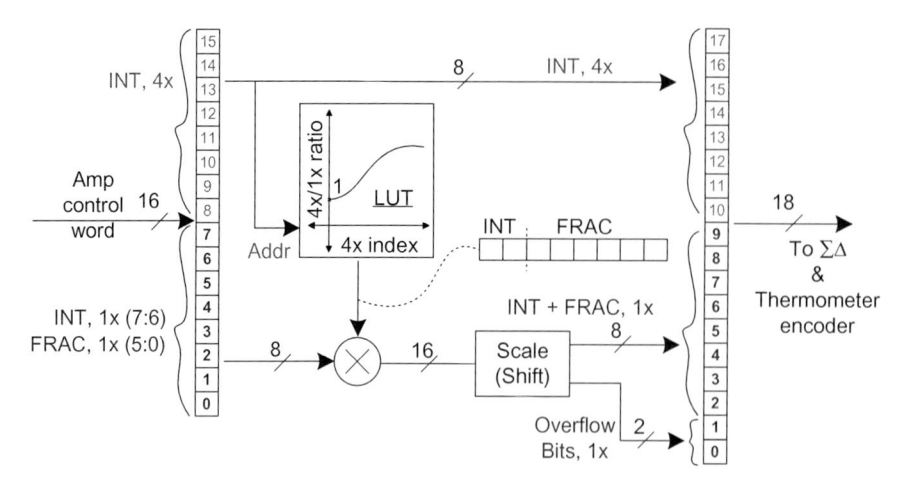

FIG. 4.16

1×-4× amplitude mismatch predistortion.

also control 1× transistor after being modulated by the $\Sigma\Delta$ modulator. The ration between the two transistors can be found by using the data shown in Fig. 4.8 where the 1× to 4× ration starts with nearly 1/4 but as the compression takes effect, the ratio changes to about $(1/4) \times (16/11) = (1.45/4)$. Thus the ratio of 1.45 should be used for the data shown in Fig. 4.8.

Thus, the correction factor, found from the look-up-table (LUT), is multiplied by the 1× part of the code to provide first-order correction for the 1×-4× distortion. Any overflow resulting from the multiplication operation is handled by the dedicated 1× transistors (marked as overflow bits) placed near the regular 1× transistors. Spare 1× transistors shown at the bottom of Fig. 4.12 are used for the overflow handling.

The phase portion of the 1×-4× distortion is extremely difficult to correct. The regular AM-PM predistortion (Section 3.2) is performed by applying correction value to the phase/frequency signal used by the ADPLL for modulation. The phase correction that can be performed using ADPLL-based AM-PM predistortion is limited by the bandwidth of this correction path. For instance if the bandwidth of this correction path is 10 MHz then it can correct of any distortion within 10 MHz. The 4× transistor roughly toggle at the same rate in the worst case scenario. Thus the AM-PM predistortion can be corrected by using this path.

But if this path had to be used for the 1×-4× phase mismatch correction, the phase adjustments have to be applied at the rate of CKV/8 (around 200 MHz) to the ADPLL input based on the rate of change of integer 1× transistors. This rate is extremely fast and the phase path response is not fast enough to allow 200 MHz correction signal to pass through to the DCO. Thus the option to correct 1×-4× phase distortion through phase path is ruled out.

An alternative option is to achieve the delay by fine adjusting the phase of the clock in the DPA. This would have the same impact as having DCO phase shifted.

This could be achieved by passing the DCO clock through a chain of invertor-based buffers and taping the output at certain points at the invertor chain based on the desired phase shifting. The problem with this proposal is that it would require passing the phase modulated RF clock through a chain of invertors which would degrade the phase noise of the clock. Thus $1\times$-$4\times$ is very difficult to correct using digital means and thus careful layout matching along with appropriate transistor sizing has to be used to ensure the phase contributions from both sizes of the transistors are aligned. This also highlights the reason for not using more than 2 size of transistors in the DPA structural design because it would create more possibilities of mismatches as explained in Section 4.3.3.

4.5.2 SIMULATION-BASED SPECIFICATIONS

The simulation model (Appendix A.3) is used to analyze the tolerable limits of the mismatches of the above categories. It is critical to quantify the impact of various mismatches on the RF spectrum and prioritize the mismatches that cause dominant spectral degradation. Fig. 4.17 shows the results of the system-level simulation modeling the mismatches in appropriate way. The specification line is shown to be with 6 dB margin over ETSI specifications. $\Sigma\Delta$ modulation is assumed with realistic sampling rates to see the impact of the mismatches on the $\Sigma\Delta$ noise shaping. Curve A shows the spectrum without any mismatch between the DPA transistors resulting in around 9 dB of margin over the internal specifications. Mismatches are added in cumulative fashion on curve A simulation starting with unit transistor ($1\times$) amplitude mismatch shown in curve B. Similarly curve C shows $1\times$ mismatch (curve B) combined with systematic amplitude mismatch between $1\times$-$4\times$ transistors. Curve E has all the mismatches combined showing about 2 dB of margin over the internal specification (resulting in combined 8 dB margin). As it can be seen (and also highlighted in Section 4.5.1), $1\times$-$4\times$ mismatch is the most dominant mismatch. Especially $1\times$-$4\times$ phase mismatch is of the greatest concern because it cannot be corrected by using a predistortion mechanism in contrast to the $1\times$-$4\times$ amplitude mismatch. The impact of MSB ($4\times$) transistor mismatch has been found to be less dominating on the overall spectrum. Once the mismatches shown in Fig. 4.17 are included in the analysis, varying $4\times$-$4\times$ mismatch does not impact the spectrum noticeably. Hence this mismatch has been excluded in the simulation plots presented here for the sake of simplicity.

A second example of such result highlighting the process for deriving the specifications is shown in Fig. 4.18. $1\times$-$4\times$ phase mismatch has been used as a sweep parameter. The figure is focused around 20 MHz offset where the transmitter has the lowest margin. The first curve (in the legend) contains all the other mismatches ($1\times$ amplitude and phase mismatch with $1\times$-$4\times$ amplitude mismatch) with 0° $1\times$-$4\times$ phase mismatch. Using this curve as the baseline, $1\times$-$4\times$ phase mismatch is incremented in offsets of 1° showing progressive degradation of the spectrum. It can be seen from the results that having $1\times$-$4\times$ phase mismatch greater than 3° results in internal specification failure. Thus the specification has been selected

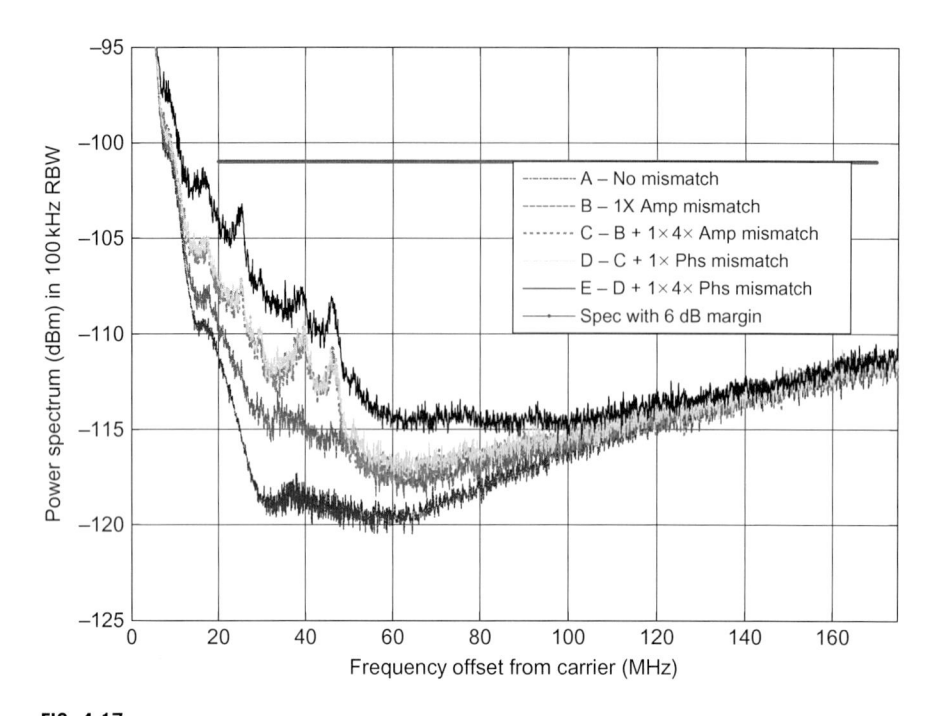

FIG. 4.17

Contribution of individual mismatches on the spectrum.

to be 3°. Similar process has been used to run simulations across the mismatch combinations to identify the tolerable limits on other mismatches as well. These results are summarized in Table 4.1 ensuring 6 dB of margin for the presented transmitter.

4.5.3 DYNAMIC ELEMENT MATCHING

DEM has been used extensively in the field of DAC and ADC designs as a way to minimize the impact of mismatches on a specific spectral region. Although the DEM cannot eliminate the spectral energy resulting from the mismatches, it can either spread it or shape it in such a way as to move it out of the band of interest [109]. For the DEM to work properly, a few conditions have to be satisfied [110, 111]. The primary condition requires the scrambling pattern be random and uncorrelated with the data. Although this can be satisfied with a reasonably random sequence, the generation of the scrambling pattern and the repetitive access of the transistors at the scrambling pattern frequency results in spectral spurs present at the RF output. In addition, DEM can only correct for random mismatches so it cannot improve the 1×-4× systematic mismatch.

If an oversampled DEM is used to correct for the 1×-1× mismatch, it might interfere with the $\Sigma\Delta$ quantization noise shaping. Additionally, the use of higher-

4.5 Key categories of mismatches and DEM

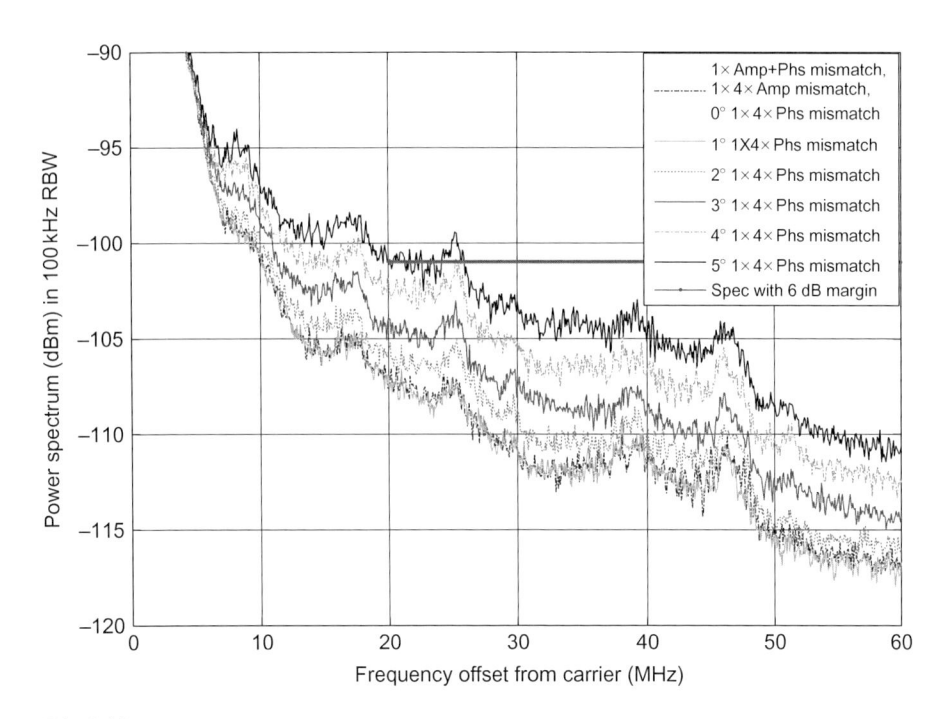

FIG. 4.18

Contribution of 1×-4× phase mismatch on the spectrum.

Table 4.1 Simulation-Based Mismatch Specifications

Mismatch Category	Transistor Types	Mismatch Specifications
Amplitude mismatch	4×-4×	8%
	1×-1×	10%
	1×-4×	10%
Phase mismatch	4×-4×	3°
	1×-1×	3°
	1×-4×	3°

order $\Sigma\Delta$ modulation would require more complex DEM schemes, which are difficult to synthesize at nearly 1 GHz speed without excessively increasing the power consumption.

To analyze the effects of DEM of the RF spectrum, a sine wave is used to amplitude modulate the RF carrier without any intentional phase modulation. Fig. 4.19 shows the impact of DEM on the RF spectrum with the sine wave-based amplitude modulation with $f_c = 1784.8$ MHz and -2 dBm output power. The effects

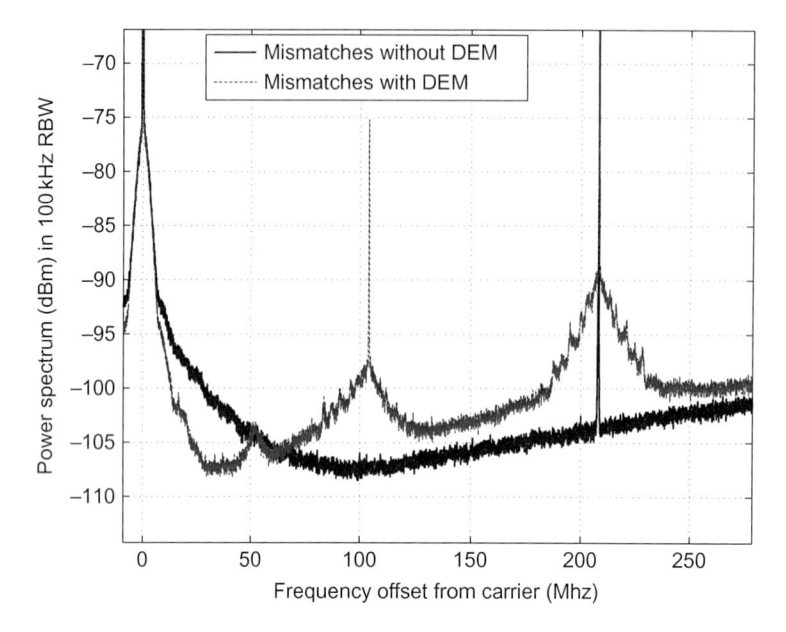

FIG. 4.19

Impact of DEM on RF spectrum.

are shown with both amplitude and phase mismatches present. The 1×-4× mismatch is not enabled here because it is not affected by the DEM. Although the DEM improves the performance at lower frequency offsets, the replicas and degradation in spectrum suffered at higher frequencies makes it an impractical solution for the presented transmitter.

Another negative effect of the DEM is the increase in the switching activity, which was neglected by the above simulations. The increase in switching activity increases the likelihood of digital high-frequency noise coupling to the sensitive RF circuits, resulting in further spurs and increased noise floor. In addition to above critical disadvantages of the DEM, minor factors, such as increased loading on the clock to drive the DEM transistors, increased area, higher number of interface lines between digital and analog boundaries, etc., make it rather unattractive for the RFDAC design. Based on these factors, DEM is not used in the active mode of operation of the presented transmitter.

4.5.4 MEASUREMENT RESULTS

The mismatch between individual transistors is systematically characterized in a production version of the IC fabricated in the 65 nm CMOS technology. The measurements of individual strengths of the transistors are carried out using both the

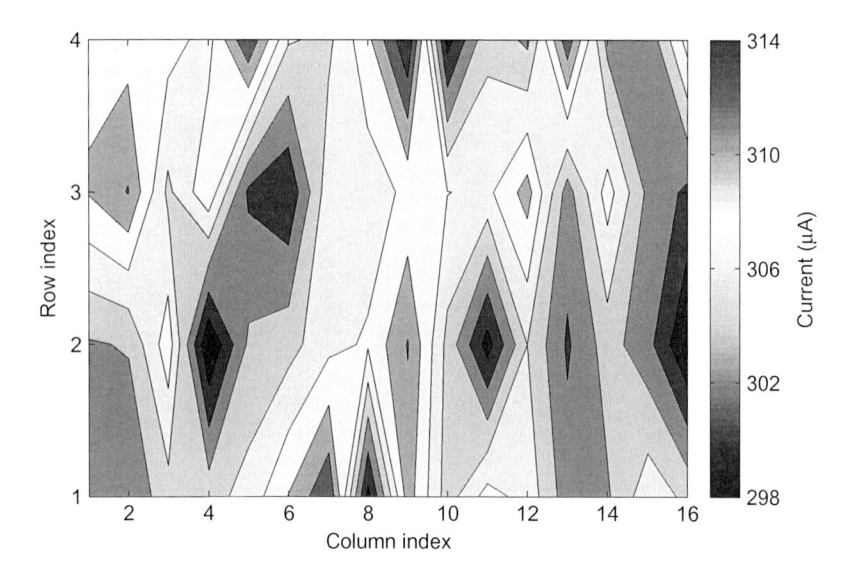

FIG. 4.20

Measured DC amplitude mismatch between the 4× transistors in the first quadrant.

on-chip receiver-based approach and with a spectrum analyzer. One such example of the measurements is shown in Fig. 4.20, where the first quadrant of the DPA is shown. Various colors show the dc current output of 4× transistors in μA under 1.4 V V_{DD} supply. The transistors are enabled individually and their load current is measured after significant averaging to reduce the measurement noise. As it can be seen from the contour plot, there is no systematic gradients in the quadrant. Additionally the standard deviation of the mismatch between the transistors is found to be only 1.3% which is significantly better than the specification mentioned in Section 4.5. The spread of the mismatch is also contained to within ±3%. Similarly the mismatch for other three quadrants was found to be 1.28%, 1.33%, and 1.16%.

Similar measurements are performed over all the quadrants of various process corner devices and the mismatch is found to be within the specifications mentioned in Section 4.5.

Individual mismatches between a set of 16 1× transistors are shown in Fig. 4.21. These measurements are obtained by the same procedure used to characterize the 4× transistors. The 16 transistors measured are from the Integer 1× bank shown in Fig. 4.12. Here the mismatch is directly shown to highlight the spread and the standard deviation (STDEV) of the mismatch. The STDEV of the mismatch is 2.7% which is well within the specifications.

Average 1×-4× amplitude mismatch is found to be 4% from measurements. The 1×-4× phase mismatch is measured to be 1° which meets the specifications with margin.

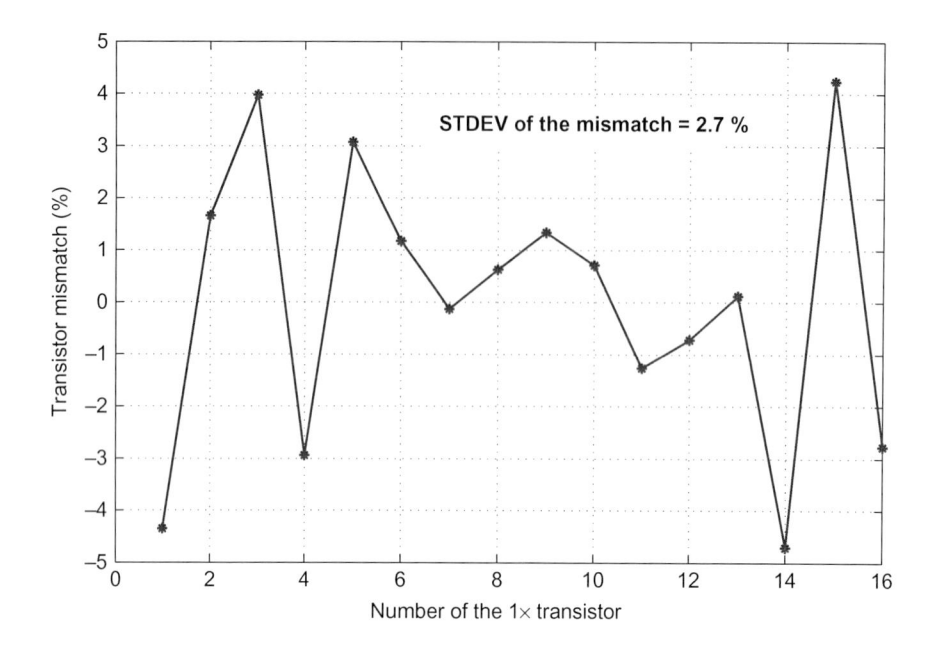

FIG. 4.21

Measured DC amplitude mismatch between the 1× transistors.

4.6 CLOCK DELAY ALIGNMENT

4.6.1 EXPLANATION OF THE PROBLEM

As shown in the conceptual schematic of the DPA in Fig 4.1, the data input and clocks are passed thorough an AND gate before controlling the gate of the DPA output transistors. Although the amplitude signal before thermometer encoding changes in "staircase" fashion containing binary values, the thermometer encoder output is a binary value controlling each of the transistors individually as shown in Fig. 4.6. The rate at which the value of the thermometer encoder output changes will depend on the type of the transistor (4× integer or 1× integer). The same also applies to the $\Sigma\Delta$ transistor control signals connected to 1× fractional transistor.

For integer transistor, the rate of change will also depend on the power level and the position of a particular transistor in the thermometer encoding. For instance, if the DPA is transmitting near maximum power, the transition placed at location 1 of the thermometer encoder will always be ON during the active transmission. But the transistor near position 1024 will be rarely turned ON. And there will be transistors near the peak power DPA code that will be toggling at nearly CKV/8 rate.

The other signal connected to the input of the AND gate is the clock which is operating from 0 to $+V_{DD}$ with approximately 50% duty cycle. The clock is running at the center frequency of RF carrier which is around 2 GHz for the highband transmission.

If the data input to the AND gate changes while the clock is zero then the data input gets registered and stays constant during the active phase of the clock resulting in intended data latching. But if the data changes while the clock is active, the change gets propagated to the output of the transistor and thus the DPA spectrum gets switching noise resulting in spectral degradation. The intended operation is to make sure the data input to the AND gate is stable while the clock is high. And all the data input transitions occur during the negative phase of the clock. Thus, the data transitions need to be aligned such that they always happen during the inactive period of the clock as shown by CLK_C and CLK_D interaction in the timing diagram of Fig. 4.22. This might sound trivial for a typical digital design but for the DPA running at the RF clock speed, there are challenges involved in guaranteeing this condition

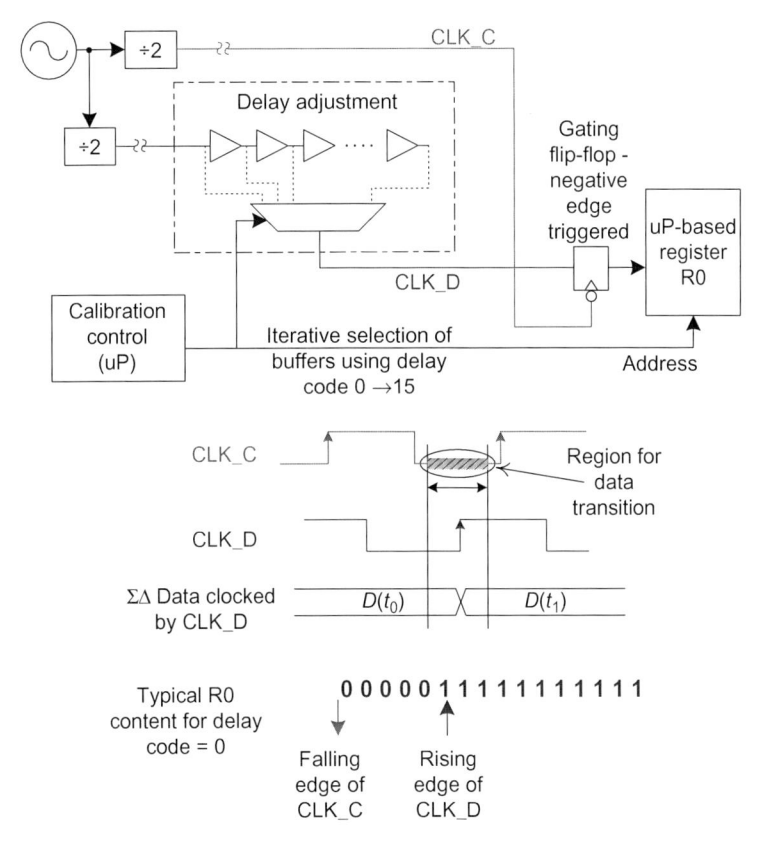

FIG. 4.22

Conceptual details of delay alignment calibration.

over process, voltage, temperature, and frequency corners. The propagation of the clock from the DCO to the input of the DPA AND gate could vary over these corners. Similarly the data propagation delay could have variations. Additionally there could be buffer sizing variations on the DCO output or the data input (coming from $\Sigma\Delta$ or thermometer encoder outputs). Hence it is not possible to predict and align the data and clock inputs in advance and it requires calibration and compensation mechanism to align the transitions for every IC.

4.6.2 SELF-CALIBRATION AND COMPENSATION MECHANISM

The delay alignment adjustment is achieved by using a chain of delay buffers, implemented by using pairs of inverters, to align the delay between data and clock as shown in Figs. 4.1 and 4.22. CLK_C contains the phase modulated RF clock which is mixed with the data clocked by CLK_D as shown in Fig. 4.1. Thus the phase noise or jitter on the CLK_C directly translates into the output phase noise. If the clock CLK_C is passed through a delay buffer chain, the phase noise at the output of the transmitter will degrade which is undesirable. On the other hand, if the data signals are to be delayed, a large number of buffer chains would have to be implemented for every thermometer encoder output—total of 265 buffer chains. A solution is devised to this problem by delaying the clock that is used by the thermometer encoder output generation—CLK_D. CLK_D is the main clock used by the thermometer encoder and the $\Sigma\Delta$ module. Thus the CLK_D directly controls the transitions on the data inputs of the DPA AND gates. And the data transitions are calibrated to happen only during the OFF period of the CLK_C and thus do not propagate through the AND gate. Hence the phase noise or jitter on CLK_D is far less crucial for the DPA output phase noise. Delaying CLK_D will effectively delay the data signal because the data signal generation timing is directly proportional to $\Sigma\Delta$ clock. The delay resolution of the compensation scheme depends on the delay of the inverters which in turn depends on the PVT conditions.

The optimum delay relationship is when the CLK_D has a rising edge during the middle of the CLK_C OFF period as the data signals change on the rising edge of the CLK_D. The details of calibration are shown in Fig. 4.22 to achieve this relationship. The inverter-based delay chain outputs are multiplexed to select the generation of CLK_D. A total of 16 controls are used to cover the full dynamic range of the delay values. Each invertor pair is controlled by one delay code thus making the output noninverted. Thus a "invertor delay code" is defined as a delay of two invertors in the given process technology.

Fig. 4.23 shows the pseudo-code for the self-calibration mechanism. On-chip microprocessor is used extensively throughout the calibration process to reduce the use of dedicated hardware. Once the DCO has settled to the desired carrier frequency, the microprocessor-based calibration control starts stepping through the 16 delay codes. Based on the delay code generated by the microprocessor, corresponding buffer output is connected to CLK_D by the multiplexer—shown using *SetDelayMux()* procedure. This is repeated by the microprocessor to eventually

4.6 Clock delay alignment

Algorithm 1 Invertor delay code selection

```
 1: procedure PROGDELAY(lowband)           ▷ lowband = 1 if band of operation is
    LB
 2:     PowerUpTX()
 3:     ProgramTXFreq()
 4:     LockADPLL()
 5:
 6:     indexSelected = FindOptimumDelay(lowband)
 7:     if ((indexSelected == 16) && (lowband == 1)) then
 8:         EnableClockInverter()              ▷ Try inverted clock in lowband
 9:         indexSelected = FindOptimumDelay(lowband)
10:     SetDelayMux(indexSelected)            ▷ Final selection of the mux
11:     return

12: procedure FINDOPTIMUMDELAY(lowband)
13:     indexSelected ← 16
14:     for i ← 0, 15 do
15:         SetDelayMux(i)
16:         R0(index) ← GetFlipFlopOutput()   ▷ Read the flipflop output
17:
18:     for j ← 0, 15 do
19:         if lowband == 1 then
20:                                            ▷ R0(j:j+12) is a circular operation
21:             if R0(j : j + 12) == 0b000000111111 then
22:                 indexSelected ← j
23:         else
24:             if R0(j : j + 6) == 0b000111 then   ▷ R0(j:j+6) is a circular
    operation
25:                 indexSelected ← j
26:
27:     return indexSelected                  ▷ The optimum index found
```

FIG. 4.23

Algorithm 1.

select all the buffer output to drive CLK_D as shown in Fig. 4.22 and shown by the "for" loop in Fig. 4.23.

The selected CLK_D is sampled by a negative edge triggered flip-flop sampling on CLK_C. Using the falling edge of CLK_C, the distance from this falling edge to the rising edge of CLK_D is derived in terms of the invertor delay code length. This is graphically shown in the waveforms of Fig. 4.22. For each value of invertor delay code, the sampling value at the flip-flop is stored in a 16 bit register R0 as shown by *GetFlipFlopOutput* procedure. The value of R0 is an indication of the relationship between the two clocks. For example, if the contents of R0 starts with a one then the rising edge of CLK_D has already taken place before the falling edge of the CLK_C. This would not be a desired because the CLK_D rising edge is occurring during the active phase of CLK_C. On the other hand, if CLK_D rising edge occurs when CLK_C is OFF, the value of R0 will contain a few zeros followed by a 1. This case is shown graphically using the waveforms and corresponding R0 contents in Fig. 4.22. As shown, first few delay codes will occur before the rising edge of CLK_D and thus the value will be 0. After the rising edge of CLK_D, the remaining delay codes will produce 1. The number of zeros to be selected in the R0 is optimized based on the nominal value of the inverter delay for the 65 nm process technology such that the number of zeros roughly equal to half the clock period for the given carrier frequency. For example, if one invertor delay is 25 ps then a buffer containing two invertors will have delay of 50 ps. If the clock pulse is 500 ps, half clock period is 250 ps. And the desired alignment would require 125 ps from the falling edge of CLK_C to rising edge of CLK_D. Based on the buffer delay of 50 ps, 3 zeros would be required before a 1 in the content of R0 as shown in the binary code for highband in Fig. 4.23. The indexing operation on R0 in Fig. 4.23 is circular in nature. Thus if the value of j is 12, $j + 6$ would be equal to $18 - 16 = 2$. In this example, if the content of R0 was found to be 1111_0000_0111_1100, the code which would provide optimum clock relationship would be 7 which would allow 3 zeros before a 1 in the code. The delay alignment procedure takes 10 μs to execute and the procedure is performed on every burst to ensure compliant performance.

For lowband, the clock frequency is nearly half of the highband clock frequency. Hence the clock period is nearly double of the highband clock period. For the same example mentioned above, the clock period would be 1 ns. Hence, instead of looking for 3 zeros in the buffer R0, 6 zeros will be searched for as shown in line 24 of Fig. 4.23.

The calibration algorithm is executed before transmission of every data burst because the frequency or the temperature change could alter the delay alignment value. The execution takes less than 15 ms with 26 MHz processor clock speed. Measured data on silicon for $f_{rf} = 1784.8$ MHz is shown in Fig. 4.24. Three key frequency offset locations (20 MHz, 57 MHz, and 105 MHz) from the RF spectrum are selected for the performance comparison across various invertor delay codes. The points below −106 dBm represents the points highlighted for CLK_C in Fig. 4.1 where data transition is optimum. Around 8–9 dB of degradation can be observed in the spectrum if the data changes outside the "safe" window of CLK_C. In other words, if the data changes while the CLK_C is high, degradation of the performance is sever. The algorithm is used in the silicon to perform self-calibration using the scheme mentioned in the previous paragraph. The code selected by the self-calibration algorithm is highlighted in the figure as well. As it can be seen, the delay code selected (8) is the most optimum point in the middle of CLK_C low period. The performance has been evaluated over PVT conditions and similar optimum behavior has been observed using the self-calibration algorithm.

A similar performance plot is shown for lowband in Fig. 4.25. One interesting difference between the two plots is the percentage of the clock period covered by the invertor chain. For highband plot (Fig. 4.24), the clock period is 560 ps corresponding to 1784.8 MHz clock frequency. On the other hand Fig. 4.25 has the period of 1135 ps corresponding to 880.2 MHz frequency. The inverter-based buffer delay is approximately 50 ps. Thus, 11 buffers would cover the full clock cycle of the

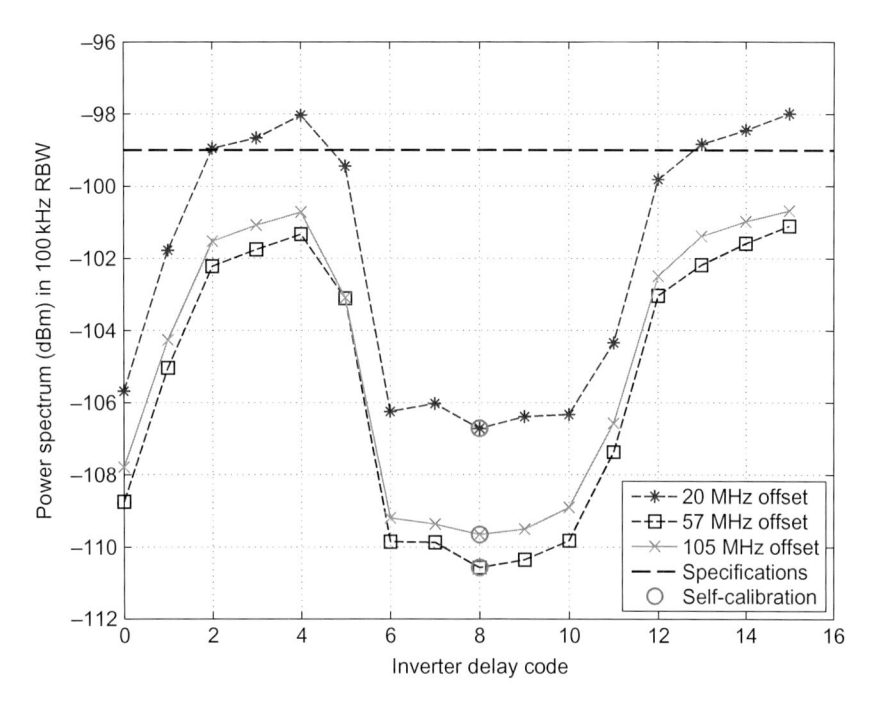

FIG. 4.24

HB WBN performance over delay and selection of optimum delay by calibration mechanism.

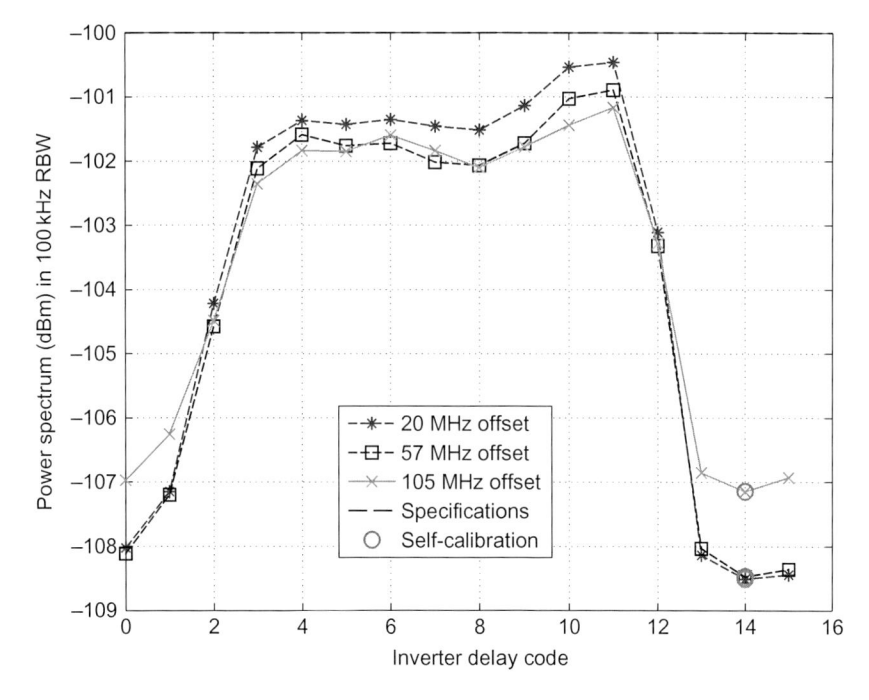

FIG. 4.25

LB WBN performance over delay and selection of optimum delay by calibration mechanism.

highband frequency mentioned earlier. But 23 buffers are needed to cover the full clock period for lowband. Instead of adding additional buffers, an inverted phase of the clock is used to cover the second half of the clock period. Thus, an additional multiplexer selects between either in-phase clock or inverted-phase clock. This clock is passed through the invertor buffer chain mentioned earlier. Using this approach, only 12 buffers are needed to cover half period of the clock even for LB. This ensures optimum delay code is always covered within the range of inverter delay codes over PVT and frequency conditions.

4.7 ANALYSIS OF PARASITIC COUPLING

4.7.1 POSSIBLE COUPLING PATHS

High-speed digital logic containing $\Sigma\Delta$ and the DPA are two of the most critical blocks in the system running significant amount of logic at nearly 1 GHz speed. The $\Sigma\Delta$ logic and the thermometer encoder functionality are combined in a single digital module interfacing with the DPA. The combined digital activity with fast rise and fall time causes generation of significant energy that could couple into the DPA transistors. Since the number of lines interfacing between the two modules is large, they have to be located within close proximity of each other in the chip layout. The proximity further increases the likelihood of signals coupling from $\Sigma\Delta$ module, with first and higher-order $\Sigma\Delta$, to sensitive analog circuits of the DPA as shown in Fig. 4.1.

One of the strong coupling paths identified in postlayout parasitic-extracted SPICE simulation is shown in Fig. 4.26. The $\Sigma\Delta$ activity from the high-speed digital logic controls the gate terminal of the $\Sigma\Delta$ transistors. But this activity also couples into the source terminal of the 256 4× integer transistor via parasitic path

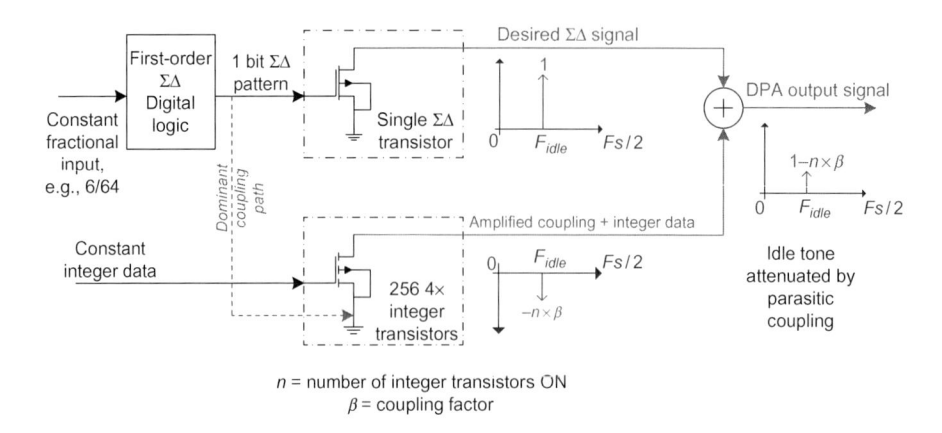

FIG. 4.26

Basics of $\Sigma\Delta$ idle tones and coupling.

of the shared power supply ground terminal. The $\Sigma\Delta$ 1× DPA transistor acts as inverting amplifier but the parasitic coupling signal is amplified through the integer DPA transistors in a noninverting phase. This results in the parasitic coupling being out of phase with the original signal at the output. As more integer transistors are switched-in, the parasitic coupling strength grows and eventually exceeds the original $\Sigma\Delta$ 1× contribution. In addition to the main coupling path, there are additional coupling paths which might be inverting or noninverting with minor delay compared to original path. This results in severe distortion of the $\Sigma\Delta$ noise shaping. It is necessary to characterize the coupling in the silicon to better quantify its strength and the effect of this coupling on the RF spectrum. The following section proposes a new method for doing the same.

4.7.2 A NOVEL METHOD OF CHARACTERIZING $\Sigma\Delta$ PARASITIC COUPLING USING IDLE-TONES

The possibility of $\Sigma\Delta$ signal with higher activity coupling into lower activity integer signals is present in any architecture using $\Sigma\Delta$ with integer/fractional partitioning to achieve finer resolution. This section first highlights one way of characterizing this coupling in silicon without needing sophisticated high precision voltage or current measurement equipments. For a given *constant* integer and fractional input to the DPA (or any DAC using $\Sigma\Delta$), the DPA output contains constant voltage (or current) proportional to integer code, along with a digital $\Sigma\Delta$ signal toggling at a fixed duty cycle based on the fractional input code. The rate of toggling is periodic and it is directly proportional to the constant fractional input for first-order single loop $\Sigma\Delta$. This results in frequency domain tones, called idle channel tones or simply idle tones, appearing at the fraction of the carrier frequency [102, 112].

For a first-order single-loop $\Sigma\Delta$ modulator, the relationship between a constant fractional value presented at the input and the frequency of the first idle tone can be derived by using closed form equations due to the simpler operation of first-order $\Sigma\Delta$ [112]. The digital $\Sigma\Delta$ is considered to be based on N bit wide accumulator. The fractional input bit-width is assumed to be N bit without loss of generality. The fractional input value can be represented by the ratio of two integers $\frac{x}{y}$ where $x < y \le (2^N - 1)$. Reducing the $\frac{x}{y}$ to its lowest terms (or irreducible form), it can be written as $\frac{X}{Y}$. Based on the $\Sigma\Delta$ operation, the accumulator overflow will be periodic with repetition rate of N samples. Thus the $\Sigma\Delta$ output bitstream repeats at every Y bits. Any binary sequence that repeats itself with a period of Y contains the fundamental spectrum at the frequency of sampling rate divide by the parameter Y. Hence the first idle tone frequency is given by,

$$f_{idle\_tone} = \frac{f_C}{Y} \qquad (4.3)$$

where f_C is the frequency of operation for $\Sigma\Delta$ conversion. As an example, for the presented transmitter, $N = 6$ bits and, for a carrier frequency of 1664 MHz, $f_C = 1664$ MHz. Lets assume a DC offset of $\frac{x}{y} = \frac{6}{2^6}$ which results in

$$\frac{X}{Y} = \frac{3}{32} \text{ and,}$$

$$f_{idle\_tone} = \frac{1664 \text{ MHz}}{32} = 52 \text{ MHz} \qquad (4.4)$$

Thus with a fixed fractional DC offset code of 0x6 (out of the 6 bits $\Sigma\Delta$ code - max value 0x3F), 52 MHz idle tone can be observed in the spectrum with constant level irrespective of the number of integer transistors turned ON. Here the "integer transistor" refers to either 1× or 4× integer transistor banks. The next section provides a link between the idle tones and the parasitic coupling.

4.7.3 RELATIONSHIP BETWEEN IDLE TONES AND $\Sigma\Delta$ PARASITIC COUPLING

The power spectral density of the fundamental idle tone should be constant in frequency irrespective of the corresponding integer code dialed. This behavior ensures that $\Sigma\Delta$ operation is completely independent of the integer code thus ensuring additive and independent behavior between the two. This is absolutely desired to ensure the best $\Sigma\Delta$ resolution enhancement. But, when the $\Sigma\Delta$ parasitic coupling is present (thus making $\Sigma\Delta$ output dependent on the integer code dialed), the level of the idle tone will vary (even if the fractional code is constant) when integer code is changed. As shown in the conceptual block diagram of Fig. 4.26, the $\Sigma\Delta$ fractional digital output could couple into the integer transistors through parasitic paths. This parasitic signal, after going thorough amplification and potential phase change, is combined with the desired $\Sigma\Delta$ transistor contribution at the output. If the parasitic signal couples out-of-phase compared to the original signal, the two signals combine destructively. For example, consider the strength of $\Sigma\Delta$ transistor be normalized to 1. The amount of parasitic coupling can be represented, on an average, by a coupling factor β. The total parasitic coupling coming out of integer transistors will depend on the number of integer transistors turned ON by having an active high signal on their gate. Thus, for n active integer transistors, with a coupling factor of β, and a normalized amplification factor of 1, the parasitic coupling at the output can be given by $(-n \times \beta)$. Overall contribution at the output pin is the combination of the two contributions and can be written as $[1 - (n \times \beta)]$ as shown in the idle tone strength of the simple spectrum in Fig. 4.26 with 0 to $Fs/2$ span on x-axis. If the coupling factor is high enough, the $(-n \times \beta)$ factor can be greater than 1 and it can overcome the desired $\Sigma\Delta$ in-phase signal as n increases.

This scenario is observed in the lab measurements during one of the test chips leading to the final production silicon. Fig. 4.27 shows the measured result where the out-of-phase signal nearly cancels the desired $\Sigma\Delta$ transistor contribution at around integer code of 100. For the rest of the power control range with digital code greater than 100, the parasitic coupling path dominates the desired $\Sigma\Delta$ modulation.

4.7 Analysis of parasitic coupling

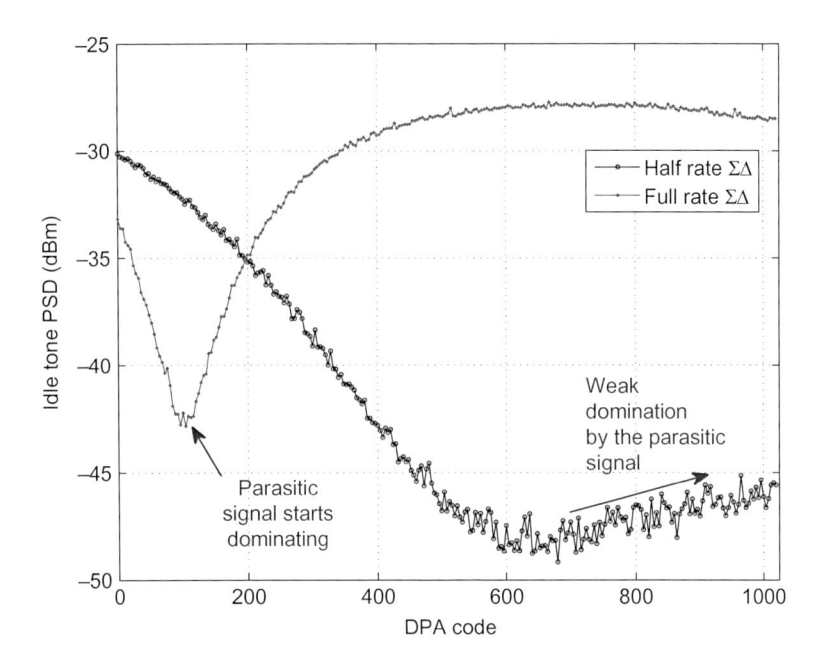

FIG. 4.27

Measured impact of $\Sigma\Delta$ switching additivity on idle tones.

It results in nonmonotonous amplitude response where increasing the fractional code results in decreased amplitude output because the $\Sigma\Delta$ contribution is 180° out of phase with the integer contribution. The parasitic coupling path is found to be frequency dependent where slowing down the $\Sigma\Delta$ clock frequency by half resulted in significant reduction of the above mentioned coupling as shown by the half rate $\Sigma\Delta$ curve in Fig. 4.27.

This type of parasitic coupling can create severe spectral degradation due to the nonmonotonous nature of the $\Sigma\Delta$ response. With modulated amplitude code, the fractional input is considerably more random resulting in increased noise floor and reduction of effective number of bits of resolution. The simulated effect of the coupling (Fig. 4.27) on the modulated spectrum, at DPA output, is shown in Fig. 4.28. Strong presence of spurs can be seen due to the coupling of low frequency integer transistor signal to the $\Sigma\Delta$ transistor.

For the presented design, this phenomenon is thoroughly characterized. Special care is taken to minimize any coupling by isolating the power supplies for SAM and DPA modules along with large amount of decoupling capacitors. SPICE simulation with full extraction is used with carefully planned stimulus to verify the coupling level at layout level. The measured performance on the final silicon shows almost negligible amount of coupling from the $\Sigma\Delta$ to the integer transistors. The measurement results are covered in Chapter 5.

FIG. 4.28

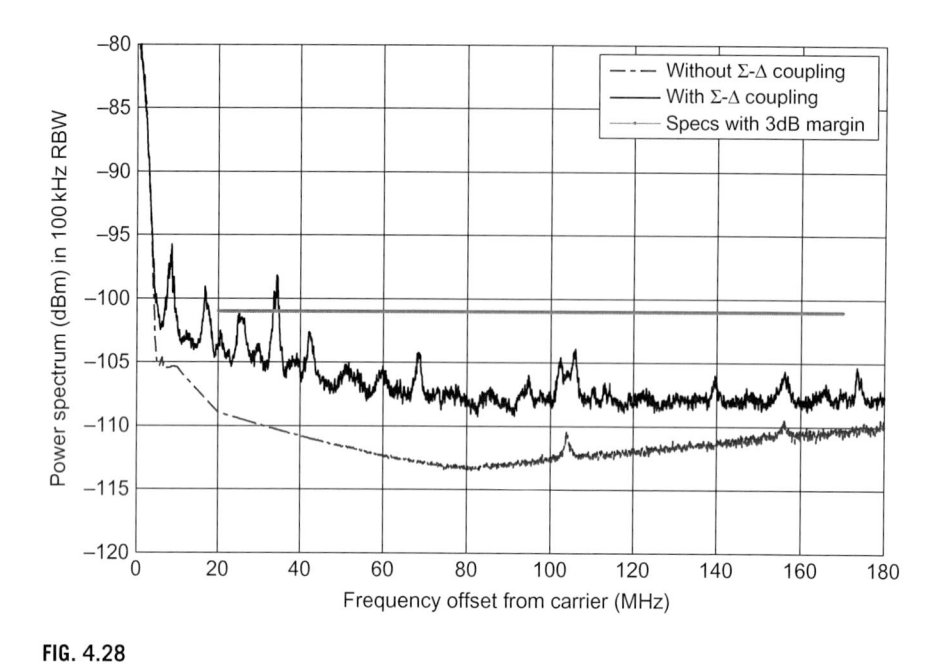

Simulation spectrum showing the effect of $\Sigma\Delta$ leakage on the RF spectrum.

CHAPTER 5

Simulation and measurement results of the polar transmitter*

CHAPTER OUTLINE

5.1 Simulation Results 117
5.2 Measurement Results... 118
    5.2.1 Predistortion 118
    5.2.2 Transmitter Close-In Performance 118
    5.2.3 Transmitter Wideband Noise Performance 123
    5.2.4 Performance Comparison 125
5.3 Conclusion 125

This chapter summarizes the simulation and silicon-measured results for the polar transmitter. Simulation and measurement results are shown in the individual sections where necessary to highlight the design methodology and justifications for individual modules. This chapter covers the results not covered in the individual sections. It also covers the transmitter level results over multiple parameters such as PVT, frequency, power level, etc.

5.1 SIMULATION RESULTS

Fig. 5.1 shows the performance of the simulation model contrasted with the lab measurement in the identical transmitter setup ($f_{carrier}$ = 1784.8 MHz, output power = −2 dBm, first-order $\Sigma\Delta$). As it can be seen, the simulation data closely correlates with the lab measurement—especially for the $\Sigma\Delta$ noise shaping. The fluctuations in the measured spectrum are mostly due to the measurement accuracy and calibration-related issues. This correlation serves to validate the correctness of the simulation model, including the modeling of various impairments such as $\Sigma\Delta$ coupling, various transistor mismatches, DPA nonlinearity, etc. More details of the simulation model are provided in Appendix A.3.

---

*The authors acknowledge substantial contributions from Prof. Dr. Poras T. Balsara (UT Dallas) and Dr. Oren Eliezer (Texas Instruments).

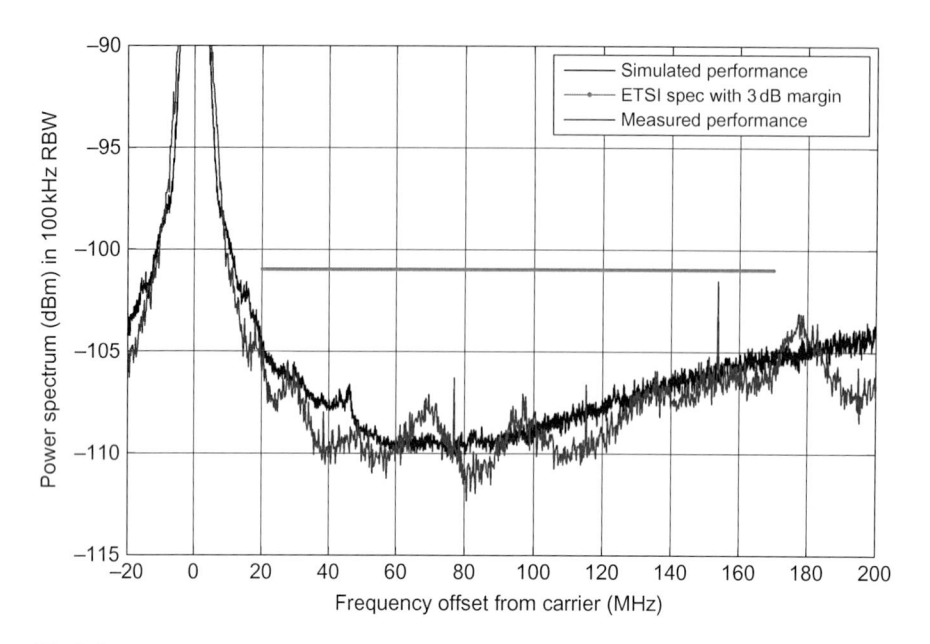

FIG. 5.1

Comparison of simulation results with lab measurements.

5.2 MEASUREMENT RESULTS

5.2.1 PREDISTORTION

An exhaustive set of measurements were performed to fully characterize the impact of various innovations in the predistortion module on the RF spectrum and EVM performance. Fig. 5.2 shows one such example measurement comparing the close-in spectrum of the transmitter output signal with and without predistortion. For this measurement, the transmitter output power is −2 dBm in high band mode with $f_{carrier} = 1710.2$ MHz in DCS band. As specified in Section A.2, the measurement is performed with 30 kHz resolution bandwidth (RBW). At 400 kHz offset from the carrier, the spectral power is −56.1 dB without use of predistortion, which is only 2 dB better than the specification of −54 dB. With the use of predistortion, the spectrum improves consistently above 300 kHz offset from the carrier. At 400 kHz offset, the improvement is about 9 dB resulting in −65 dB spectral power, which provides 11 dB of margin over the ETSI specifications in Section A.2.

Measurements of AM-AM and AM-PM distortions are covered in Section 3.2.

5.2.2 TRANSMITTER CLOSE-IN PERFORMANCE

Fig. 5.3 shows the time mask of the transmitted EDGE signal. The specifications are shown with a red envelope. The transmitter complies with the time mask specified

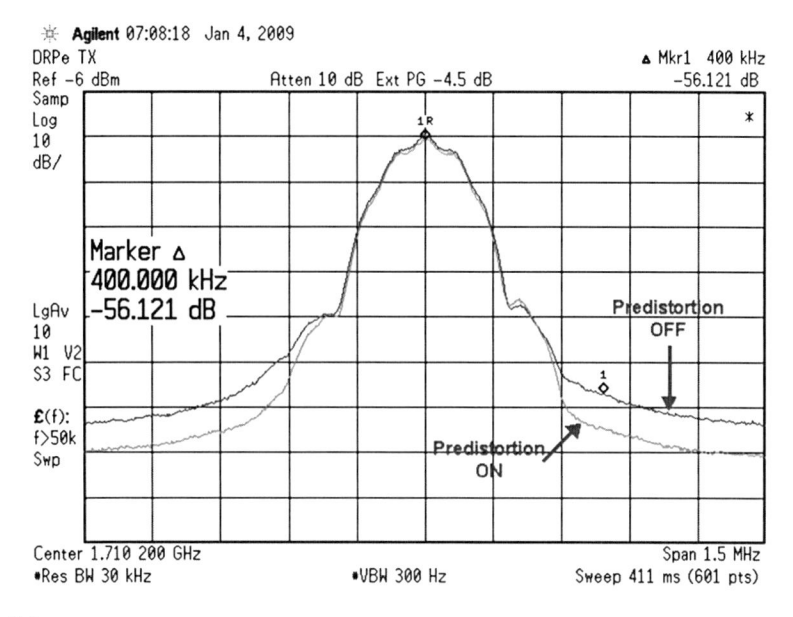

FIG. 5.2

Measurement results of predistortion performance.

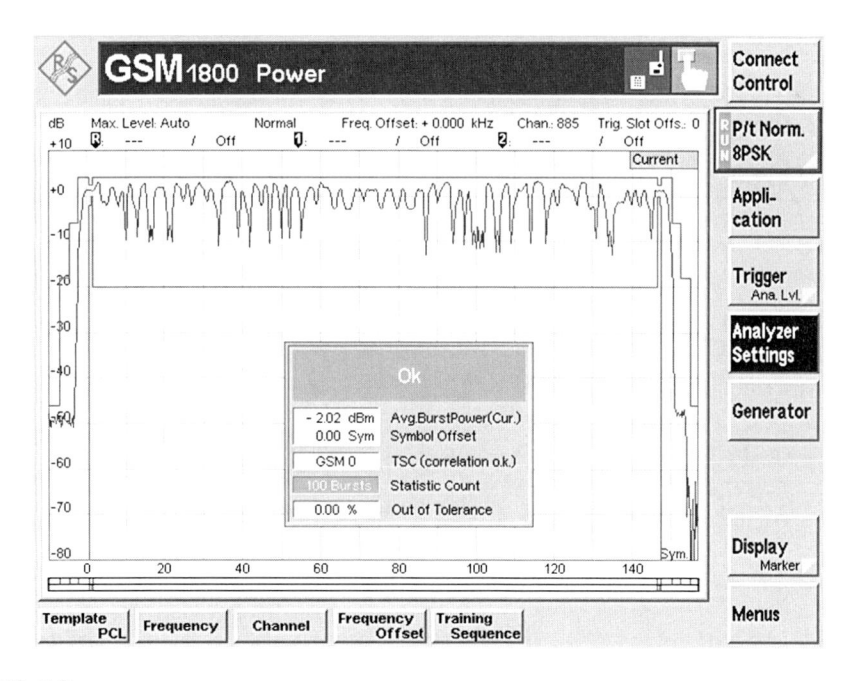

FIG. 5.3

A typical EDGE burst with time mask.

by the ETSI EDGE specifications [113]. The same compliance is observed across frequencies and power levels. These measurements are performed using Rhodes & Schwartz CMU 2000 instrument designed to test EDGE burst transmission compliance.

Fig. 5.2 shows the EDGE transmitter spectrum in continuous mode of operation without the burst mode of operation. Fig. 5.4 shows the spectral performance of the transmitter in burst mode of operation using the CMU 2000. The spectrum shown is extracted by averaging the useful portions of the burst transmission containing data. The power ramp-up and ramp-down are removed before measuring the spectrum as specified in Ref. [113]. The 400 kHz offset is shown to be -61.7 dB on the left side (lower sideband) of the carrier and -61.8 dB on the right side (higher side band) of the carrier.

Fig. 5.5 shows a measured constellation at the transmitter SoC output. It characterizes the key constellation specifications mentioned in Section A.2. The RMS EVM (max) is 1.78% while the peak EVM is 5.34% compared with the 10% and 30% specifications, respectively. Other parameters such as magnitude and phase error are shown as well for reference.

The following figures show the performance of a typical transmitter device across frequency bands of operation. Fig. 5.6 shows the measured EVM across the frequencies of low band. Over the frequency spread, the worst-case EVM is found to be 2.6%. Similarly Fig. 5.7 shows transmitter performance in the high band.

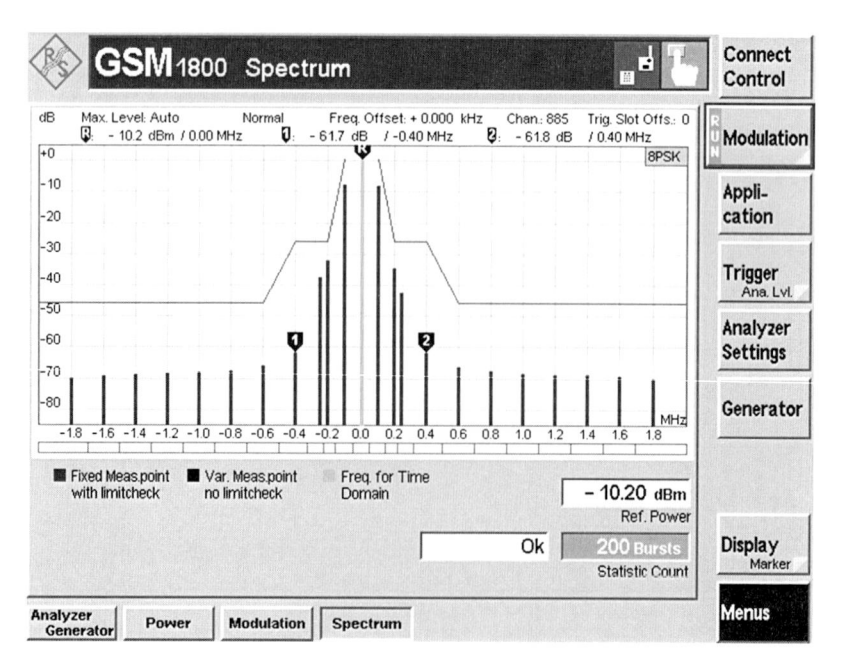

FIG. 5.4

EDGE burst mode spectral performance.

5.2 Measurement results

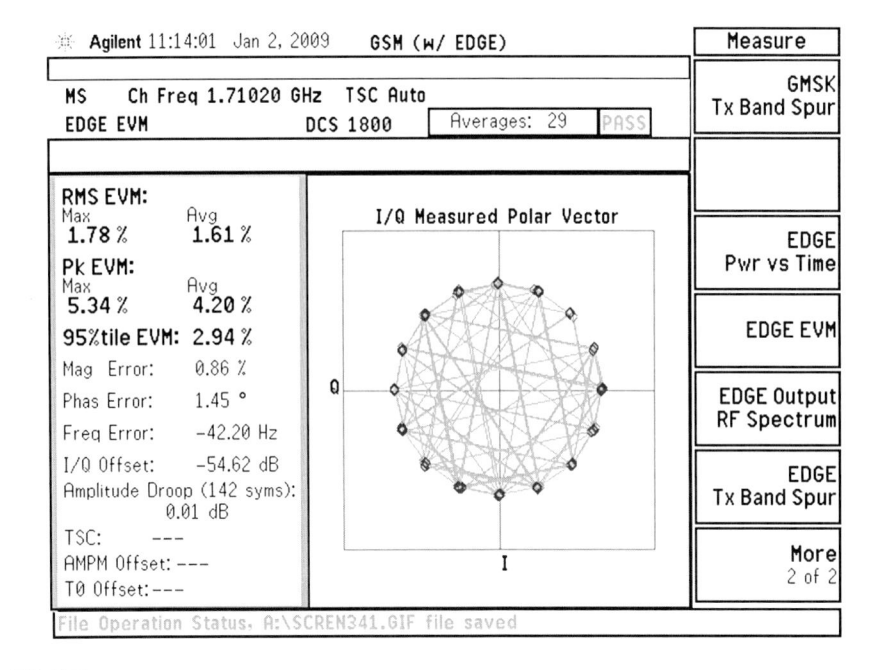

FIG. 5.5
EDGE EVM and constellation.

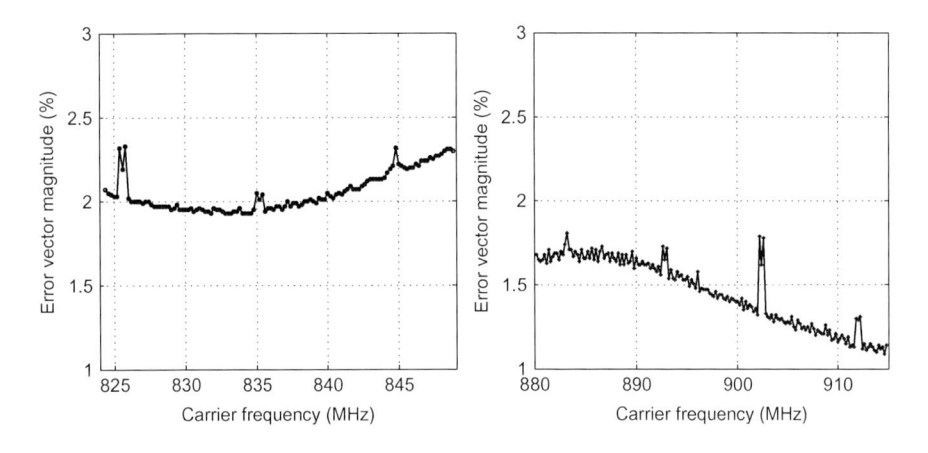

FIG. 5.6
EDGE EVM across low band frequencies.

Occasional spikes in the EVM over frequency are due to the integer channel problem covered extensively in Ref. [84].

Figs. 5.8 and 5.9 show the 400 kHz offset and 600 kHz offset spectral performance across low and high band frequencies. Similar integer channel-related distortion can be observed on certain high band carrier frequencies [84]. The low band

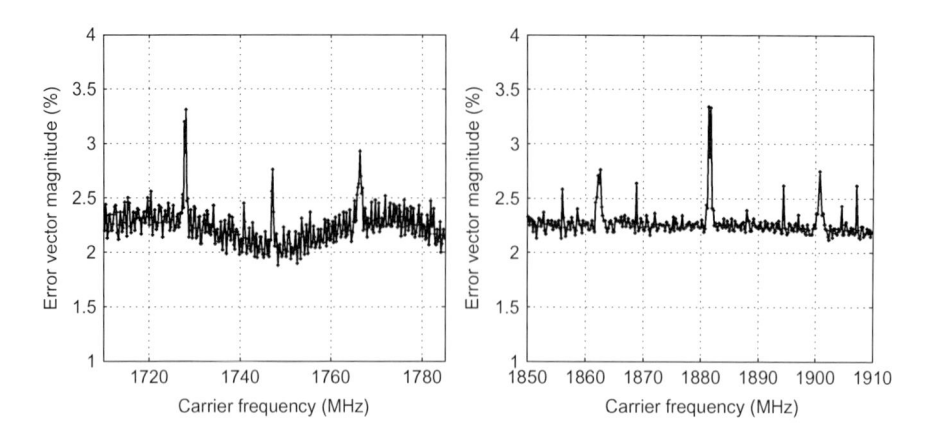

FIG. 5.7

EDGE EVM across high band frequencies.

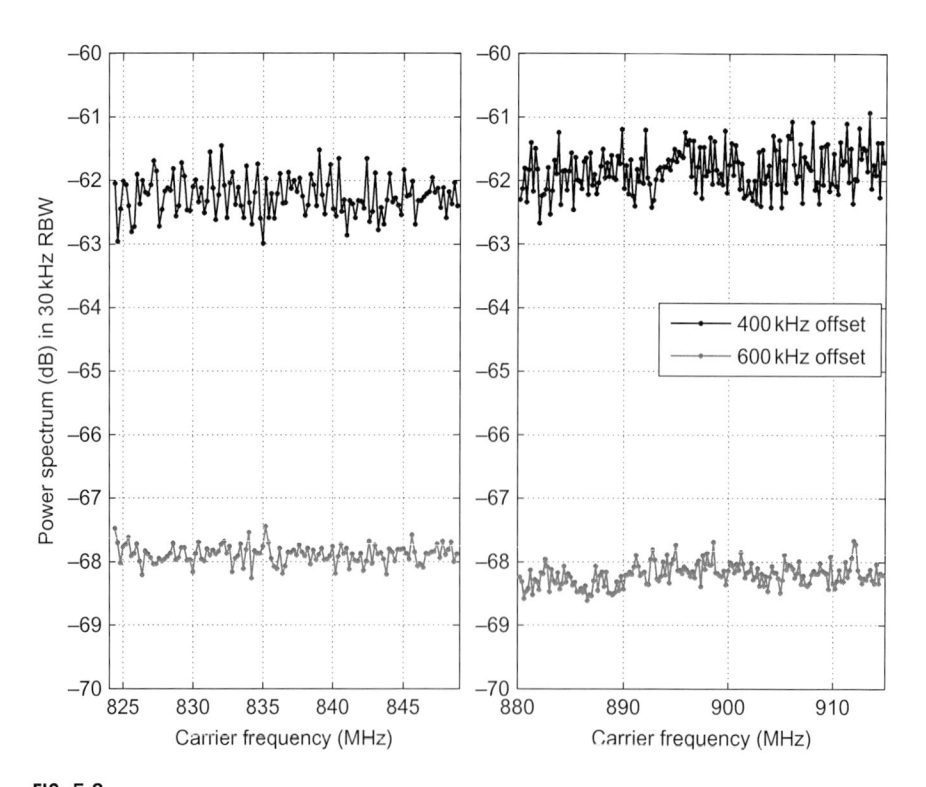

FIG. 5.8

EDGE 400 kHz and 600 kHz offset spectral performance across low band frequencies.

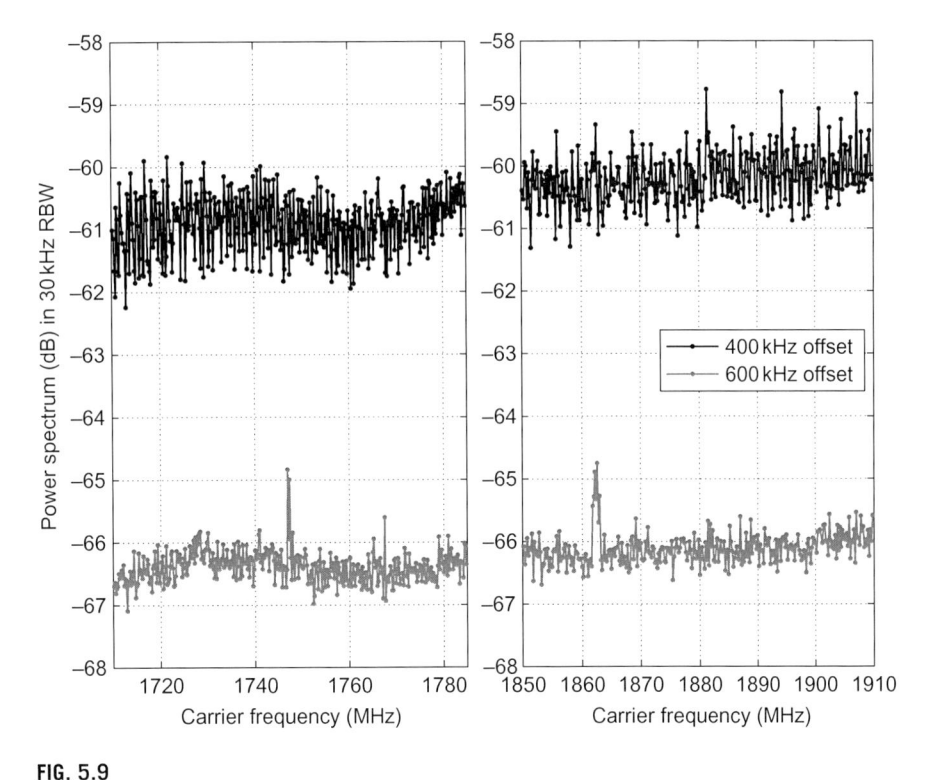

FIG. 5.9

EDGE 400 kHz and 600 kHz offset spectral performance across high band frequencies.

400 kHz offset spectral performance is better than −61 dB, which provides 7 dB margin over ETSI specifications of Fig. A.5. Similarly 600 kHz offset spectral performance on LB is better than −67 dB, which provides 7 dB margin over ETSI specifications. Worst-case high band performance at 400 kHz offset is −59 dB, which provides 5 dB margin over the specifications. And the high band 600 kHz offset performance of −65 dB provides 5 dB margin as well.

5.2.3 TRANSMITTER WIDEBAND NOISE PERFORMANCE

Fig. 5.10 shows wideband noise performance for $f_{carrier} = 914.8$ MHz in low band with +2 dBm output power. The rise in the noise floor is caused by the noise shaping of the $\Sigma\Delta$ modulator. The two major replicas are spaced at 114.35 MHz from the carrier, which is equal to the CKV/8. These are present because there is no interpolation on the integer path after the interpolative filter. Since the replicas are far away from the desired RX band as shown by the red ETSI specification line, they do not cause any violations of specifications. There is more than 3 dB margin over the internal specifications (containing an additional 3 dB of margin over ETSI specifications) throughout the spectrum. A similar replica can be seen in the high band spectral performance in Fig. 5.11 at $f_{carrier} = 1784.8$ MHz and −2 dBm output

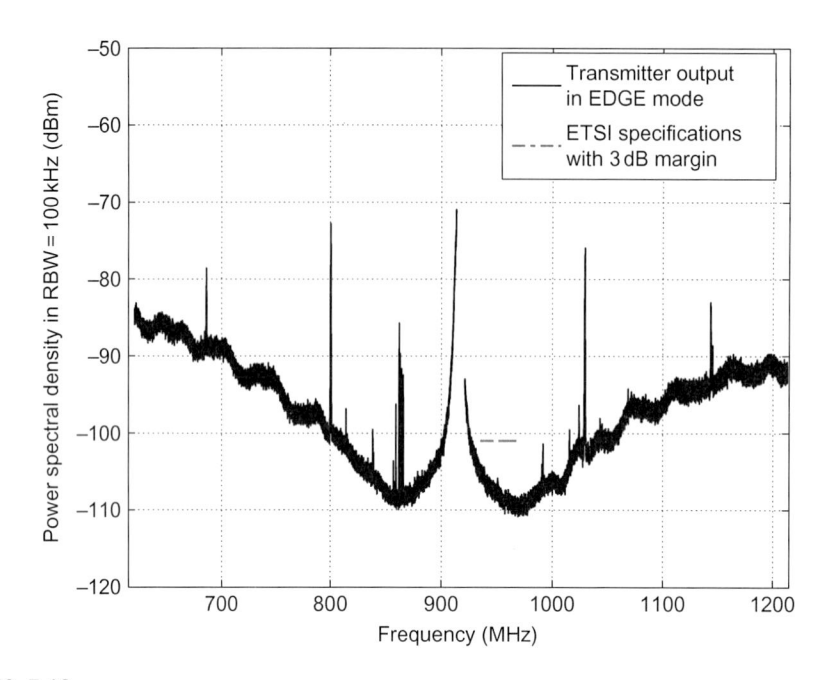

124 CHAPTER 5 Simulation and measurement results of the polar transmitter

FIG. 5.10

Wideband noise measured for a low band channel.

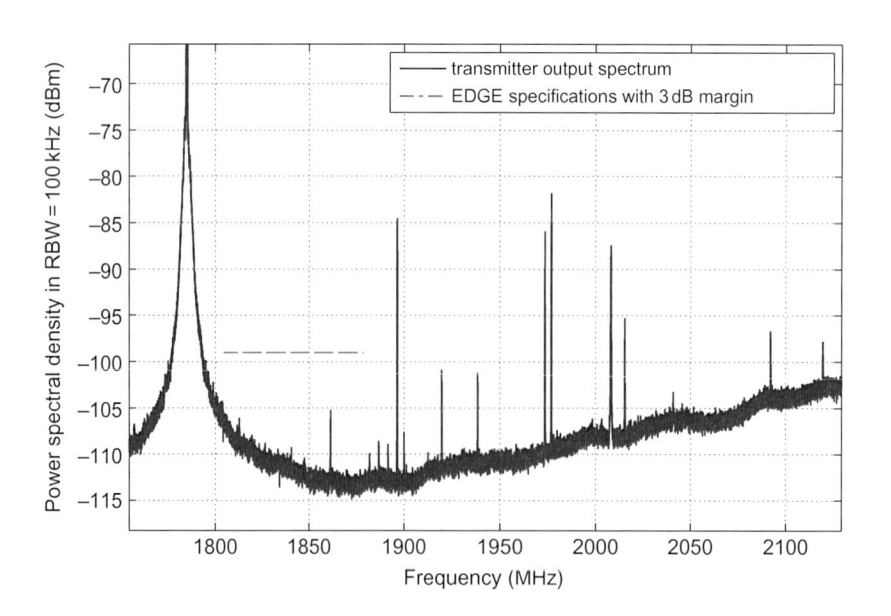

FIG. 5.11

Wideband noise measured for a high band channel.

power. The spacing of the first major spur is 111.55 MHz, which is CKVD/8, from the carrier. The high band spectrum also contains more than 3 dB margin over all the frequencies. Similar performance is observed over all the transmit frequency bands.

5.2.4 PERFORMANCE COMPARISON

Table 5.1 compares the performance of the presented transmitter with existing EDGE transmitters—ADI [40], Skyworks [90], and RFMD [42]. In most cases, the performance has been found to be at par or better than the existing solutions. The key difference between these transmitters is the implementation area and power consumption. The presented transmitter is more than $4\times$ smaller in area than the existing transmitters. Additionally, the power consumption is less than $3\times$ compared with the transmitter used in Ref. [90]. In addition to these three transmitters, the transmitter used in Ref. [114] had a power consumption of 320 mW (with 3.5 mm^2 area), which is inferior to the presented transmitter. Overall the presented transmitter has the same or better performance than the existing transmitters but with significantly smaller area and power consumption. This achievement can be attributed to highly digital nature of the transmitters. The authors believe the presented transmitter is far superior in its calibration and compensation time for each device but a numerical comparison is difficult due to the proprietary nature of such numbers.

5.3 CONCLUSION

An RFDAC-based polar GSM/EDGE transmitter was introduced where the RFDAC achieves 10 bit integer and 3 bits fractional resolution using $\Sigma\Delta$-based noise shaping. A fully digital architecture was proposed for the transmitter, which meets all GSM/EDGE specification by solving critical issue inherent to the digital architecture. A new interpolation filter was introduced which achieves two orders of higher filtering than the conventional CIC filter with roughly the same complexity. The same filter also provides inherent delay alignment capabilities for the amplitude and phase path. Clock skew alignment—a problem unique to digital architectures—was presented and a built-in calibration and compensation mechanism is shown to calibrate the right delay value over PVT corners. Parasitic coupling was shown to be one of the issues affecting the design due to high-frequency digital switching. A new way to characterize the parasitic coupling was shown using $\Sigma\Delta$ idle channel tones.

A mismatch centric design of the RFDAC was also presented for a digital polar EDGE transmitter. A new concept of phase mismatch was presented in the context of high-frequency upconverting RFDAC and specifications for the same are derived based on the wideband noise (WBN) specifications of the transmitter. Digital predistortion was designed for precompensation of the systematic $1\times$-$4\times$ mismatch by characterizing the mismatch using on-chip receiver and storing the distortions in the look-up-table. Additional specifications were derived for tolerable random clock skew in the array of transistors. Finally silicon-based measurements of individual

Table 5.1 Performance Comparison With Existing Solutions

8PSK TX Parameter	Presented TX		ADI		Skyworks		RFMD		3GPP Spec		Units
	LB	HB	LB	HB	LB	HB	LB	HB	LB	HB	
RMS EVM	2	2.5	<3	<3	2.9	3.8	2	3		9	%
Peak EVM	3	6	<9	<9	4.8	6	5	8		30	%
400 kHz mask	−62	−61	<−60	<−60	−60	−60	−59	−57		−54	dBc
600 kHz mask	−68	−66	<−64	<−64	−65	−64	−63	−62	−60	−58	dBc
20 MHz noise	−158	−154	−159	−153					−156	−146	dBm/Hz
SoC output power	2	1	2	1					N/A		dBm
Current consumption	94	105	75	72					N/A		mA
Supply voltage	1.2				2.7						V
Power consumption	110				390						mW
TX area	0.8		7.8		3.7						mm^2
Technology	65 nm digital CMOS		0.5 μm SiGe BiCMOS		130 nm digital CMOS		250 nm digital CMOS				N/A

transistors were used to confirm the compliance of derived mismatch specifications for the RFDAC over PVT conditions.

In addition, a new predistortion scheme for a digital polar EDGE transmitter was proposed, which realizes a compensation mechanism of low area and low complexity for the distortions created in the amplitude modulator of the transmitter, as well as for possible additional distortions contributed by an external PA. The predistortion is based on look-up-tables whose sizes have been optimized based on the analysis of the impact of the resultant quantization error on the transmitter's modulated spectrum. The look-up-tables holding the measured distortions are used in conjunction with an interpolation scheme to dynamically derive the necessary predistortion for each instantaneous level of desired output amplitude, thus eliminating the need for a computation of the inversion of the distortion characteristics. The revealed scheme maintains the performance of the transmitter across the desired dynamic range.

The transmitter is able to achieve significant margins over ETSI specifications for GSM and EDGE standards with very small area of 0.8 mm^2. The generic method of the RFDAC design can be used for any DAC design to derive the physical mismatch specifications from high-level performance specifications. The challenges faced and their solutions will work as a guide for the future designer using digitally intensive architectures in multi-GHz radio designs.

The small-signal digital polar architecture proposed in this research can be extended to many standards requiring frequency modulation, which is comparable to the EDGE or Wideband CDMA (WCDMA) standards. Future extensions of this transmitter can also be used to implement low-cost versions of various voice communication transceivers such as Project P25 [115]. The direct use of this architecture in LTE and future radio standards is difficult, due to challenges highlighted in Section 3.5. Even though a different architecture is suggested for LTE and beyond, the core blocks proposed in this work (such as the interpolative filter, predistortion, DPA, etc.) can be used along with Cartesian-based architectures to support the newer standards. Similarly the design methodology highlighted for the RFDAC design (including the methods for mismatch modeling, self-calibration for clock-skew, characterization of coupling, etc.) are valuable tools for the design of any transceiver.

CHAPTER 6

Idea of all-digital I/Q modulator*

CHAPTER OUTLINE

6.1 Concept of Digital I/Q Transmitter 129
6.2 Orthogonal Summing Operation of RFDAC 131
6.3 Conclusion 141

As discussed in the preceding chapters, for wide modulation bandwidths, due to their direct linear summation of the I and Q signals and thus the avoidance of the bandwidth expansion, Cartesian [16, 57–66] modulators are proved to be a better choice than their polar [13, 14, 17, 19, 46–49] or outphasing [51, 54–56] counterparts. Reference [16] proposed a digitally controlled I/Q modulator that utilizes current sources to isolate the orthogonal I and Q paths. The exploitation of the current sources, however, worsens the far-out noise. Additionally, in order to produce the required RF output power, that approach employs an external power amplifier. Later, an I/Q direct digital RF modulator is introduced in Ref. [63] in which an FIR-based quantization noise filter is embedded in order to filter the quantization noise in the receiver frequency band. Implemented in 130 nm CMOS, it also employed numerous current sources to isolate the orthogonal paths as well as to establish the proper coefficient value for the FIR filtering operation. However, its drain efficiency, at 15.4 dBm output power, does not exceed 13%. Moreover, the related noise floor is not better than −152 dBc/Hz at 20 MHz offset. To alleviate the foregoing concerns, a new digital I/Q modulator concept is introduced. Section 6.1 explains the principal concept behind the proposed digital I/Q modulator. Section 6.2 discusses the various types of orthogonal summing in digital intensive or all-digital I/Q transmitters. Finally, Section 6.3 summarizes this chapter.

6.1 CONCEPT OF DIGITAL I/Q TRANSMITTER

Fig. 6.1 illustrates the concept of the digital I/Q modulator. The desired IQ is constructed by vectorial summing of their composite I and Q digital vectors. Their

---

*The authors acknowledge substantial contributions from Prof. Dr. Leo de Vreede (TU Delft).

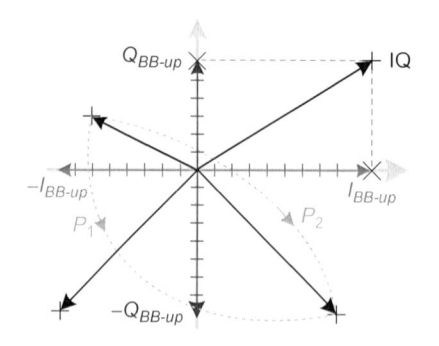

FIG. 6.1

Digital I/Q modulation concept; Its related IQ constellation vectors.

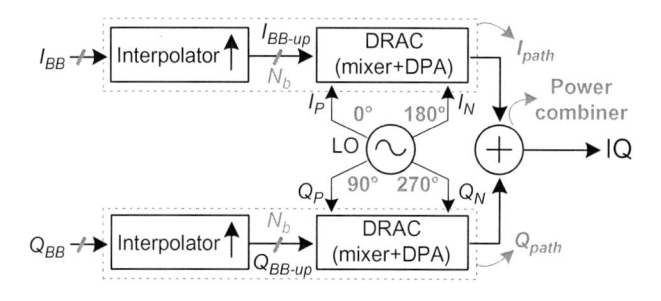

FIG. 6.2

Digital I/Q modulation concept; Its related idealized block diagram.

code resolution (N_b) must be high enough to cover all I/Q points of the corresponding trajectory connecting the symbols [22].

This indicates that, for supporting only an m-symbol constellation diagram, the resolution of the digital I/Q modulator should be at least[1]

$$N_b \geq \log_2\left(\sqrt{\frac{m}{4}}\right) \quad (6.1)$$

In addition, N_b also affects the subsequent quantization noise, which is discussed in more detail in the following chapters. A significant issue related to any transmit modulator is its agility to traverse from one I/Q point to another. As graphically depicted in Fig. 6.1 by P_1 and P_2 paths, traversing along P_2 trajectory instead of P_1 makes the complex baseband modulation faster and, consequently, the modulator must manage a wider bandwidth as well as a higher sampling rate. To do so, based on the idealized block diagram in Fig. 6.2, the I_{BB} and Q_{BB} digital baseband signals are

---

[1] Of course, the resolution based on Eq. (6.1) is not enough to support the corresponding IQ trajectories.

upsampled as $I_{BB\text{-}up}$ and $Q_{BB\text{-}up}$. This process ensures that the spectral images will be attenuated and located far away from the carrier and thus can easily be filtered out. The $I_{BB\text{-}up}$ and $Q_{BB\text{-}up}$ are $2 \times N_b$-bit (N_b for in-phase as well as N_b for quadrature component) upsampled digital signals, which should be directly upconverted to their continuous-time reconstructed RF output signal. As a result, these signals are applied to a pair of DRACs, comprising an array of 1-bit unit cell mixers and 1-bit unit cell DPAs.

The DRACs are clocked in tact of differential quadrature upconverting clocks I_P, I_N, Q_P, and Q_N.[2] According to Fig. 6.1, the four quadrants of the constellation diagram must be addressed by the modulator. The switching between quadrants can be achieved by swapping between I_P/I_N or/and between Q_P/Q_N according to the sign bits of $I_{BB\text{-}up}$ and $Q_{BB\text{-}up}$. The DRAC outputs are connected to a power combiner that facilitates the conversion of the upconverted digital signals into the reconstructed RF output. In fact, the digital I/Q modulator represents an RFDAC. In this approach, however, the primary challenge is related to the orthogonal summing of the I and Q DRAC outputs in order to reliably reconstruct the modulated RF signal [26, 116–118].

6.2 ORTHOGONAL SUMMING OPERATION OF RFDAC

From the digital communications theory, in order to maintain high bandwidth efficiency, the baseband information is generally represented by two orthogonal streams, that is, I and Q signals, each modulated by the corresponding orthogonal carrier signal (i.e., basis function), and they are subsequently summed. Based on that, the I/Q RFDAC of Fig. 6.2 has two signal paths, namely, the I_{path} and the Q_{path} performing the following operations:

$$I_{path}(t) = (I_P(t) - I_N(t)) \times I_{BB\text{-}up}(t) = 2 \times I_P(t) \times I_{BB\text{-}up}(t) \quad (6.2)$$

$$Q_{path}(t) = (Q_P(t) - Q_N(t)) \times Q_{BB\text{-}up}(t) = 2 \times Q_P(t) \times Q_{BB\text{-}up}(t) \quad (6.3)$$

The final modulated IQ signal is generated by vectorial summation of Eqs. (6.2), (6.3) and mathematically expressed as:

$$\begin{aligned} IQ(t) &= I_{path}(t) + Q_{path}(t) \\ &= 2 \times \{I_P(t) \times I_{BB\text{-}up}(t) + Q_P(t) \times Q_{BB\text{-}up}(t)\} \end{aligned} \quad (6.4)$$

The transmitted RF signal is a band-pass filtered version of $IQ(t)$.

$$RF_{out}(t) = \text{filter}[IQ(t)] \quad (6.5)$$

Eq. (6.4) reveals two aspects of the quadrature modulation: orthogonality and summation. The summing operation must be orthogonal, and there should be no

---

[2] These clocks comprise four phases separated by 90°.

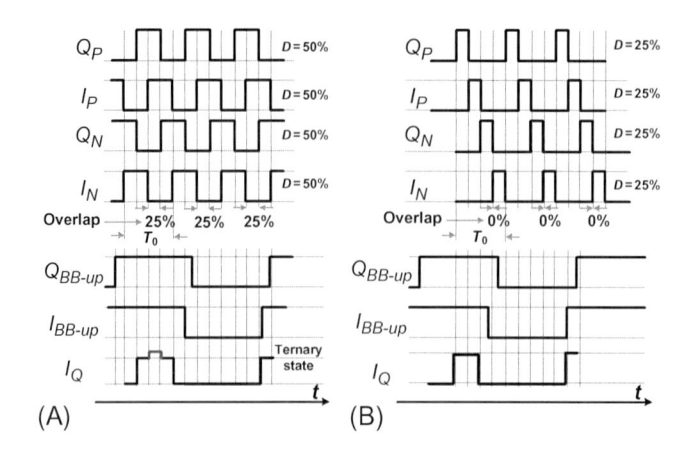

FIG. 6.3

Conceptual diagram of I/Q signals: (A) $D = 50\%$; (B) $D = 25\%$.

interaction or correlation between I_{path} and Q_{path}, or the EVM, BER, and spectral regrowth will arise. The digital carrier signals are typically rectangular pulses with a $D = 50\%$ that toggle between ground and supply. If duty cycle of the upconverted clock is 50% [66], there invariably exists an overlap between I_P/I_N and Q_P/Q_N. Mathematically, their orthogonality can be verified by employing a dot product operation. Fig. 6.3A is a conceptual illustration of the carrier signals. Based on mathematical principles, two signals are orthogonal with respect to each other if an integral of their inner product over the interval of one period is zero. The orthogonality of the carrier signals ($D = 50\%$) is examined as follows:

$$\frac{1}{T_0} \int_0^{T_0} [(I_P - I_N) \cdot (Q_P - Q_N)] \, dt = 0.25 \tag{6.6}$$

where T_0 is the clock period, and the clocks are assumed of unity amplitude. Based on Eq. (6.6) the carrier signals, I_P/I_N and Q_P/Q_N, are not orthogonal. Fig. 6.3A intuitively confirms Eq. (6.6) and clearly demonstrates that, when I_P and Q_P are both simultaneously one, provided that I_{BB-up} and Q_{BB-up} are also one, then the resulting IQ signal of a practical circuit implementation exhibits a third state. This signifies that the I_{path} and Q_{path} are correlated, and the receiver cannot determine the component to which this unsolicited state belongs. Let's consider an idealized digital I/Q modulator depicted in Fig. 6.4. Employing the $D = 50\%$ clocks, the foregoing circuit is simulated. According to its SPICE simulated constellation diagram of Fig. 6.4B, its related EVM at 16 dBm RF output power is −21 dB. Hence, to improve linearity, a sophisticated DPD algorithm would be required [66]. In addition, the drain efficiency of its composite DPA is deficient due to the fact that the maximum conduction angle is 75% of the RF clock cycle.

6.2 Orthogonal summing operation of RFDAC

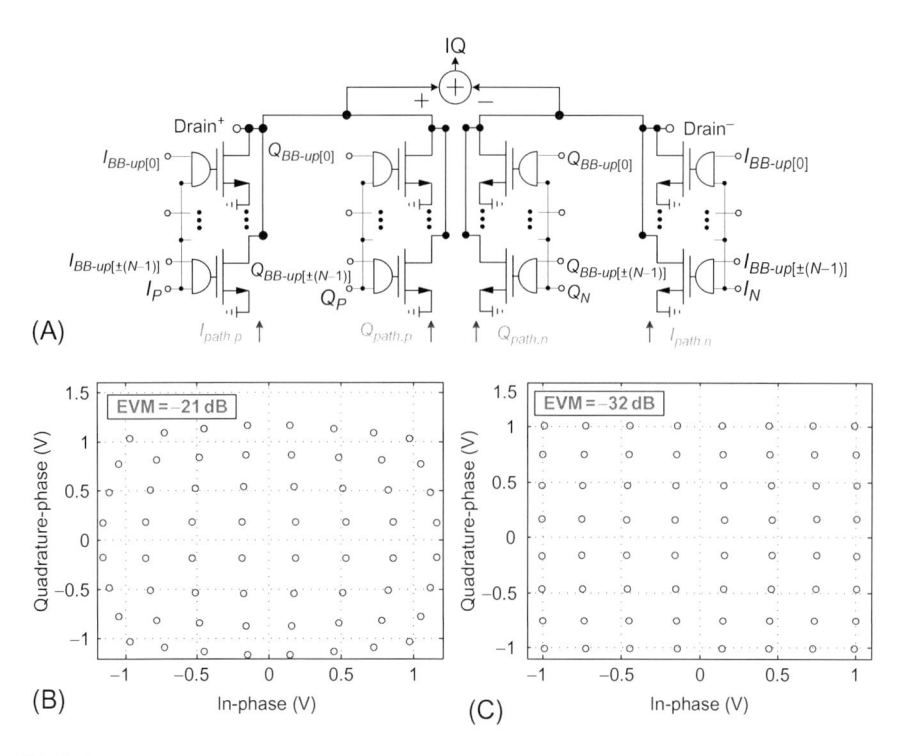

FIG. 6.4

(A) Idealized schematic of digital I/Q modulator; Its related SPICE simulated constellation diagrams for (B) $D = 50\%$; (C) $D = 25\%$.

To make the carrier signals orthogonal, their overlapping part should be eliminated. As a result, to perform orthogonal summation, the duty cycle of upconverting clocks is selected at 25% to deter any interaction between the I_{path} and Q_{path}. Based on Fig. 6.3B, the overlap between I_P/I_N and Q_P/Q_N is now zero and the resultant IQ signal comprises only two states. The orthogonality of the new carrier signals can be expressed as:

$$\frac{1}{T_0} \int_0^{T_0} [(I_P - I_N) \cdot (Q_P - Q_N)] dt = 0 \qquad (6.7)$$

Thus, this also mathematically confirms the orthogonality of the component carriers. The solution could be considered as a TDD in which the linear addition of the time-shifted I and Q paths is accomplished by allocating individual time slots to enter the I/Q information into the system. Employing the aforementioned $D = 25\%$ upconverting clocks of the digital I/Q modulator of Fig. 6.4A, the circuit level simulated constellation diagram of Fig. 6.4C is achieved. Its corresponding EVM at 16 dBm RF output power is -32 dB. As a result, this system only needs

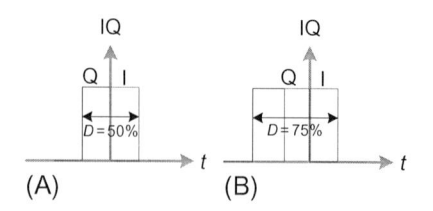

FIG. 6.5

Illustration of full power combining of I_{path} and Q_{path} employing upconverting clock for (A) $D = 25\%$; (B) $D = 50\%$.

Table 6.1 Comparison Between First, Second, and Third Harmonic Components: $D = 50\%$; $D = 75\%$

IQ Duty Cycle	Fundamental (dB)	Second (dB)	Third (dB)
50%	−2.34	−327	−11.88
75%	−5.35	−8.36	−14.89

a very simple DPD [119, 120] and, more significantly, the related drain efficiency of its composite DPA is higher due to the 50% maximum conduction angle. Note that, according to Fig. 6.4, the I/Q RFDAC can cover the entire four-quadrant constellation diagram.

It should be pointed out that, at the maximum power of operation, since both I_{path} and Q_{path} are activated, the subsequent drain voltage waveform is a square-shaped signal with $D = 50\%$ providing that the duty cycle of their related upconverting clocks is 25% (see Fig. 6.5A). In contrast, if the duty cycle of the upconverting clock is chosen at 50%, which is depicted in Fig. 6.5B, their subsequent IQ signal is a square-shaped waveform with the duty cycle of 75%. Table 6.1 summarizes their corresponding fundamental, second, and third harmonic frequency components. According to it, the fundamental component of the 50% IQ waveform is $\sqrt{2}\times$ higher than with the 75% one. Moreover, the 50% IQ waveform, ideally, creates a zero second harmonic frequency component which facilitates the design of the following transformer balun in the power combiner. On the other hand, the preceding clock generator circuits of the 50% duty cycle differential quadrature signals are less complicated than their 25% counterparts.

As stated earlier in (6.4), an additional aspect of the quadrature modulation is the summation. There would be at least four different ways of adding the orthogonal base function signals.

In the first approach [16, 58], as depicted in Fig. 6.6, the summation is performed by electrically connecting unit-weighted ideal current sources and adding them according to the input data. The controlling signals of the current sources are $dI^+_{0,1,\ldots,N-1}$ and $dQ^+_{0,1,\ldots,N-1}$ and are expressed as:

6.2 Orthogonal summing operation of RFDAC

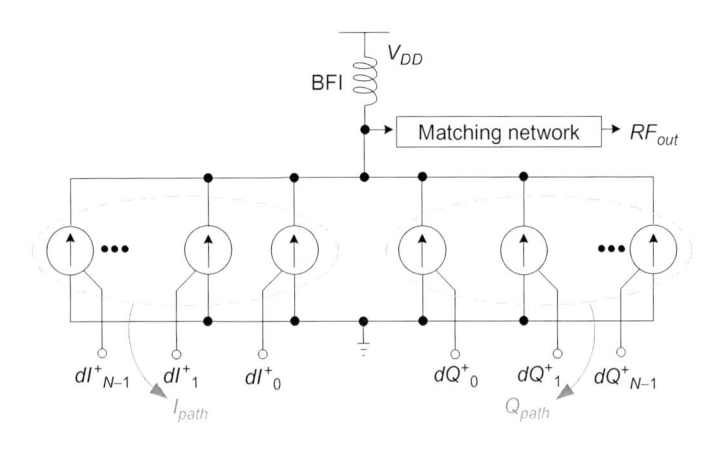

FIG. 6.6

Analog current source arrays summing.

$$dI^+_{0,1,\ldots,N-1} = I_P \times I_{0,1,\ldots,N-1} \tag{6.8}$$

$$dQ^+_{0,1,\ldots,N-1} = Q_P \times Q_{0,1,\ldots,N-1} \tag{6.9}$$

which indicates the implicit mixer operation. For maintaining the orthogonality of the I/Q signal, the I_{path} and Q_{path} must be uncorrelated such that one signal path output does not affect the operation of the other. This might require resorting to a current source impedance boost technique, such as cascoding. Unfortunately, stacking of the MOS transistors in a cascode structure is difficult in the modern low-voltage CMOS technologies and further produces an excessive amount of leakage and noise. In summary, numerous disadvantages exist related to this approach. First, this structure comprises MOS current sources which continuously work in the saturation region, therefore, create more noise than when operated in other regions. This structure is primarily implemented employing the Gilbert mixer topology. The thermal output spot noise of a Gilbert cell mixer can be approximated as [60]:

$$\overline{V^2_{out-n}} = 4KT\gamma \frac{G^2}{g_m} + KT\pi \frac{G}{g_m} \tag{6.10}$$

where $K = 1.38 \times 10^{-23}$ J/K denotes the Boltzmann constant, T is the absolute temperature, the parameter γ has a value of 2/3 in saturation region for long channel devices, g_m is the device transconductance, and $G \propto g_m R_{out}$ is the voltage gain of the Gilbert mixer (R_{out} is the equivalent mixer output resistance). Since the gain of the mixer is constant due to linearity constrain, in order to reduce the noise, g_m should be increased. Thus, as a second disadvantage, the power efficiency of this approach is minimal, as is apparent in Refs. [16, 58]. Third, the linearity of this structure is not promising due to the voltage to current conversion in the Gilbert cell.

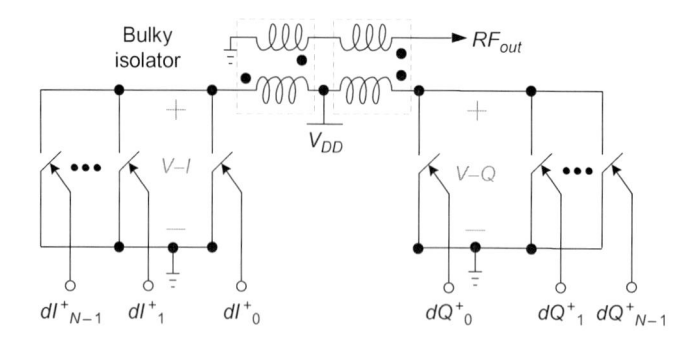

FIG. 6.7

Two separate digital switch arrays using transformer summing.

The second approach illustrated in Fig. 6.7 might also be realized as a pair of digitally controlled RF-modulated resistor structures. This approach could benefit from a less complicated circuit, the lack of stacked devices, and the elimination of noisy current sources. Since the MOS switch operates either in the off or in triode state, less noise is introduced. In addition, because of the switched-mode behavior, it has a potential to produce higher power efficiency [50]. Unfortunately, the final I/Q signal summation is difficult to accomplish in this method since the individual I and Q outputs are not currents but, instead, voltages. Due to the fact that the individual voltages of the RF outputs of the matching networks must be included, bulky microwave-type isolator/combiner would be required, otherwise, the RF voltage level of the I_{path} will affect the impedance of the Q_{path}, and vice versa. Hence, the I/Q orthogonality of this structure will not be preserved.

The third approach is based on employing the quadrature passive mixer which utilizes 25% duty cycle upconverting clock, which is depicted in Fig. 6.8A [60, 61]. Note that, in this approach, the baseband input signals are converted to analog continuous-time waveforms exploiting two separate DACs and their following LPFs (see [60, 61]). To boost the subsequent upconverted signals, however, this approach requires stand alone on-chip as well as off-chip power amplifiers. Thus, its power efficiency is deficient. Moreover, this approach is a digitally intensive approach rather than a fully digital approach. In order to transform it into a fully digital structure, the upsampled baseband signals are directly applied to passive mixer arrays [63]. In this approach, the in-phase and quadrature-phase signals are orthogonally summed employing analog current source cells, which are depicted in Fig. 6.8B. Note that the aforementioned current sources also provide the required FIR filter coefficient in order to notch the generated noise of the transmitter at the receiver frequency band. Implemented in a 130 nm CMOS process, this approach requires large voltage headroom due to adopting current sources. Moreover, this structure necessitates an additional power amplifier to increase the transmitted RF power thus its power efficiency is deficient.

6.2 Orthogonal summing operation of RFDAC

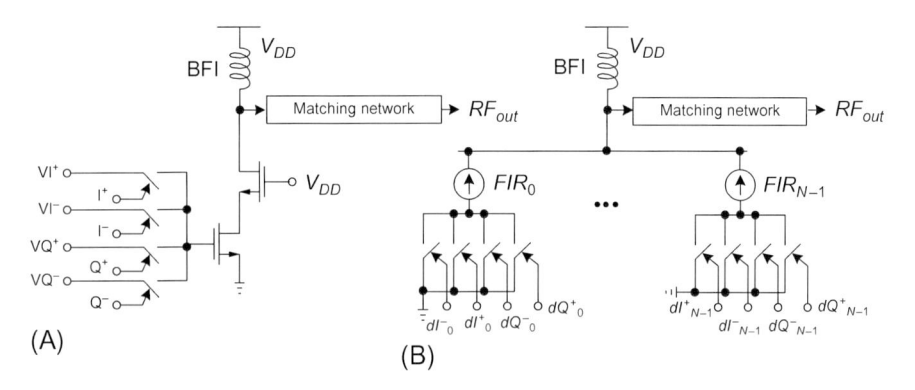

FIG. 6.8
(A) Quadrature passive upconverting mixer using 25% duty cycle upconverting clock.
(B) Analog current source FIR arrays summing.

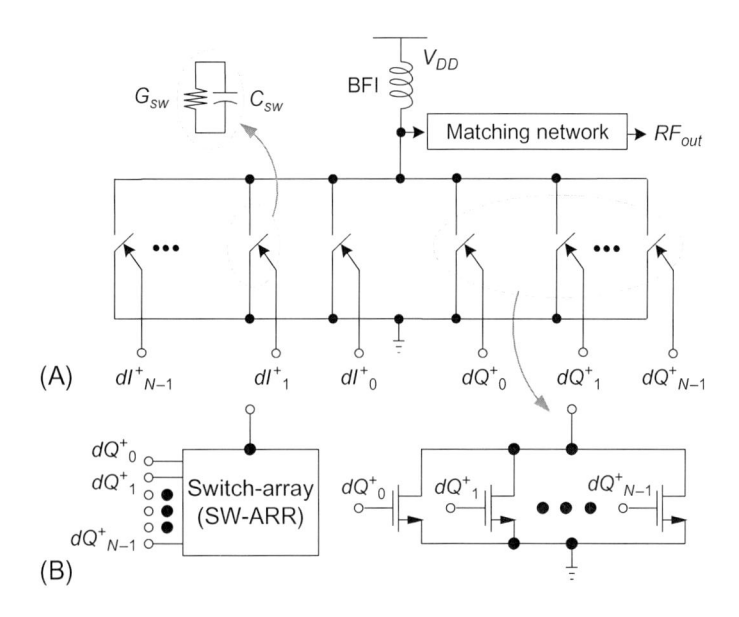

FIG. 6.9
Electric summing; (A) digital switch arrays; (B) switch array structure.

In the fourth approach, the total I/Q summation is also performed by the addition of unit-weighted digital switches (Fig. 6.9A), each contributing a finite conductance G. The difference between the second and fourth approaches is that, in the former, the I_{path} and Q_{path} are isolated and subsequently summed in the voltage domain while, in the latter, these paths are electrically summed (like in the first approach). As a

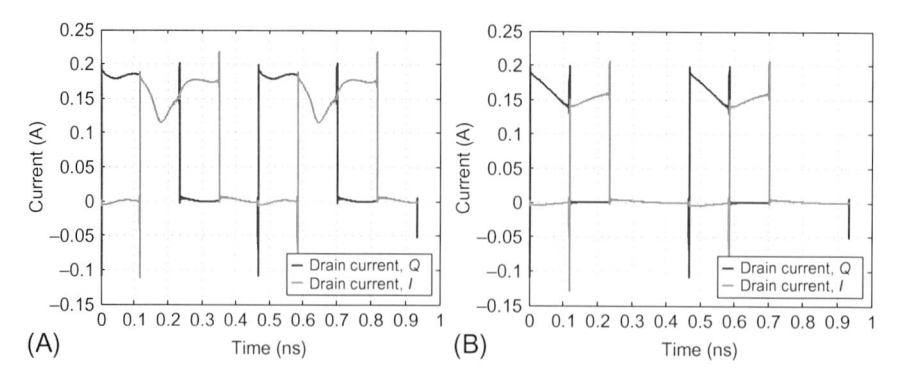

FIG. 6.10

Simulated I/Q drain currents: (A) $D = 50\%$; (B) $D = 25\%$.

result, this approach selectively exploits the best advantages of the first and second approaches.

Utilizing the Agilent ADS circuit simulator, two simulations ($D = 50\%$, and $D = 25\%$) are performed based on the approach exhibited in Fig. 6.9A. The corresponding I/Q switch array banks comprise 512 unit-weighted switches. The dimensions of each NMOS switch are $W/L = 660\,\text{nm}/100\,\text{nm}$. The frequency of operation targets the basestation WCDMA band-I at 2.14 GHz. Simulation results confirm that, for the $D = 50\%$ case in Fig. 6.10A, the orthogonal summation is not feasible. The reason is that, in the overlap region, the drain voltage further decreases and forces both I and Q switch banks to enter deep-triode, which causes a subsequent drop in the drain current. However, for $D = 25\%$ in Fig. 6.10B, there is no overlap; therefore, the summation possesses potential to be orthogonal. As depicted in Fig. 6.9A, each digital switch can be modeled as a combination of a parallel conductor (G_{sw}) and drain capacitor (C_{sw}). The off-state resistance of the switch is substantial and can be approximated as a zero conductance. The on-state conductance and resistance of the NMOS switch are:

$$G_{sw\text{-}ON} = \mu_n C_{ox} \frac{W}{L} (V_{GS} - V_{th}) \quad (6.11)$$

$$R_{sw\text{-}ON} = \frac{1}{G_{sw\text{-}ON}} \quad (6.12)$$

where μ_n and C_{ox} are electron mobility and gate oxide capacitance per unit area of NMOS transistor, respectively. In addition to the finite conductance (Eq. 6.11), C_{sw} depends on the width of the switching transistor [121].

$$C_{sw} = W \times E \times C_j + 2 \times (W + E) \times C_{jsw} \quad (6.13)$$

where E is drain length, C_j and C_{jsw} are junction capacitance per unit area, and sidewall capacitance per unit length, respectively. Note that junction capacitance is inversely related to drain voltage.

6.2 Orthogonal summing operation of RFDAC

$$C_j = \frac{C_{jo}}{\left[1 + \frac{V_R}{\Phi_B}\right]^m} \tag{6.14}$$

where V_R is reverse bias voltage to the drain terminal, Φ_B is the junction built-in potential, and m is an exponent power typically in the range of 0.3 and 0.4. As a result, C_j and, consequently, C_{sw} strongly depend on drain DC value. Based on Eqs. (6.11), (6.13), due to the fact that that they are proportional to the switch channel width, there exists a trade off between $G_{sw\text{-}ON}$ and C_{sw}. The parasitic capacitance impacts the resonant frequency of the power combining network and diminishes its quality factor Q_F. By cumulatively turning the switches on in the sequential manner (see Fig. 6.9), the total conductance of I_{path} and Q_{path} will be increased as:

$$G_{sw\text{-}I\text{-}N_I} = \sum_{i=0}^{i=N_I-1} G_{sw\text{-}I}(i) \tag{6.15}$$

$$G_{sw\text{-}Q\text{-}N_Q} = \sum_{i=0}^{i=N_Q-1} G_{sw\text{-}Q}(i) \tag{6.16}$$

where $N_I \leq N$ and $N_Q \leq N$ are the number of ON-state NMOS switches of I_{path} and Q_{path}, respectively. N is the number of NMOS switches in each individual I/Q path. Moreover, the total resistance of each switch array (I or Q) is:

$$R_{on} = \frac{1}{\sum_{i=0}^{i=N-1} G_{sw}(i)} \tag{6.17}$$

Eqs. (6.15), (6.16) indicate that, by controlling the state of the digital switches, the conductance (current capability) and, consequently, the output amplitude of the I/Q switch banks can be varied.

$$G_{sw\text{-}I\text{-}N_I} \propto I_{BB\text{-}up} \tag{6.18}$$

$$G_{sw\text{-}Q\text{-}N_Q} \propto Q_{BB\text{-}up} \tag{6.19}$$

Collectively considering Eqs. (6.4), (6.15), (6.16), (6.18), (6.19), the modulator can, therefore, cover the I/Q constellation points of a chosen digital communication standards, for example, WCDMA, Wi-Fi, or WiMAX.

$$IQ \propto I_P \times G_{sw\text{-}I\text{-}N_I} + j \times Q_P \times G_{sw\text{-}Q\text{-}N_Q} \tag{6.20}$$

To cover all four quadrants of the constellation diagram, the modulator requires the complementary (or differential) phases of quadrature I_P and Q_P clocks. According to Fig. 6.11A, I_P, Q_P, I_N, and Q_N each has a 25% duty cycle and a relative phase difference in multiples of $90°$. In fact, by swapping between I_P/I_N or between Q_P/Q_N, the sign bits of the baseband data can be reversed.

$$\begin{aligned} IQ &\propto \pm(I_P - I_N) \times G_{sw\text{-}I\text{-}N_I} \pm j \times (Q_P - Q_N) \times G_{sw\text{-}Q\text{-}N_Q} \\ &\propto \pm 2 \times I_P \times G_{sw\text{-}I\text{-}N_I} \pm j2 \times Q_P \times G_{sw\text{-}Q\text{-}N_Q} \end{aligned} \tag{6.21}$$

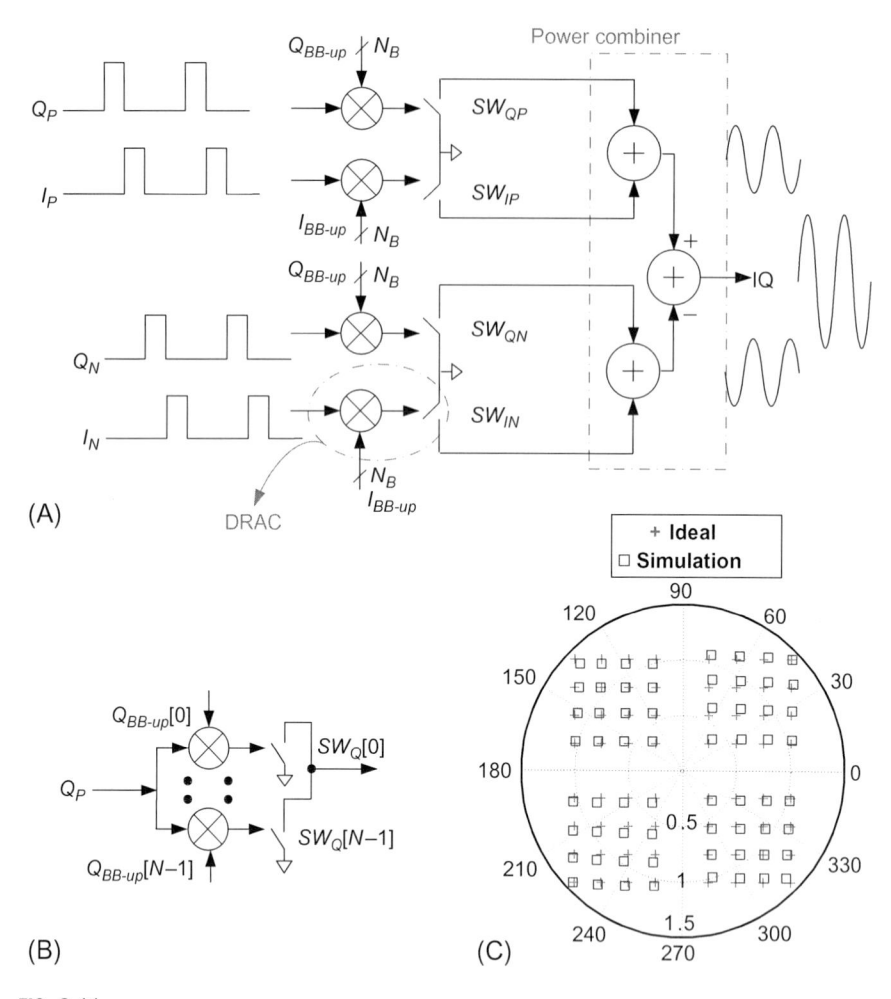

FIG. 6.11

Illustration of: (A) an RFDAC-based orthogonal digital I/Q modulator, (B) DRAC, (C) Spice postlayout simulation result of 64-point I/Q constellation diagram.

Each clock subsequently mixes with the baseband data and finally drives the corresponding switch. The combination of the mixer and switch array thus constitutes the DRAC (Fig. 6.11B), which is an individual I or Q modulator within the composite I/Q RFDAC. A simulation of a 64-point constellation of the proposed approach is demonstrated in Fig. 6.11C and indicates as follows:

1. Using the four banks of DRAC, the amplitude and phase of the resultant IQ signal are modulated.
2. By interchanging between the differential clocks, the four quadrants of the constellation diagram can be addressed.

Note that the thermal output spot noise of this structure employing upconverting clock with $D = 25\%$ is as follows:

$$\overline{V_{IQ,n}^2} = 2 \times (4KTR_{sw\text{-}I\text{-}N_I} + 4KTR_{sw\text{-}Q\text{-}N_Q}) \times (D)$$
$$= 2 \times (4KTR_{sw\text{-}I\text{-}N_I} + 4KTR_{sw\text{-}Q\text{-}N_Q}) \times (\frac{1}{4}) \quad (6.22)$$
$$= 4KTR_{on}$$

As a result, comparing to Eq. (6.10), the noise performance of the proposed modulator is superior. The modulator output spot noise with respect to the maximum power of operation is as follows:

$$P_{relative\text{-}noise\text{-}power} = \frac{\overline{V_{IQ,n}^2}}{V_{out\text{-}full}^2} \Rightarrow$$
$$P_{relative\text{-}noise\text{-}power}|dBc/Hz = 10\log\frac{\overline{V_{IQ,n}^2}}{V_{out\text{-}full}^2} \quad (6.23)$$

Considering $R_{on} = 1\Omega$, and $V_{out\text{-}full} = 1\,V$, the relative spot noise power based on Eq. (6.23) is -197.81 dBc/Hz. Note that Eq. (6.23) does not take the noise contribution from the clock signals into consideration. Otherwise stated, the thermal noise of the I/Q switches only minimally affect the eventual noise performance of the modulator. Specifically, this amount of spot noise is sufficient to meet the stringent communication standards at the receiver frequency bands.

6.3 CONCLUSION

A new all-digital I/Q RF modulator is described. Employing an upconverting RF clock with a 25% duty cycle ensures the orthogonal summation of I_{path} and Q_{path}, which avoids nonlinear signal distortion. It was clarified that electric summing of I and Q digital unit array switches is the most appropriate I/Q orthogonal summation approach. Moreover, to address all four quadrants of the constellation diagram, the differential quadrature upconverting RF clocks must be utilized. In addition, it was explained that employing switches instead of utilizing current sources leads to superior noise performance of the all-digital I/Q transmitter.

CHAPTER 7

Orthogonal summation: A 2 × 3-bit all-digital I/Q RFDAC*

CHAPTER OUTLINE

- 7.1 Circuit Building Blocks of Digital I/Q Modulator 144
  - 7.1.1 Digitally Controlled Oscillator 145
  - 7.1.2 Divide-By-Two Circuit.. 146
  - 7.1.3 25% Duty Cycle generator... 146
  - 7.1.4 Sign Bit Circuit... 147
  - 7.1.5 Implicit Mixer Circuit... 148
  - 7.1.6 2 × 3-Bit I/Q Switch Array Circuits... 148
- 7.2 Measurement Results.. 154
- 7.3 Conclusion 159

Following the orthogonal summing concept of the digital I/Q modulator in Chapter 6, this chapter presents a simple 2 × 3-bit all-digital I/Q modulator to justify the proposed solution [116, 118]. Note that, as will be explained in Chapter 8, the baseband code resolution of three bits is not an appropriate option for supporting the most stringent communication standards. However, in order to simply substantiate the proposed *orthogonal* I/Q RF modulator, its resolution is selected to be 2 × 3-bit, which is the most straightforward and feasible I/Q RF modulator. In the conventional analog I/Q approach of Fig. 1.5, the DACs and mixer tend to be of a rather bulky and power hungry. The proposed I/Q approach in Fig. 7.1 employs a pair of DRACs which comprise implicit mixers and switch array banks that directly convert the digital signal input to its RF waveform representation. Hence, the circuit functions as an RFDAC with a discrete-time complex-valued digital input and a real-valued continuous-time output. Note that, in such an approach, the traditional analog-circuit issues of calibration and timing misalignment no longer pose a problem since the digital discrete-time operation is clock-cycle accurate with modern technology sufficiently supporting sampling rates in the GHz range. This yields ultra-fast settling of the RFDAC conversion circuit. Consequently, these digital circuits can ensure fine timing accuracy that is constant (to at least within a clock cycle delay) and not subject to processing and environmental changes.

---

*The authors acknowledge substantial contributions from Prof. Dr. Leo de Vreede (TU Delft).

CHAPTER 7 Orthogonal summation: A 2 × 3-bit all-digital I/Q RFDAC

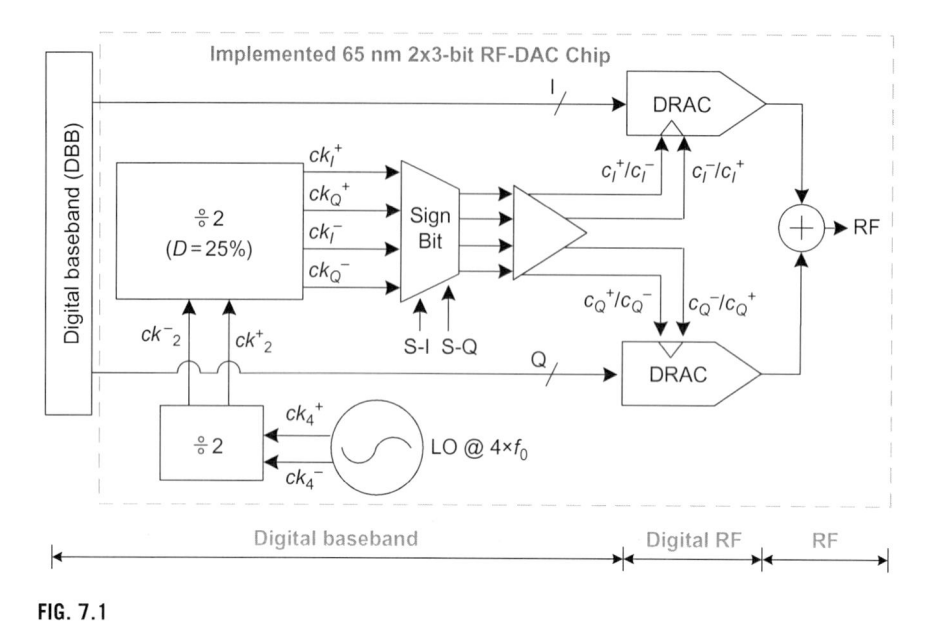

FIG. 7.1

Implementational block diagram of the 2 × 3-bit all-digital I/Q modulator.

In this specific implementation, the output clock of an on-chip LO, which operates at 4× of the desired carrier frequency, is first down-converted by a divide-by-4 circuit. Inherently, it directly provides the required I and Q clocks with their appropriate phase relationship and duty cycle. The differential in-phase (c_I^+/c_I^-) and quadrature-phase (c_Q^+/c_Q^-) clocks, at fundamental frequency of f_0, are subsequently "multiplied" by the baseband I/Q signals through the implicit mixer and drive the transistor switch arrays. The outputs of the switch arrays are connected to a power combining network that adds and converts the discrete-time pulses into a continuous-time RF output signal. As previously discussed, the proposed approach represents an RFDAC which, as such, does not require the baseband DAC and explicit mixers of Fig. 1.5. Moreover, the bandwidth of the digital modulator is only limited by the passive output power combiner and the speed of the digital circuitry. The main technical challenge is the orthogonal summing of I/Q pulses in order to reconstruct the modulated RF signal. Section 7.1 thoroughly explains the circuit building blocks of the proposed modulator including a digitally controlled oscillator, divide-by-two circuit, 25% duty cycle generator, sign bit circuit, implicit mixer, and a 2×3-bit I/Q switch array circuit. Section 7.2 addresses the related measurement results. The conclusion of the chapter is drawn in Section 7.3.

7.1 CIRCUIT BUILDING BLOCKS OF DIGITAL I/Q MODULATOR

In the remainder of this section, the modulator's building blocks will be sequentially unveiled and their circuit design techniques described.

7.1.1 DIGITALLY CONTROLLED OSCILLATOR

An on-chip 8-GHz DCO is included to generate the clock signals for the digital I/Q modulator that targets 3G cellular basestation applications operating at 2 GHz.[1] Following are explanations for the selection of the 8-GHz DCO resonating frequency:

1. The energy of the DCO resonating at 8 GHz will not substantially leak to the 2 GHz output.
2. The injection pulling of the 2 GHz RF output has a sufficiently weak 8 GHz harmonic to disturb the DCO resonant tank.
3. The 8 GHz DCO occupies less area than the 4 GHz or 2 GHz DCO due to its smaller inductor.
4. Frequency-down conversion of the 8 GHz signal automatically yields the proper fundamental I and Q differential clock signals in which their related duty cycle is 25%.

Fig. 7.2A demonstrates the DCO core circuit which is a conventional cross-coupled LC-tank DCO with a PMOS current source. Fig. 7.2B exhibits the simulation results of the differential output nodes ck_4^+ and ck_4^-. In order to be able to construct the constellation diagram and calculate the EVM of the modulator as it was illustrated in Fig. 6.11C, the modulated phase must be measurable. Therefore, the modulator requires a stable phase reference. To achieve that, the DCO is injection-locked to an external 16 GHz oscillator through the tail current source M_3. Since the source nodes of the cross-coupled transistors M_1 and M_2 exhibit 2× the frequency of their drain

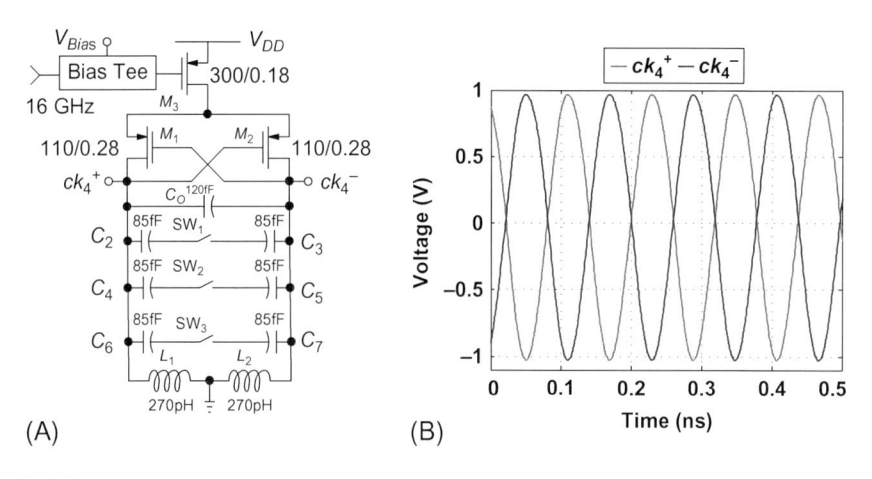

FIG. 7.2

DCO: (A) circuit schematic; (B) simulation results.

---

[1] Courtesy of my group colleague Akshay Visweswaran. The DCO and the first divider have been designed and laid-out by him.

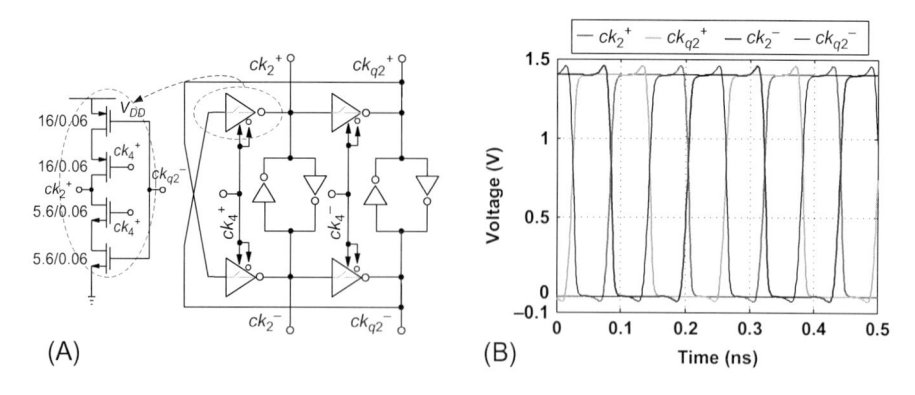

FIG. 7.3

First divider: (A) schematic; (B) simulation results.

nodes, this node is forced to lock to the injected signal (16 GHz reference clock) [122]. It should be noted that, for injection-locking of the DCO, the bias voltage node (V_{Bias}), and three-bit switched-capacitor (varactor) bank should be tuned. The varactor elements force the DCO into the vicinity of 8 GHz so that it can be easily injection-locked.

7.1.2 DIVIDE-BY-TWO CIRCUIT

The differential signals of ck_4^+ and ck_4^- are applied to the first divide-by-two circuit, which is depicted in Fig. 7.3A. For the frequency down conversion, as demonstrated in Fig. 7.1, the chip employs two divide-by two circuits. The divider comprises four gated inverters, namely C²MOS latches [123, 124], that latch the input clocks. The back-to-back inverters [14] guarantee that the prohibited state will not occur. This divider provides a very low jitter and the speed increases with process scaling. Moreover, the divider output frequency is 4 GHz, and it creates quadrature outputs, which are exhibited in Fig. 7.3B. For the subsequent stage, only one differential node (i.e., ck_2^+ and ck_2^-) of the divider is required.

7.1.3 25% DUTY CYCLE GENERATOR

As a result, the ck_2^+ and ck_2^- nodes drive the second divider which subsequently produces the four clock phases with the targeted duty cycle of 25% at 2 GHz [125] (see Fig. 7.4A). The clock nodes (ck_2^+ and ck_2^-) run at 2× the fundamental frequency (4 GHz), while ck_I^+, ck_Q^+, ck_I^-, and ck_Q^- represent the 2 GHz clock signals with a 25% duty cycle. The simulation result of the second divider is illustrated in Fig. 7.4B. The divide-by-two and 25% duty cycle signal generation are explained as follows. In each input clock cycle (i.e., the differential nodes of ck_2^+ and ck_2^-), two of the PMOS transistors (e.g., M_5 and M_6) are activated, which makes one of the drain nodes high

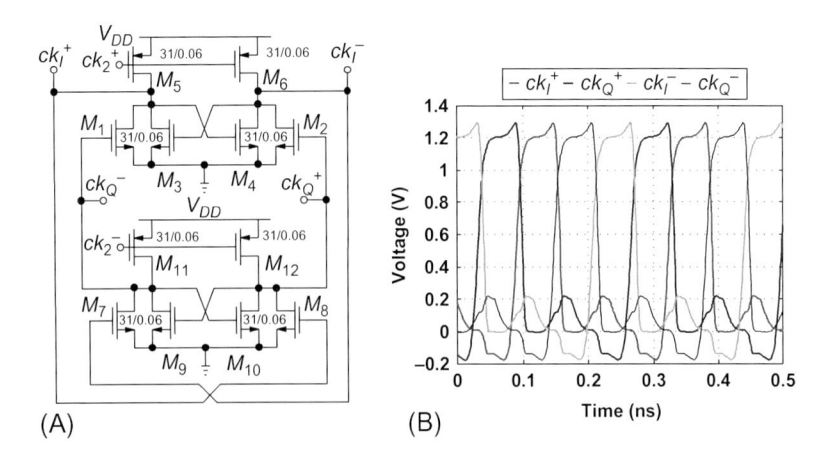

7.1 Circuit building blocks of digital I/Q modulator 147

FIG. 7.4

Second divider: (A) schematic; (B) simulation results.

(e.g., ck_I^+) and the other (cross-coupled differential drain node, ck_I^-) low. In this case, the previous state of the nodes ck_I^+, ck_Q^+, ck_I^-, and ck_Q^- are low, high, low, and low, respectively. It should be mentioned that ck_I^- also has a tendency to be high (since M_6 is also active) but, because of the positive feedback from the cross-coupled transistors of M_3 and M_4, it eventually returns to a low level (see in Fig. 7.4B, the small ripple of each node around the ground). Simultaneously, the other two (e.g., M_{11} and M_{12}) are off thereby generating floating nodes that allow the ck_Q^- to remain low, and ck_Q^+ decreases from a high to a low level because ck_I^+ is now in high level state. Therefore, $ck_I^+, ck_Q^+, ck_I^-, ck_Q^- =$ "1000." This indicates that, in each half cycle of the input clock, the high level output voltage of the circuit rings from one node to another node and, consequently, in two cycles of the input clock (divide-by-2), the high level voltage passes through all of the output nodes (i.e., the circuit generates a 25% duty cycle).

7.1.4 SIGN BIT CIRCUIT

The slightly distorted, four differential quadrature clock signals of ck_I^+, ck_Q^+, ck_I^-, and ck_Q^- are "cleaned" by the subsequent clock buffer stages and applied to the sign bit circuit. To address the four quadrants of the I/Q constellation diagram, as mentioned in Chapter 6, the modulator additionally requires two sign bits. Fig. 7.5A demonstrates the sign-bit circuit for the in-phase *I* clocks, which proves to be a multiplexer. According to the simulation result of the sign bit circuit (see Fig. 7.5B), when the sign bit is high, then ck_I^+ and ck_I^- pass directly through the transmission gate switches to the c_I^+ and c_I^- nodes. When the sign bit is low, the inputs ck_I^+ and ck_I^- are swapped through the pass-gate switches to reach the c_I^- and c_I^+ nodes, respectively.

FIG. 7.5

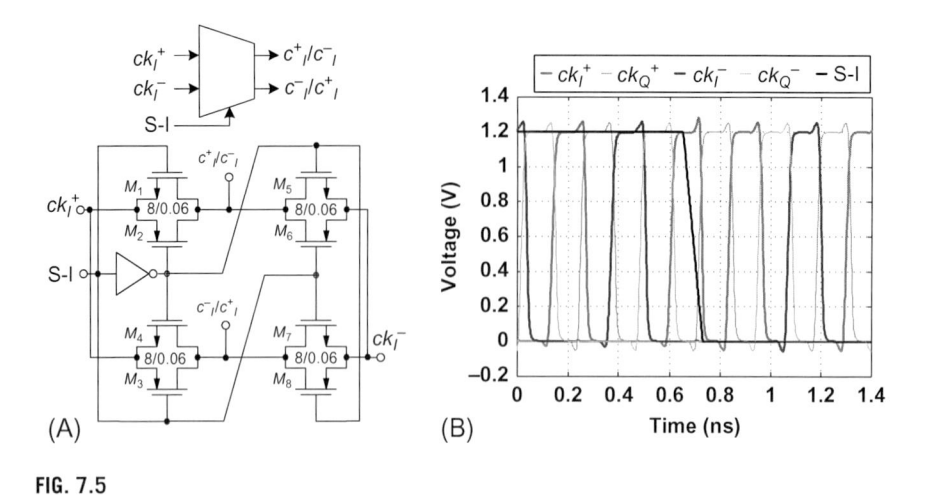

Sign bit selector: (A) schematic; (B) simulation results.

The same circuit is employed for the Q clock, thus, in total, the modulator requires two multiplexer circuits.

7.1.5 IMPLICIT MIXER CIRCUIT

The four clock signals transit through clock buffers to arrive at the implicit mixer stages. Fig. 7.6A indicates one of the unit cells of the implicit mixer which is realized as a transmission gate-based AND gate. The clock signal c, which could be any of the c_I^+, c_Q^+, c_I^-, or c_Q^- signals of Fig. 6.11, passes through the transmission gate if the baseband data bit I or Q is asserted, and the resulting signal is termed dI or dQ, which is an upconverted version of the I or Q, respectively. The pull-down NMOS transistor (M_3) is utilized for suppressing carrier leakage. The M_3 transistor should be appropriately sized in order to shunt the carrier energy at the gate of the switch array transistor in the unit's off-state without excessive loading in the unit's on-state.

7.1.6 2 × 3-BIT I/Q SWITCH ARRAY CIRCUITS

Fig. 7.6B demonstrates the corresponding equivalent transistor switch array of Fig. 6.11, which possesses 2 × 2 bits of resolution. Considering the sign bits, it is indeed a 2 × 3-bit modulator. The switch arrays are implemented in a pseudo-differential configuration. The differential design of transistor switch arrays affords several advantages over the single-ended version.

1. It dramatically reduces the second harmonic distortion. As an example, based on simulation results for the single-ended power cells, the second harmonic distortion is approximately −24 dB while, for the differential counterpart, it is better than −60 dB.

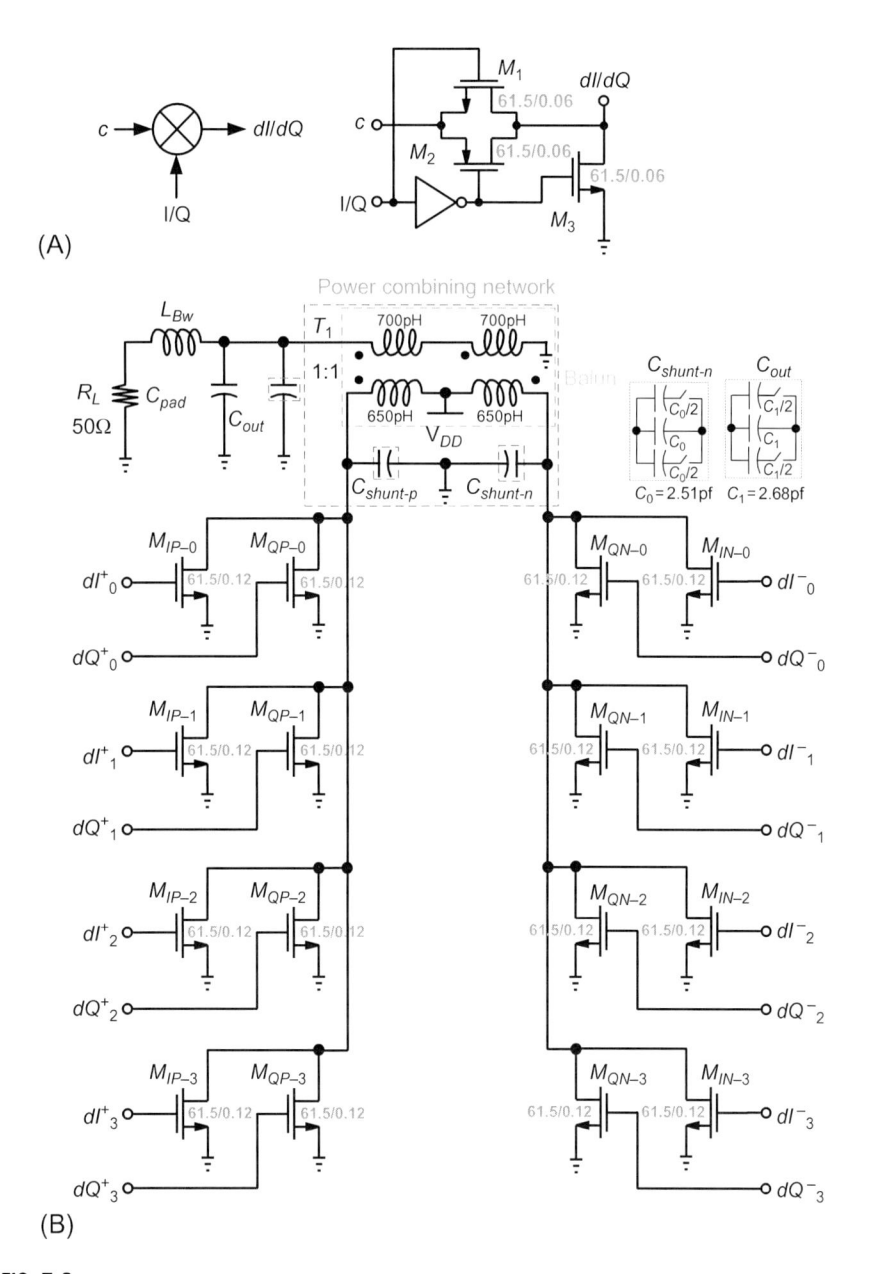

7.1 Circuit building blocks of digital I/Q modulator

FIG. 7.6

Schematic of (A) implicit mixer unit; (B) 2 × 2-bit switch array banks.

2. It doubles the output power. Moreover, due to differential generation of quadrature clocks employing differential DCO and dividers, if they are implemented as single-ended digital power cells, then the I/Q modulator wastes more power. As a result, the system efficiency will be diminished.
3. It is also more tolerable to common-mode noise through the supply and substrate coupling. Furthermore, the supply and ground bond-wire inductance have less impact on the power cells. Consequently, the ripple and instability are substantially reduced. In addition, the supply and ground bond-wire would not affect the power combining network. Most importantly, the differential configuration eases the oscillator's injection pulling and pushing phenomena.
4. It eliminates the requirement of bulky on-chip DC current choke. The reason lies in the fact that the differential drain currents are pulled from the common supply voltage in which they tend to cancel each other. As a result, the inductance of the passive current source does not require to be very high.

The dI and dQ signals drive the gates of the switch array transistors. When the I (or Q) code is kept constant while the Q (or I) code is varying, the conductance of the switches changes and, consequently, the amplitude and phase of the composite RF output also change. In addition, when both the I and Q codes change, the amplitude and phase of the RF output change due to the orthogonality. The size of the transistor switch is $W/L = (64 \times 0.96\,\mu m)/(0.12\,\mu m)$. The length of the switches are designated as twice the minimum feature length of the corresponding process technology to ensure good matching. The disadvantages are diminished output power, larger clock buffers, and more occupied area.

To confirm the correct operation of the modulator, three simulations are performed for three different I/Q codes which include $(I, Q) = (1, 0.5), (0.5, 1)$, and $(1, 1)$, respectively. Fig. 7.7 exhibits simulated drain currents of $M_{IP\text{-}0...3}$ and $M_{QP\text{-}0...3}$ transistors of Fig. 7.6 with corresponding output voltages and the related constellation points of the RF output. The corresponding phasors of the constellation points are $1.43\angle 63.06°$, $1.46\angle 27.22°$, and $1.81\angle 44.5°$ versus the expected $1.43\angle 63.44°$, $1.43\angle 26.57°$, and $1.81\angle 45°$, thus demonstrating the appropriate basic operation of the digital I/Q modulator. In addition, recall that Fig. 6.11C illustrates the postlayout simulation results of the 64-point digital modulator. The highest output power is 14.26 dBm and the related calculated EVM with no digital predistortion is 4.83%, according to Eq. (1.3).

The transistor switch arrays are connected at the output to the power combining network, which will be thoroughly explained in Chapter 9. It comprises the balun transformer T_1 with input $C_{shunt\text{-}p}$, $C_{shunt\text{-}n}$ and output C_{out} tuning capacitors. As will be clarified in Chapter 9, the shunt input and output capacitances of the power combiner are employed to fine-tune the amplitude and the phase relationship of the I/Q signal as well as the output power at 2 GHz. These capacitances consist of a fixed component part (with values of C_0, C_1) and a tuning component part (with values of $C_0/2$, $C_1/2$), which are tuned utilizing NMOS switches. The size of NMOS switches should be carefully selected since it affects the quality factor and, consequently, the insertion loss of the power combining network. The shunt input and output

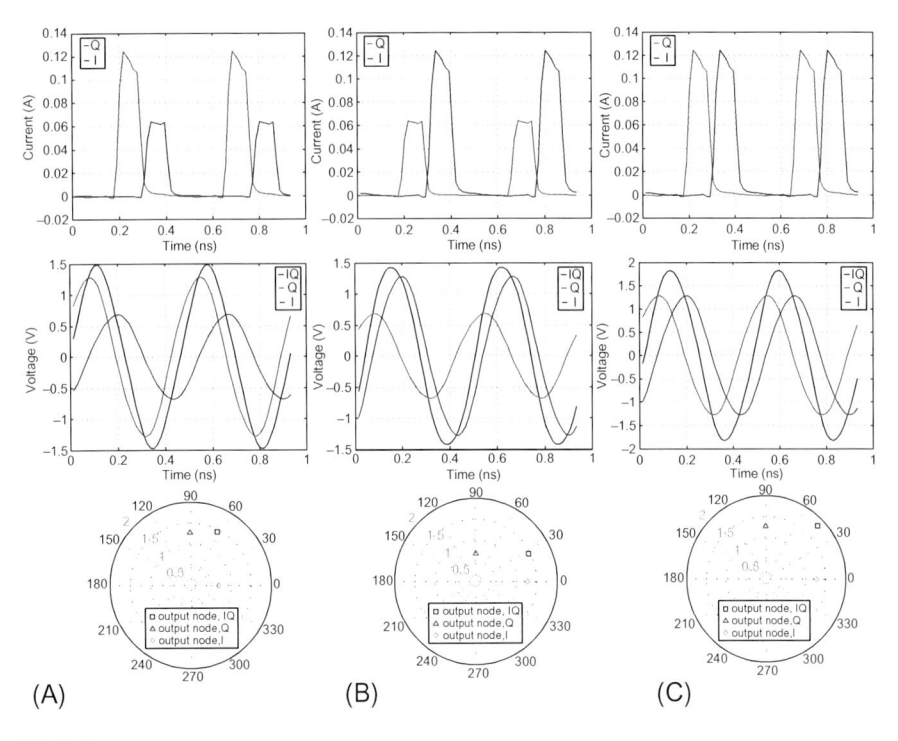

FIG. 7.7

Plots of drain current (*top row*), output voltage (*middle row*), and related constellation points (*bottom row*) for the codes of: (A) $(Q, I) = (1, 0.5)$; (B) $(Q, I) = (0.5, 1)$; (C) $(Q, I) = (1, 1)$.

capacitances are realized as fringe interdigitated capacitors which use metal layers one through seven of our selected technology to improve the capacitor density per unit area.

The layout of the balun transformer T_1 is depicted in Fig. 7.8A. T_1 is utilized to convert the differential signals to a single-ended output. It impacts the modulator performance and, hence, requires special attention to its layout. The balun transformer uses metal layers 6 and 7 as well as an aluminum layer for decreasing losses due to the series resistance. The transformer traces are 12 μm wide for maximizing the current handling capability and with 3 μm gaps between them which imposes by metal density rule of the process technology. The total size of the transformer is 500×500 μm² with a 1:1 turns ratio (with center tap). The size is selected to be large enough to address the frequency range of 1 GHz to 3 GHz. The transformer layout has been simulated employing ADS Momentum. The resulting S-parameter model has been converted to an equivalent lumped circuit model [126] that is used in the time-domain circuit simulations. However, the equivalent lumped circuit is a

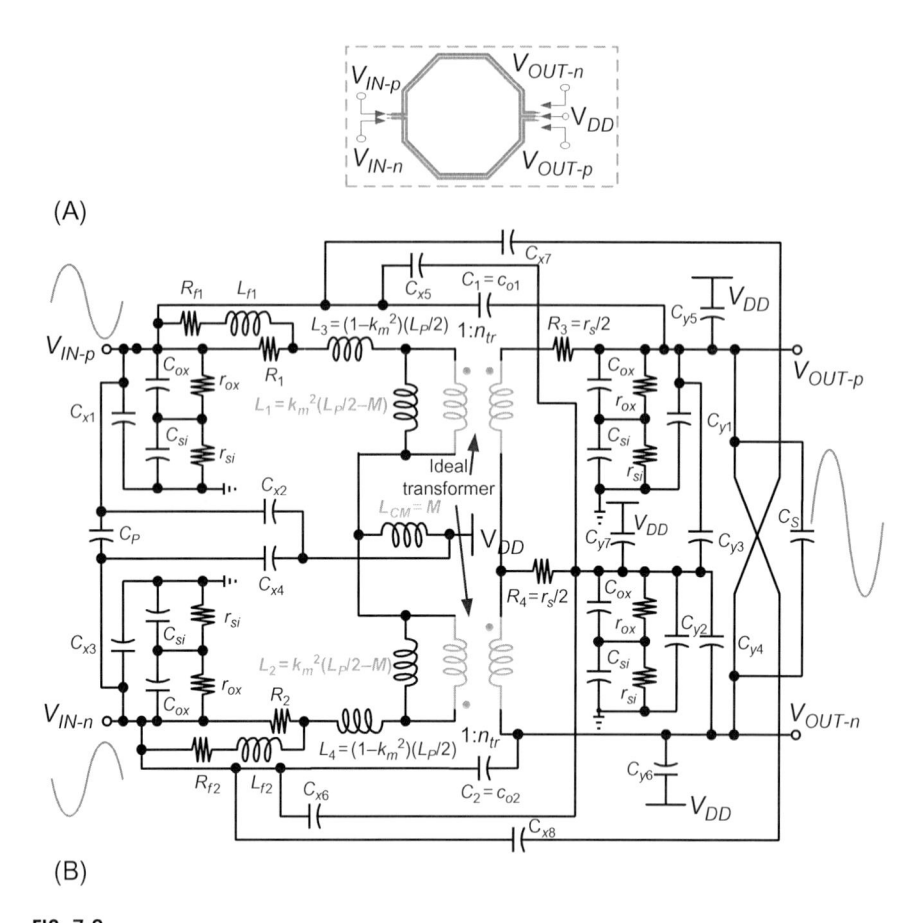

FIG. 7.8

(A) Layout of balun; (B) corresponding lumped equivalent circuit.

modified version of [126], which is exhibited in Fig. 7.8B. In this modified equivalent lumped model, the mutual inductance between two primary differential legs of the balun is included. Therefore, by using the T-section model for the primary side, the magnetizing inductor is:

$$L_{pm} = k_m^2 \times \left(\frac{L_p}{2} - M\right) \tag{7.1}$$

where M is mutual inductance, L_P is self-inductance of the primary winding [126], and the related coupling factor of the transformer is $k_m = 0.65$. Moreover, in this model, all of the parasitic capacitances between the six ports (two inputs, two outputs, V_{DD}, and ground) of the balun are included. In addition L_{f1}, R_{f1}, and L_{f2}, R_{f2} are parallel to R_1, and R_2, respectively, which could model the metal skin effect

FIG. 7.9

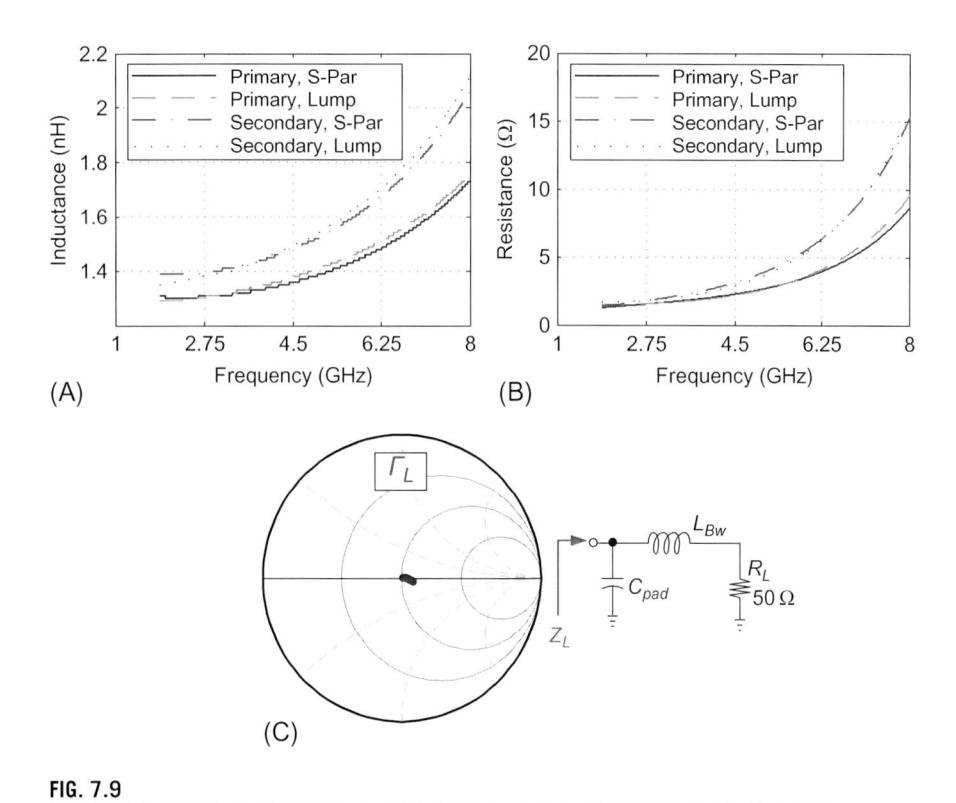

(A) Primary and secondary inductance; (B) primary and secondary loss resistance; (C) load reflection coefficient of Z_L in 1 to 5 GHz frequency range.

or frequency-dependent substrate losses. Fig. 7.9A and B compares the simulation results between the Momentum S-parameter and the equivalent lumped model for the primary and secondary inductance and series loss resistance versus frequency of the balun transformer. According to Fig. 7.9A and B, the equivalent lumped model coincides favorably with the Momentum simulation results. Based on simulation results, the insertion loss of balun is approximately 1.7 dB which causes the drain efficiency of the modulator to drop from almost 33% to about 21%. It should be mentioned that, in Fig. 7.6B, the R_L is connected to the power combining network via a bond wire inductor (L_{BW}). In addition, the effect of capacitance of the RF pad (C_{pad}) should be considered. The effect of these extra passive components is simulated for 1–5 GHz frequency range, and the corresponding load reflection coefficient (Γ_L) is shown in Fig. 7.9C. Based on simulation results, if $L_{Bw} = 1\,nH$ (which is a typical value for 1-mm-long bond-wire) and $C_{pad} = 376\,fF$, then $Z_L = 53\angle 0\,\Omega$, as a result, the corresponding Γ_L, the related VSWR, and its return loss (RL) are 0.029, 1.06, and 30.71, respectively. This slight deviation from the ideal 50 Ω has a very minimal effect on the functionality of the power combining network. Thus, the modulator enjoys the small VSWR, and its performance of the RF power, efficiency, and modulation accuracy will not be diminished. However, as will be discussed in Chapter 9, in the design steps of power combining network, C_{pad} is considered as parts of C_{out}.

7.2 MEASUREMENT RESULTS

As an experimental validation of the proposed orthogonal I/Q combining concept, a 2×3-bit (including one sign bit) direct-digital I/Q modulator circuit is fabricated in 65 nm CMOS technology. Fig. 7.10A exhibits the micrograph of the implemented chip. The total chip area is $1.2 \times 2\,\text{mm}^2$ with an active part of $0.6 \times 1.15\,\text{mm}^2$. The chip is mounted on an FR4 PCB board (see Fig. 7.10B) and all of the pads, including two RF pads (16 GHz input clock reference and 2 GHz RF output), are wire bonded. The test board is designed such that very short bond-wires could be utilized for all RF signals as well as for supply and ground connections.

The chip height is approximately 600 μm, and it is placed into a designated hole on the FR4 board which results in even shorter bond-wires (see Fig. 7.10C). For the correct operation of ESD[2] protection circuits, digital control bits are surrounded by ground and supply pads. In addition, as depicted in Fig. 7.10A, SW_1, SW_2, and SW_3 are the digital control frequency bits of the DCO. Also $C_{shunt-p}$, $C_{shunt-n}$, and C_{out} of Fig. 7.6B are used for tuning the power combining network. Moreover, the I/Q baseband digital control bits are $S-Q$, $S-I$, Q_1, I_1, Q_2, I_2, Q_3, I_3, Q_4, and I_4. During the measurement process, these digital baseband bits are purposely and statically switched on and off in order to address a desired constellation point.

At the start of the measurements, the chip is first injection locked to a 16 GHz reference signal source. Fig. 7.11A and B illustrates the resulting phase noise spectrum of the modulator output while the DCO is injection-locked to 16 GHz reference signal source. The measured phase noise are demonstrated at two different output power levels of 12.32 dBm and 5.63 dBm. For both cases, the measured phase noise at 1 MHz offset is approximately -140 dBc/Hz. The DC drain current of the transistor switch arrays is 60 mA at the maximum measured output power of 12.32 dBm. Note that after de-embedding the 0.3 dB loss of the SMA cable as well as board losses, the output power is actually 12.62 dBm. Consequently, the related drain efficiency is 20% ($V_{DD} = 1.4$ V), which is considered excellent from the I/Q RF modulator perspective. Fig. 7.11C indicates the highest output power and carrier leakage superimposed. The carrier leakage is -71 dBc below the peak output power. In Fig. 7.11D, the variation of the output power versus the I/Q codes is plotted, which indicates the 10.5 dB dynamic range of the modulator. Note that

---

[2]Courtesy of our group's technical supporter, Atef Akhnoukh. The ESD and related circuits have been designed and laid-out by him from scratch.

FIG. 7.10

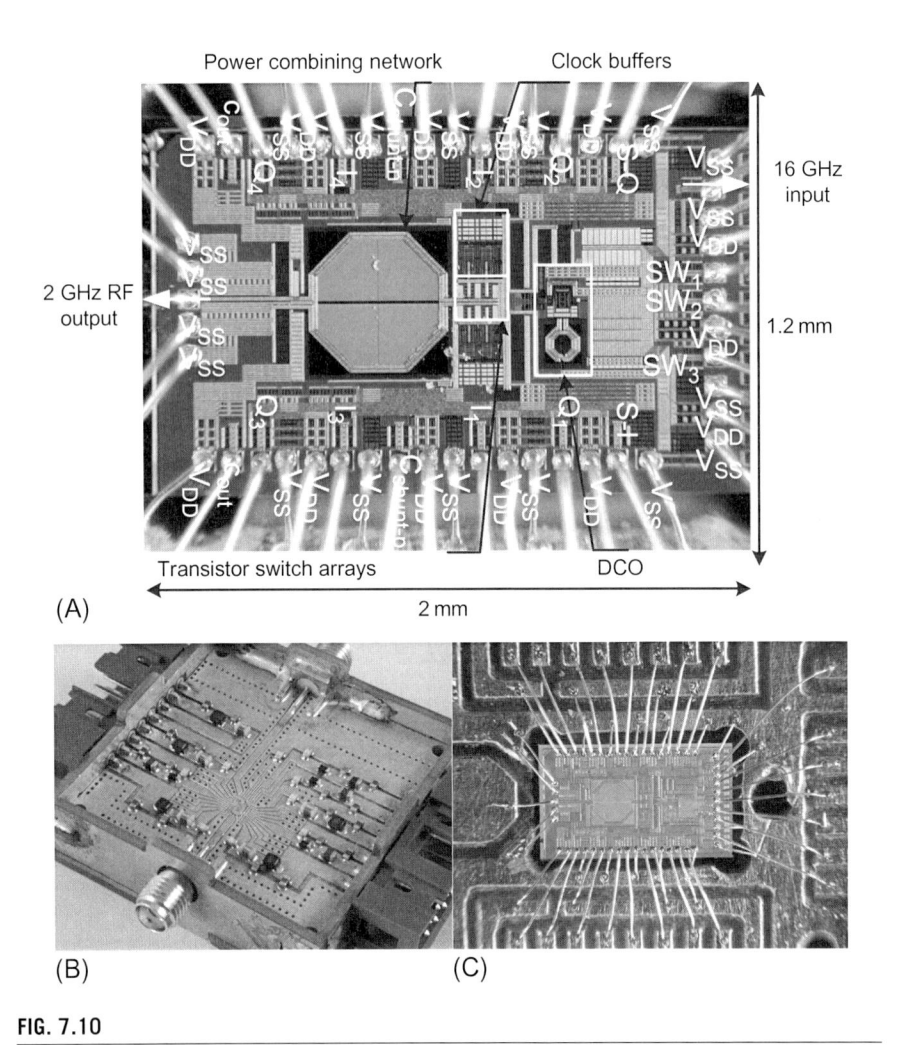

(A) Micrograph of the 2 × 3-bit-bit, 2 GHz all-digital I/Q modulator including 8 GHz DCO;
(B) FR4 PCB board; (C) mounted chip.

this value is in accordance with the modulator resolution of 2 bits when the supply is kept constant. Since the carrier leakage is less than −58 dBm, it affords the opportunity to increase the resolution of the modulator to a higher number of bits in a rather unambiguous manner. In fact, the dynamic range is now exceeding 70 dB (dynamic range = maximum power (12.62) − minimum power (−58)). Therefore, the resolution could be extended to more than 11 bits ($11 \times 6 = 66 \leq 70$). Note that the post layout simulated carrier leakage was approximately −70 dBm. The difference

FIG. 7.11

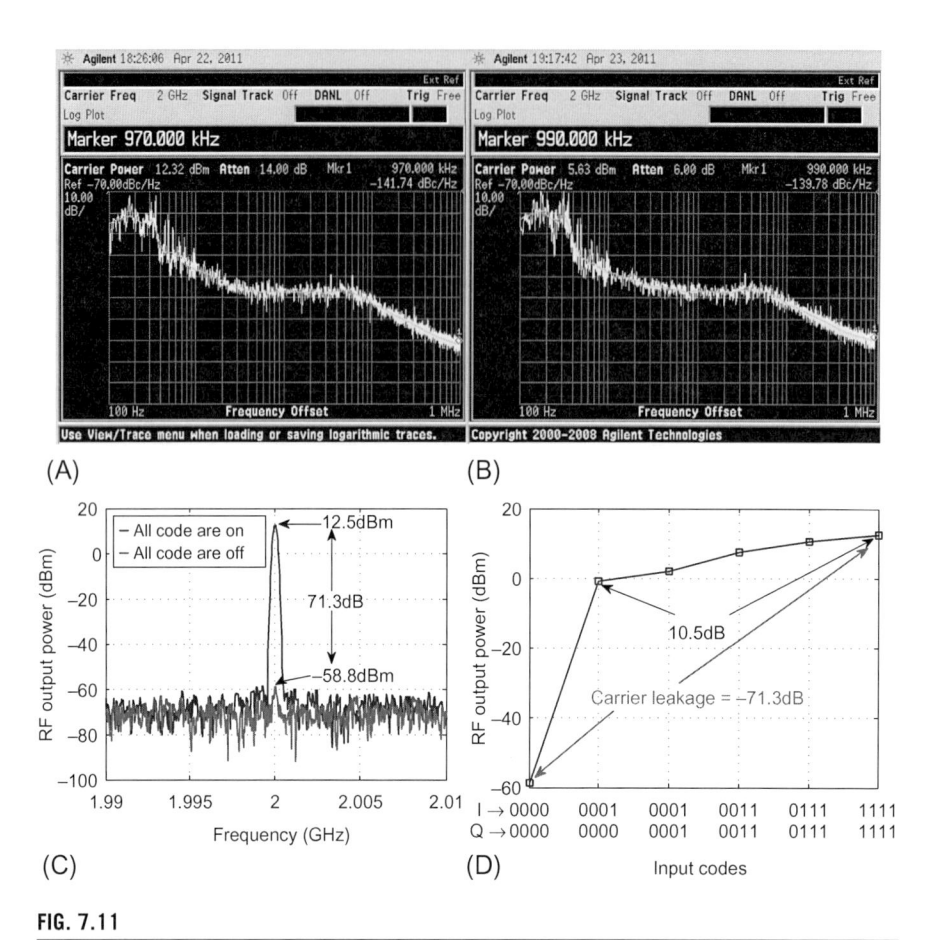

Phase noise of injection-locked RF output at 2 GHz; (A) at 12.32 dBm; (B) at 5.63 dBm. Measured (C) output power spectrum; (D) output power versus input codes.

between the simulation and measurement could be a result of coupling between the clock traces, the drain nodes, and the digital/RF ground plane.

The implemented modulator is tested in two different constellation modes, specifically, in QPSK and 64-point QAM. For both of these measurements, time domain RF output signals are captured and saved. FFT of these signals is subsequently calculated, and the amplitudes and phases are plotted. In the QPSK mode, by statically changing the sign bits, the phase of the RF signal can be adjusted by 90° (see Fig. 7.12A). In addition, by increasing the I/Q code from the lowest to the highest, the modulator amplitude is also changed, which is exhibited in Fig. 7.12B. The corresponding constellation points and the related EVM are shown in Fig. 7.12C and D, which indicate that, for the output power less than 6 dBm the EVM is less than 1.5% in QPSK modes of operation. Note that, to show the functionality of the

FIG. 7.12

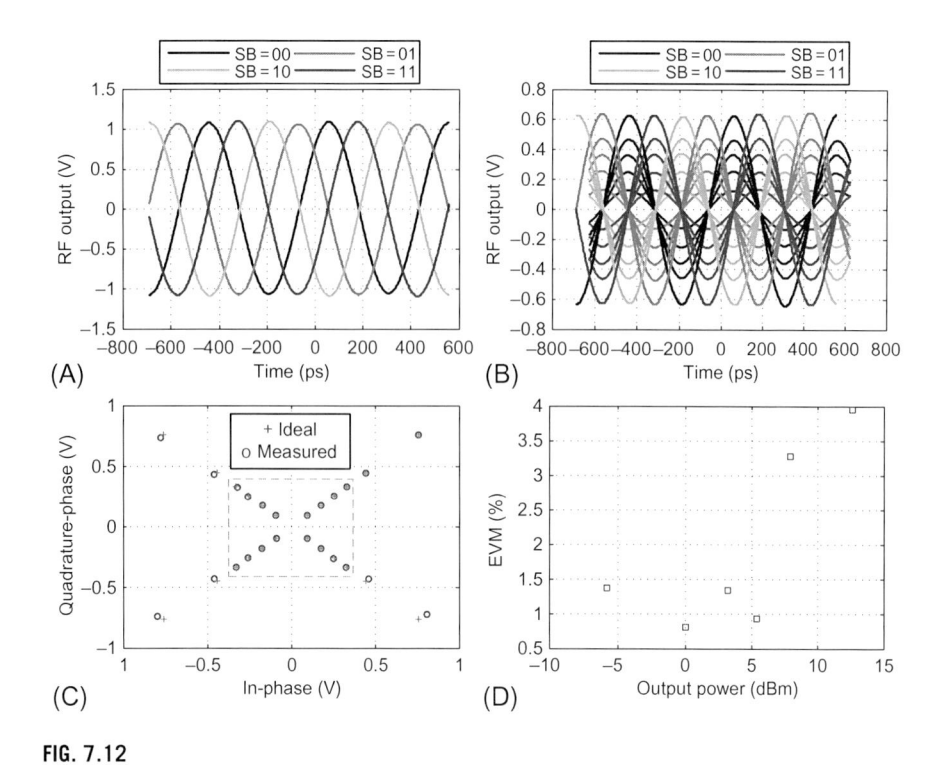

Measured time-domain 4-QPSK: (A) phase changing; (B) amplitude changing; (C) corresponding constellation points; (D) corresponding EVM.

proposed digital I/Q RF modulator, with this measurement, only six different QPSK points within each quadrant are measured.[3]

As a final point, Fig. 7.13 shows the modulator performance in 64-QAM constellation mode. The measured time-domain of three different points in each of the four quadrants (12 points in total) are demonstrated in Fig. 7.13A–D. These points are $(I, Q) = (1, 4)$, $(I, Q) = (2, 2)$, and $(I, Q) = (4, 1)$, respectively. The corresponding constellation points are depicted in Fig. 7.13E, which are designated by a white "x" that establishes the correct operation of the modulator. The entire 64-QAM constellation diagram is indicated in Fig. 7.13E. The maximum output power in this case is 6 dBm while $V_{DD} = 1$ V and the corresponding maximum drain efficiency is 10%. The related EVM is 2.36% or, equivalently, -32.53 dB, without using any predistortion, which is a favorable number. It should be mentioned that the constellation diagram will be easily improved in more advanced implementations when exploiting a higher output resolution (see Chapter 10). In this case, the

---

[3] For the output power higher than 6 dBm, $V_{DD} = 1.4$ V and for the remaining, $V_{DD} = 1$ V.

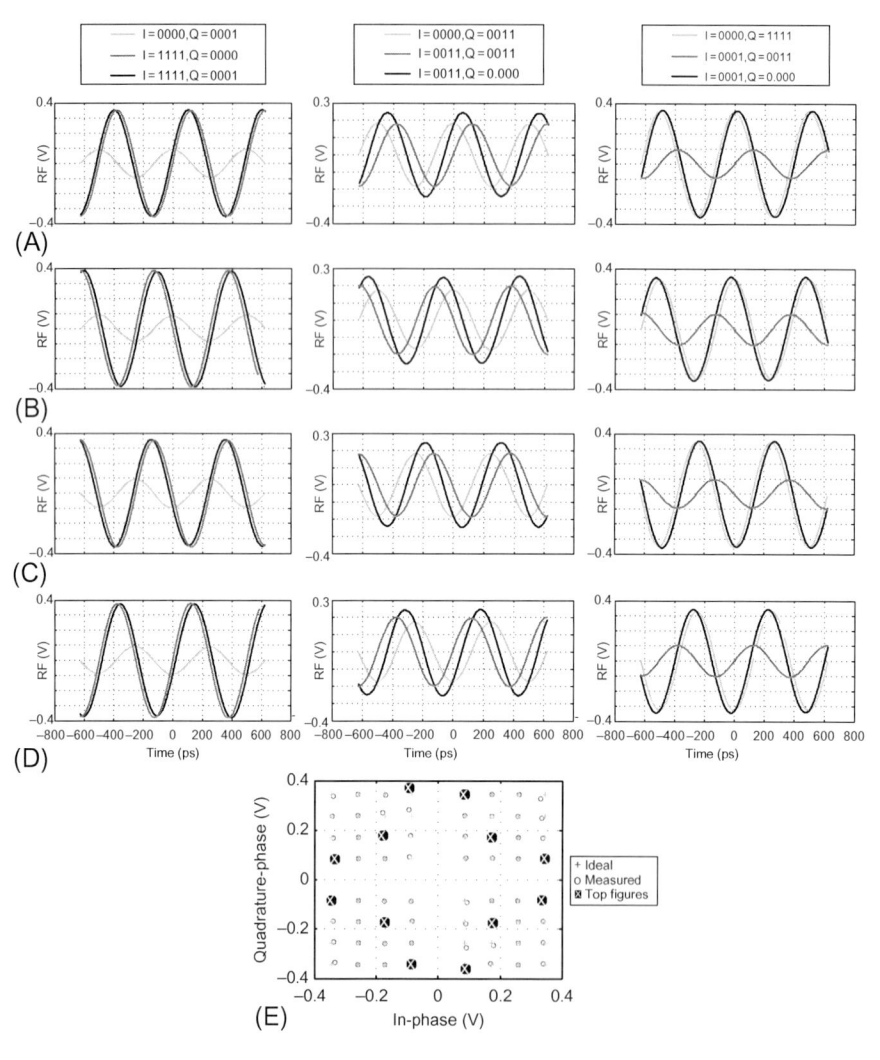

FIG. 7.13

Measured time-domain: (A) first; (B) second; (C) third; (D) fourth quadrant; (E) corresponding measured 64 points I/Q constellation diagram at 2 GHz.

predistortion techniques can be implemented in an uncomplicated manner. Fig. 7.13E reveals that the first and third quadrants of the constellation diagram exhibit improved behavior over the other two quadrants. This is due to the mismatch between the clock paths of the four transistor switch arrays.

The measurement results of 2×3-bit digital I/Q RF modulator are summarized in Table 7.1. Note that, as will be discussed in Chapter 9, Section 9.2, the peak

Table 7.1 2 × 3-Bit Digital I/Q RF Modulator

	Process (nm)	Output Power (dBm)	Carrier Leakage (dBc)	Drain Efficiency (%)	EVM (%)
All-Digital I/Q RFDAC [118]	65	5.4	−71	10	0.94[a]
		12.6		20	3.95[b]
		5.4		10	2.36[c]

[a] Only 4-point (QPSK) is considered, which is based on Fig. 7.12C.
[b] Only 4-point (QPSK) is considered, which is based on Fig. 7.12C.
[c] Only 64-point (64-QAM) is considered, which is based on Fig. 7.13E.

drain efficiency of the proposed I/Q RFDAC while operating at full RF power should be more than 40%. Based on Table 7.1, however, the drain efficiency of the contemporary I/Q RFDAC, that is, [118], at 12.6 dBm RF power is approximately 20%. The foregoing efficiency drop is regarded to the fact that the transistor length of the switch arrays is selected 120 nm, which entails more power losses. Moreover, the designated power combining network could not transform the optimum loading condition at the drain nodes of the related switch arrays. In order to improve the efficiency, in the next I/Q RFDAC test chip, which will be elaborated in more detail in Chapter 10, these oversights are addressed. In addition, employing an ultra thick metal layer, the insertion loss of the balun transformer is reduced. Thus, the achievable drain efficiency is increased up to 40%.

7.3 CONCLUSION

In this chapter, a novel 2 × 3-bit all-digital I/Q (i.e., Cartesian) RF transmit modulator is implemented, which operates as an RFDAC. The modulator performs based on the concept of orthogonal summing, which is introduced and elaborated in Chapter 6. It is based on a TDD manner of an orthogonal I/Q addition. By employing this method, a very simple and compact design featuring high-output power, power-efficient, and low-EVM has been realized. The resolution of the experimental RFDAC presented in this work is only 3-bit (including one sign bit), but it will be demonstrated in the following chapters that the resolution can be increased to 8–12 bits in an unequivocal manner for utilization in multistandard wireless applications.

CHAPTER 8

Toward high-resolution RFDAC: The system design perspective*

CHAPTER OUTLINE

8.1 System Design Considerations 162
8.2 Conclusion 169

The all-digital *orthogonal* I/Q modulator concept of Chapter 7 [116–118] substantiates that the direct orthogonal summation of the in-phase and quadrature digital RF vectors can generate the composite IQ RF vectors. According to Chapter 7, a 2×3-bit static I/Q implementation could achieve a maximum RF output power and drain efficiency of 12.6 dBm and 20%, respectively, while the corresponding RF supply voltage is 1.4 V. Moreover, its related static EVM of "4QPSK" symbols is −28.6 dB at maximum RF power. In addition, the measured static constellation diagram of Fig. 7.13E demonstrates certain mismatches between differential quadrature paths of Fig. 6.11A due to either mismatches between the 25% duty cycle differential quadrature clocks of I_P, Q_P, I_N, and Q_N or an insufficient symmetrical layout. Furthermore, the differential drain nodes of the switch array structure of Fig. 7.6B would not tolerate more than 1.3 V differential voltage swing employing low-power 65-nm CMOS technology and, as a result, their corresponding NMOS transistor switches would definitely break. Note that the maximum RF power using the switch array structure of Fig. 7.6 is 14 dBm. Likewise, the digital I/Q modulator has only been tested at an RF frequency band of 2 GHz. The ultimate universal digital I/Q transmitter would operate in multiple RF bands to support multimode/multiband communication standards.

The above disadvantages of the digital I/Q modulator of Chapter 7 must be addressed. Most importantly, the dynamic performance of the proposed digital I/Q modulator employing the actual wideband complex baseband signals has yet to be examined. Since the effective modulating sample resolution is an utmost important parameter as it directly impacts the achievable dynamic range, linearity, EVM, noise floor, and out-of-band spectral emission, we recently proposed an all-digital I/Q RFDAC with 2×13-bit resolution that can provide output power beyond 22 dBm

---

*The authors acknowledge substantial contributions from Prof. Dr. Leo de Vreede (TU Delft).

[119, 120]. Due to its high efficiency, wide bandwidth, high functionality, and fine resolution, while requiring only limited chip area, the proposed solution is a very promising candidate for future multimode/multiband transmitters. As stated in Chapter 1, Section 1.3.2, the dynamic performance of the RFDAC can be evaluated employing EVM for in-band performance and spectral purity as well as noise floor at RX frequency bands for out-of-band performance. In order to address the aforementioned issues, in this chapter, the system design perspective of the 2×13-bit RFDAC is elaborated in more detail. The circuit-level design considerations as well as digital calibration along with associated digital predistortion techniques, and measurement results will be addressed in the following chapters (Chapters 9–11). Section 8.1 systematically explains the proposed high-resolution RFDAC and investigates its system design parameters. The conclusion of the chapter is drawn in Section 8.2.

8.1 SYSTEM DESIGN CONSIDERATIONS

The dynamic performance of the all-digital I/Q RFDAC strongly depends on the interpolation rate of the I_{BB-up}/Q_{BB-up} signals and their resolution. Since, in this prototype, the digital signal processing and the I/Q baseband interpolations are performed in MATLAB and subsequently uploaded into two on-chip SRAM via UART, see also Appendix B.1, the memory length (SRAM capacity) also affects the RFDAC performance. Fig. 8.1 illustrates the system-level simulation setup structure that reflects the dependency of these parameters on its dynamic performance. First, I_{BB} and Q_{BB} are interpolated in software by a CK_R clock, which is generated by an integer-N division of the RF carrier LO clock. Thus, the CK_R clock and the baseband upsampled signals are synchronized to the LO clock. Next, I_{BB-up} and Q_{BB-up} are quantized and then uploaded into the SRAM memory. Subsequently, the SRAM memory is read out employing a CK_R clock and directly fed to the DRAC block. Since the CK_R is slower than the LO clock, the DRAC performs as a ZOH to balance the speed of baseband upsampled signals with the LO clock. For the sake of signal processing clarity, ZOH is exhibited as a separate block between the memory and the DRAC. Note that all subsequent simulations based on Fig. 8.1 are performed

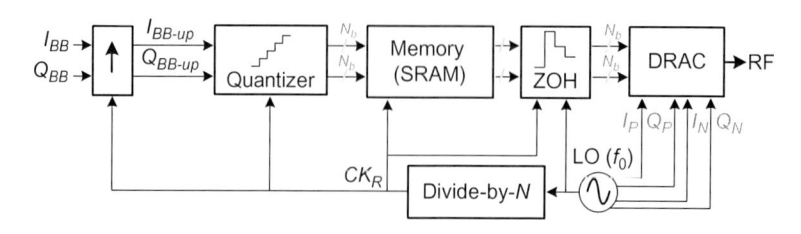

FIG. 8.1

System-level simulations at $f_0 = 2.4$ GHz: the block diagram of the test-bench.

under an assumption that the DRAC resolution is identical to that of the quantizer; the carrier frequency (f_0) is 2.4 GHz. As a result, three yet-to-be-defined variables in the RFDAC system of Fig. 8.1 are:

1. CK_R frequency (f_{CKR})
2. DRAC resolution (N_b)
3. memory length (l_{mem})

which should be appropriately selected. The f_{CKR} lower limit is determined by the highest operational bandwidth of I_{BB}/Q_{BB}. At present, the bandwidth of baseband communication signals does not exceed 160 MHz. On the other hand, the f_{CKR} upper limit could be as high as f_0. Note that, in this case, the divide-by-N would be redundant. In reality, running the CK_R at the LO rate would consume an overabundance of power, thus reducing the overall system efficiency. In fact, according to Eq. (1.2) in Chapter 1, the relationship between the system efficiency and f_{CKR} is as follows:

$$\eta_{system} = \frac{P_{RF\text{-}out}}{P_{DC\text{-}drain} + P_{DC\text{-}4\times} + P_{DC\text{-}Buffer} + P_{DC\text{-}CKR}}$$
$$= \frac{P_{RF\text{-}out}}{P_{DC\text{-}drain} + P_{DC\text{-}4\times} + P_{DC\text{-}Buffer} + V_{DD}^2 \times A \times C_{par} \times f_{CKR}} \quad (8.1)$$

where A is the circuit activity factor and C_{par} is the related total parasitic capacitances of the baseband processing circuit. Moreover, $P_{DC\text{-}4\times}$ and $P_{DC\text{-}Buffer}$ are the DC power consumption of a clock generation circuit and clock buffer which drive digital switch arrays of the RFDAC, respectively. As a result, the system efficiency is inversely proportional to f_{CKR}. For example, in this aspect, the power consumption of the SRAM at 300 MHz is 12 mW. The $P_{DC\text{-}4\times} = 36$ mW and $P_{DC\text{-}Buffer} = 84$ mW at 2.4 GHz. Moreover, $P_{DC\text{-}drain} = 396$ mW and the corresponding RF power is 22 dBm. Consequently, $\eta_{system} = 30\%$. If SRAM hypothetically works at 2.4 GHz, then its related power consumption will be 96 mW while $P_{DC\text{-}Buffer} = 150$ mW. Note that the power consumption of $P_{DC\text{-}Buffer}$ is due to the contribution of the CK_R in DRAC structure. Hence, $\eta_{system} = 23.3\%$.

Fig. 8.2 exhibits the simulations for which f_{CKR} is swept from 150 to 600 MHz in increments of 150 MHz while I_{BB}/Q_{BB} are 64-tone/80 MHz signals. The subsequent RF power spectrum is shaped by the Sinc function of the ZOH interpolation:

$$Sinc = \text{sinc}^2\left(\frac{f - f_0}{f_{CKR}}\right) \quad (8.2)$$

The ZOH operation creates spectral replicas at multiples of sampling frequency f_{CKR} away from the f_0 carrier:

$$f_n = f_0 \pm n \times f_{CKR} \quad (8.3)$$

where $n = 1, 2, \ldots$. In conclusion, the upsampling and synchronization operations represent a ZOH that performs as a sinc-filter with its corresponding zeros located at f_n. As such, the spectral images are notched by the ZOH operation. Note that doubling f_{CKR} not only reduces the out-of-band emissions but also lowers the spectral replicas

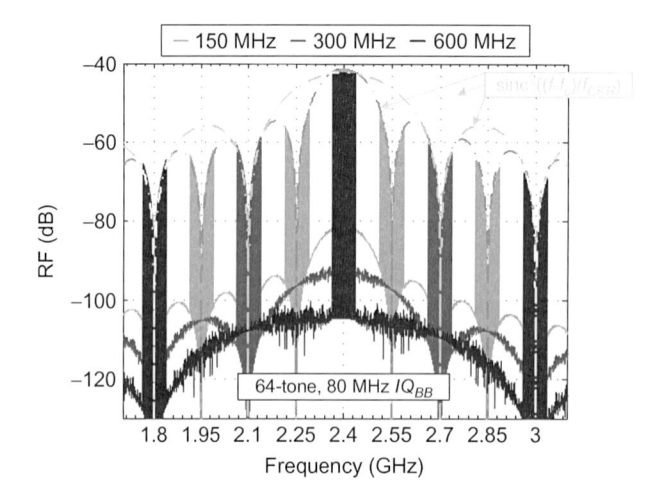

FIG. 8.2

System-level simulations at $f_0 = 2.4$ GHz: sweeping over upsampling clock frequency f_{CKR} for 64-tone, 80 MHz I_{BB}/Q_{BB} baseband signals.

by 6 dB. If f_{CKR} is 150 MHz, then it would be impractical to support the 160 MHz baseband signals. On the other hand, 600 MHz clock consumes twice as much power than at 300 MHz. Furthermore, SRAM in a low-power 65 nm CMOS process would not be feasible at 600 MHz. Therefore, f_{CKR} is selected 300 MHz that is generated employing a $\div 8$ divider.

Another simulation is performed by sweeping the bandwidth (two-tone frequency spacing) of I_{BB}/Q_{BB} from 20 MHz to 80 MHz. According to Fig. 8.3A, the "wider band" signals produce higher out-of-band spectra, while the spectral replicas are larger (6 dB/octave). This is merely the limitation of the present implementation and is entirely due to the limited sample-storing memory relative to the signal period.

Fig. 8.3B further illustrates that doubling l_{mem} improves the noise floor, although this would not be a limitation in practical transmitters. Since, in this work, the upsampled baseband signals residing in SRAM are fed to DRAC, this configuration performs as an FFT executor. Consequently, the higher number of FFT points result in the lower out-of-band spectrum. In this work, however, l_{mem} is selected 8-kword (every word is 16 bits) to save the chip area. We should emphasize that the SRAM storage of modulating samples was chosen over a real-time reception of the baseband data in order to emulate the environment of contemporary single-chip radios in which the RF transceiver is integrated with the digital baseband. This provides the benefit of avoiding contamination of the sensitive RF spectrum from the wideband modulating digital data through bond pads, bond-pad wires, and an ESD ring. Nonetheless, the

FIG. 8.3

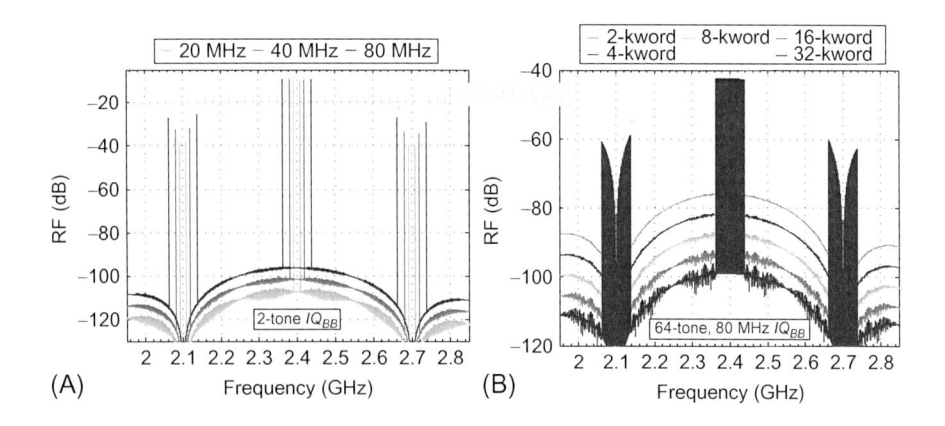

System-level simulations at $f_0 = 2.4$ GHz: (A) signal bandwidth (two-tone frequency spacing); (B) SRAM memory length ($l_{mem} = 8$-kword).

limitations of utilizing on-chip SRAM in combination with MATLAB in comparison with an actual situation can be explained as follows:

The SRAMs must be large enough to preserve the necessary baseband data. The memory length, l_{mem}, affects the following dynamic performances of the proposed I/Q RFDAC:

1. As previously stated, utilizing SRAM with larger memory length improves the out-of-band spectrum. It also affects the far-out noise performance.
2. In order to produce accurate DPD profiles, the SRAM should be at least 16 k-word. This will be explained in Chapter 10, Section 10.9.4.

As a result, l_{mem} collectively with the corresponding upsampling clock, CK_R, of the SRAM are two important design parameters. The reason can be attributed to the subsequent resolution bandwidth of RFDAC depending on l_{mem} as well as the CK_R. Since, in each period of CK_R clock, one data point is fed into the RFDAC, consequently, the total resolving time of the baseband data as well as its related resolution bandwidth are expressed as follows:

$$T_{resolve} = l_{mem} \times \frac{1}{f_{CKR}} \tag{8.4}$$

$$RBW = \frac{1}{T_{resolve}} \tag{8.5}$$

For example, in this work, since the carrier frequency, f_0, is 2.4 GHz, then the frequency of the upsampling clock, f_{CKR}, is 300 MHz. Also, $l_{mem} = 8$ k-word. Thus, the resolving baseband time and its corresponding resolution bandwidth are 27.3 μs and 36.6 kHz, respectively. Hence, the resolution bandwidth of the measured out-of-band spectrum as well as the far-out noise performance of the current RFDAC is limited to 36.6 kHz. Stated differently, to measure the far-out noise performance with the resolution bandwidth of 1 Hz while employing 300 MHz clock, the subsequent memory length should be 300 M-word. In fact, this hypothetic SRAM would occupy a prohibitively immense chip area. Moreover, it consumes exorbitant amount of power and, thus, it is avoided. In conclusion, in order to measure RFDAC out-of-band spectrum as well as far-out noise performance while its related resolution bandwidth is low enough, the most beneficial manner is to perform the measurement in the real-time test setup.

As discussed earlier, the lower limit of N_b is determined by Eq. (6.1). However, it should be much higher than that in order to meet the quantization noise requirements of practical communication standards. Thus, as with any DAC converter, increasing N_b improves the dynamic range of RFDAC. Note that the dynamic range of the RFDAC for a single-tone baseband signal can be expressed as the SNR expression of the DAC [127–129] that is as follows:

$$DR(f)_{ST}|_{f_{CKR}/2} = 1.7609 + 6.0206 \times N_b + 20 \log_{10}\left(\text{sinc}\left(\frac{f - f_0}{f_{CKR}}\right)\right) \quad (8.6)$$

in which Eq. (8.6) indicates the dynamic range of RFDAC in the Nyquist band from 0 to $f_{CKR}/2$. Note that the final term in Eq. (8.6), that is, "sinc function" is a result of the ZOH operation of the RFDAC. Moreover, the dynamic range of a single-tone baseband signal in 1-Hz bandwidth is as follows:

$$DR(f)_{ST}|_{Hz} = DR(f)_{ST}|_{f_{CKR}/2} + 10 \log_{10}\left(\frac{f_{CKR}}{2}\right) \quad (8.7)$$

Eq. (8.7) and also Fig. 8.2 indicate that doubling the sampling clock frequency, that is, f_{CKR}, improves the dynamic range by 3 dB. In addition, for a modulated or multitone signal with the PAPR, its dynamic range of in 1-Hz bandwidth could be rewritten as

$$DR(f)_{MT}|_{Hz} = DR(f)_{ST}|_{Hz} - 10 \log_{10}\left(\frac{\text{PAPR}}{2}\right) \quad (8.8)$$

where PAPR is as follows

$$\text{PAPR} = \frac{P_{RF\text{-}max}}{P_{RF\text{-}avg}} \quad (8.9)$$

where $P_{RF\text{-}max}$ and $P_{RF\text{-}avg}$ are maximum and average RF power, respectively. If the input signal is a single tone, then PAPR=2 and then Eqs. (8.7), (8.8) will be identical. Note that, throughout this book, only two complex-modulated baseband signals are employed: single-carrier M-QAM and multicarrier M-QAM OFDM signals. Under the assumption of equal probability of occurrence for all levels, the PAPR of a single-carrier 64 and 256-QAM signals are 3.7 and 4.2 dB, respectively [129, 130].[1] Moreover, the PAPR of multicarrier 64 and 256-QAM OFDM signals are 17.2 dB and 20.6 dB,[2] respectively [24, 129, 131]. According to Eq. (8.8), employing the aforementioned complex-modulated baseband signals deteriorates the dynamic range of the RFDAC. In reality, due to the fact that the maximum achievable power mode of operation is only minimally incorporated, the hard-clipping algorithm could be utilized to decrease the PAPR of the OFDM signal [132]. In this work, however, due to employing SRAM, the complex-modulated baseband signals are windowed (filtered). As a result, as will be demonstrated in Chapter 11, Section 11.3, the PAPR of applied complex-modulated signals will not exceed more than 8.6 dB.

Note that, generally, the dynamic range of modulated/multitone signal in B-Hz bandwidth can be expressed as:

$$DR(f)_{MT}|_B = DR(f)_{MT}|_{Hz} - 10\log_{10}(B) \tag{8.10}$$

The RFDAC resolution of Fig. 8.1 is swept for a single-carrier "256-QAM 80 MHz" signal and the corresponding power spectral density of the RFDAC system is depicted in Fig. 8.4. Based on Fig. 8.4 and Eq. (8.6), each extra bit improves the out-of-band spectrum by almost 6 dB. In this work, N_b is selected as 13 bits (the MSB is the sign bit) to support the most stringent communication standards.

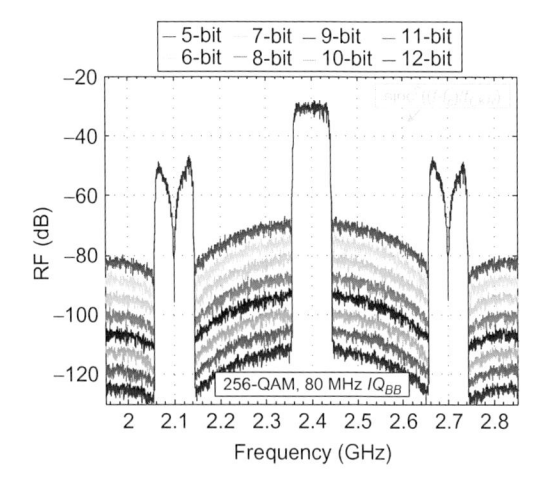

FIG. 8.4

System-level simulations at $f_0 = 2.4$ GHz: DRAC resolution ($N_b = 12$).

---

[1] Assuming the equal probability for all QAM constellation points, then its PAPR is equal to $10\log_{10} 3 + 10\log_{10}\left(\sqrt{M}-1\right)/\left(\sqrt{M}+1\right)$. Consequently, if $M \to \infty$ then PAPR $\to 10\log_{10} 3 = 4.77$.
[2] The related subcarriers (N_{sb}) are 52 and 115, respectively, and the PAPR is calculated based on $10\log_{10}(N_{sb})$.

As stated previously, the TX noise floor at RX frequency bands is another important out-of-band performance of a digital transmitter. The TX noise floor is almost dominated by the related noise of the I/Q RF modulator. The noise floor specification is dictated by the following items [6, 7, 24, 25, 60, 61, 133]:

1. How much of the TX noise floor can be leaked to RX frequency bands in order not to desensitize the receiver of the corresponding radios? Otherwise stated, the TX noise floor should be smaller (at least 9 dB) than the sensitivity of the receiver. Namely

$$N_{TX}|_{B,Hz} \leq P_{sen}|_{dBm}$$
$$\leq (10 \log_{10}(KT) + 30_{dB}) + NF + 10 \log_{10} B + SNR_{min}$$
$$\leq -173.83_{dBm/Hz} + NF + 10 \log_{10} B + SNR_{min} \qquad (8.11)$$

where KT is "available noise power" when the receiver is matched to the antenna and the temperature is 300°K, NF is the noise figure of the receiver, B is the bandwidth of interest, and SNR_{min} is the minimum SNR ratio of the receiver. As a result, the required TX noise floor with respect to the maximum RF output power, which is the negate of the dynamic range (Eq. 8.8), is as follows:

$$N_{TX}|_{dBc/Hz} = -DR(f)_{MT}|_{Hz} = N_{TX}|_{B,Hz} - (P_{TX,Max} - L_{duplexer}) \qquad (8.12)$$

where $P_{TX,Max}$ is the peak RF output power and $L_{duplexer}$ is the duplexer isolation loss. For example, with a typical number of $NF = 3$ dB and $L_{duplexer} = 45$ dB as well as $P_{TX,Max} = 22$ dBm, the required TX noise floor must be less than -148 dBc/Hz. If, in the proposed RFDAC, $N_b = 12$ bit, $f_{CKR} = 300$ MHz, and PAPR = 9 dB are selected, then these choices ensure that the TX noise floor will not desensitize the receiver.

2. The spectral emission mask of the corresponding communication standard must be satisfied so as not to corrupt the adjacent as well as alternate frequency channels of operation. For example, WCDMA standard emission mask [24] establishes that, at 12 MHz offset from the center of the transmission channel, the related emitted signal must be less than -63 dBm in a 100-kHz bandwidth which is equal to -135 dBc/Hz considering 22 dBm RF output power.

3. It is common in contemporary mobile devices to incorporate the other applications such as Bluetooth, GNSS, FM, and wireless LAN as well as other cellular standards in combination with the designated cellular system. This is referred to as co-existence of various radios on the same mobile devices. For example, in order to comply with the 3GPP co-existence standards, the TX noise at 40 MHz offset should be less than -79 and -71 dBm in a 100-kHz bandwidth in the GSM RX bands of 900 and 1800 MHz, respectively [24]. Stated differently, the TX noise floor at the foregoing RX frequency bands should be -162 and -151 dBc/Hz considering maximum RF power of 33 and 30 dBm, respectively.

As stated in Chapter 1, EVM is related to the in-band performance of the transmitter. The modulation accuracy of a transmitter is verified by EVM. Equivalently, at the receiver, the modulation error is validated employing BER. The SNR ratio of the quadrature baseband signals are inversely proportional to EVM [134]:

$$SNR_{M\text{-}QAM}|_{dB} \cong -\text{PAPR}_{dB} + 20\log_{10}\left(\frac{EVM\%}{100}\right) \Rightarrow SNR_{M\text{-}QAM} \approx \frac{1}{EVM^2} \quad (8.13)$$

Moreover, in-phase and quadrature-phase gain and phase mismatches, oscillator phase noise, and in-phase and quadrature-phase DC offset deteriorate the performance of the transmitter EVM. The proposed 2 × 13-bit RFDAC guarantees that the SNR is high enough. Furthermore, as illustrated in Fig. 8.1 and will be discussed in Chapter 10, utilizing divide-by-N as well as a master divide-by-4 ensure maintaining the quadrature accuracy of the digital I/Q modulator. In addition, employing an IQ image and leakage calibration techniques entail minimal gain/phase mismatches as well as moderate IQ DC offset. Furthermore, as will be addressed in the following chapters, the RFDAC requires a DPD algorithm that also improves EVM. As a result of IQ image and leakage calibrations, the EVM is mostly affected by the phase noise of upconverting quadrature clocks. The EVM and phase noise are approximately related to one another as follows [135, 136]:

$$EVM_{dB} \cong PN_{dBc/Hz} + 10\log_{10}(BW_{channel}) + 3.01_{dB}|_{SSB \to DSB} \quad (8.14)$$

where the right-hand side of Eq. (8.14) is, indeed, double-sideband integrated phase noise over the desired channel bandwidth. The effect of phase noise on the in-band performance, that is, EVM, as well as output-of-band spectra are depicted in Fig. 8.5. The performance of I/Q RFDAC systematically simulated employing two different phase noise profiles of Fig. 8.5A while their I/Q baseband signal is a multicarrier "44 MHz 256-QAM OFDM." Their related EVMs as well as out-of-band spectra are approximately separated by 10 dB which is, literally, the difference between the related phase noise profiles.

Finally, Fig. 8.6 exhibits the complete block diagram of the differential orthogonal 2 × 13-bit all-digital I/Q RFDAC. It comprises four paths: $I_{path,p}$, $Q_{path,p}$, $I_{path,n}$, and $Q_{path,n}$. Each path contains a DRAC comprising unit cell mixers and an array of digital power amplifiers (DPAs). The multitude of DPA outputs are connected to a differential power combing network that promotes transformation of the upconverted digital signals into a "high power" continuous-time RF output in an energy efficient manner. The represented RFDAC does not require the baseband DACs of a conventional analog I/Q transmitter. Moreover, I/Q calibration can be easily performed at baseband while its bandwidth is only limited by the speed of the digital circuitry and the passive output power combiner.

8.2 CONCLUSION

The system design considerations of the proposed high-resolution, wideband all-digital I/Q RFDAC are discussed. It is demonstrated that the upsampling clock frequency (f_{CKR}), DRAC resolution (N_b), and memory length (l_{mem}) are three important

FIG. 8.5

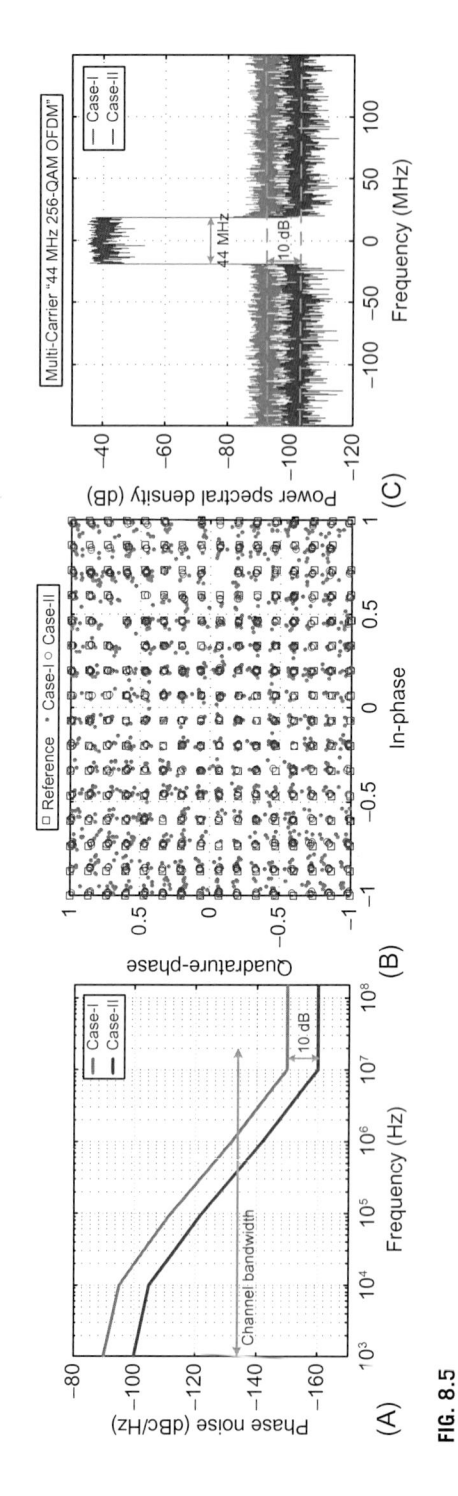

(A) Two different phase noise profiles; (B) In-band performance, $EVM_{case-I} = -23.5\,dB$, $EVM_{case-II} = -34.5\,dB$; (C) Out-of-band spectra, case-II is 10 dB better than case-I.

8.2 Conclusion

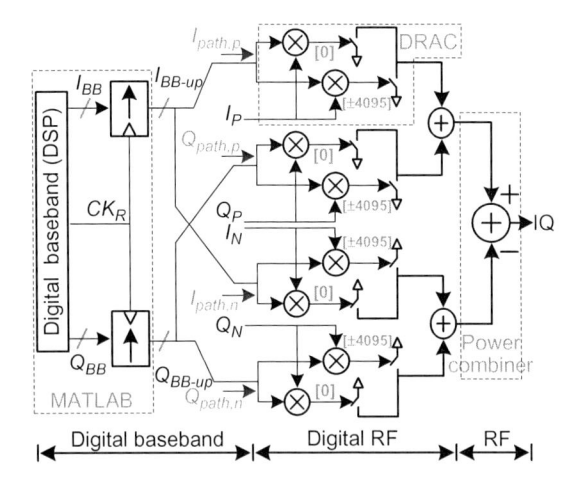

FIG. 8.6

The differential orthogonal 2 × 13-bit all-digital I/Q RFDAC.

parameters that affect the dynamic performances of the proposed RFDAC. Based on system-level simulation results and the limitation in implementing the RFDAC test-chip, they are designated as $f_{CKR}=300$ MHz, $N_b=12$ bit, and $l_{mem}=8$ k-word. The effect of these parameters on the in-band as well as out-of-band performance of RFDAC is investigated. It is concluded that exploiting 12 bits of resolution for quadrature baseband signals is sufficient to meet the most stringent communication requirements.

CHAPTER

Differential I/Q DPA and power-combining network*

9

CHAPTER OUTLINE

9.1 Idealized Power Combiner With Different DRACs 174
9.2 A Differential I/Q Class-E-Based Power Combiner 179
9.3 Efficiency of I/Q RFDAC 187
9.4 Effect of Rise/Fall Time and Duty Cycle 190
9.5 Efficiency and Noise at Back-Off Levels 191
9.6 Design an Efficient Balun for Power Combiner 193
9.7 Conclusion 200

Within the past decade, tremendous research has been directed toward the design of power-efficient on-chip combining networks. For example, Aoki et al. proposed a DAT topology which makes it possible for watt-level power generation in CMOS bulk technology for mobile and wireless applications [8–10]. In addition, Gang Liu et al., Haldi et al., and Chowdhury et al. proposed transformer-based power-combining techniques that enhance the efficiency at power back-off levels [137–139]. Moreover, Wang et al. revealed a power amplifier that also employs transformer-based matching network to generate high power as well as to achieve broadband operation [140]. These novel, well-cited papers indicate that it is possible to implement a fully integrated power-efficient RF transmitter, which generates the desired RF power level without utilizing any external passive components. Though these power amplifiers have successfully generated the desired RF power using transformer-based power combiners, they have not been implemented along with the entire RF transmitter. Most recently, Kousai et al. unveiled an analog polar RF transmitter based on a power mixer array structure, which also uses power combining for watt-level power generation [15]. Moreover, Chowdhury et al. and Ye et al. proposed inverse class-D power combiners for polar RFDACs [19, 21, 48, 49]. Furthermore, Yoo et al. employed a class-D-type power combiner for switched-capacitor digital polar transmitter [18]. Similarly, Lu et al. also demonstrated an inverse class-D power combiner for an I/Q RFDAC [66].

Considering the advantages as well as disadvantages of the foregoing power combiners, in this chapter, the design procedures of the proposed differential I/Q

---

*The authors acknowledge substantial contributions from Prof. Dr. Leo de Vreede (TU Delft).

class-E-based power-combining network will be discussed [119, 120]. Section 9.1 introduces two DRAC topologies, namely DRAC based on an AND-Gate mixer (see Fig. 9.2A) and DRAC based on a switchable cascode structure (see Fig. 9.2B). An idealized power-combining network is revealed which can be employed in either of the two DRAC structures. Section 9.2 discusses a differential power combiner, which is designed according to the class-E matching configuration at full power. Section 9.3 discusses the related efficiency of the power-combining network. Section 9.4 explains the effect of timing constraints of upconverting clock on the RFDAC performance. Section 9.5 discloses that the RFDAC operates as a class-B power amplifier at back-off power levels. Section 9.6 explains a procedure to design an efficient transformer which is employed in the power combiner. Finally, in Section 9.7, the chapter is concluded.

9.1 IDEALIZED POWER COMBINER WITH DIFFERENT DRACs

As stated previously in Chapter 6 (Section 6.2), the orthogonality and summation of the I_{path} and Q_{path} in Fig. 6.2 are two important aspects of the quadrature modulation (see Eq. (6.4)). It was later concluded that by employing 25% quadrature clocks, the vectorial addition of the I_{path} and Q_{path} is simply performed by electrical summing (Fig. 6.9). Nonetheless, the proposed all-digital high-resolution I/Q RFDAC of Fig. 8.6 should generate the desired RF power and, most importantly, must be power-efficient. The on-chip differential power-combining network of Fig. 8.6 performs the efficient RF power combination of $I_{path,p}$, $Q_{path,p}$, $I_{path,n}$, and $Q_{path,n}$.

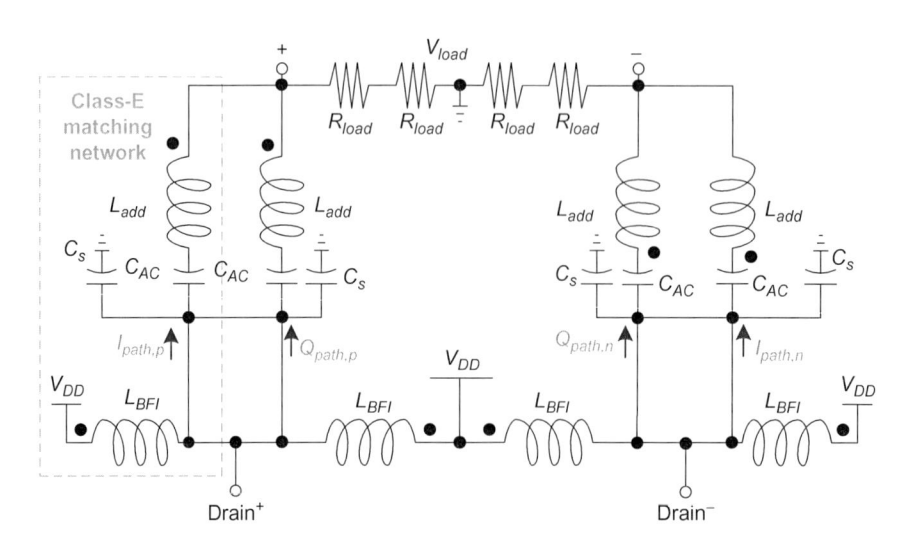

FIG. 9.1

Schematic diagram of an idealized power-combining network.

9.1 Idealized power combiner with different DRACs

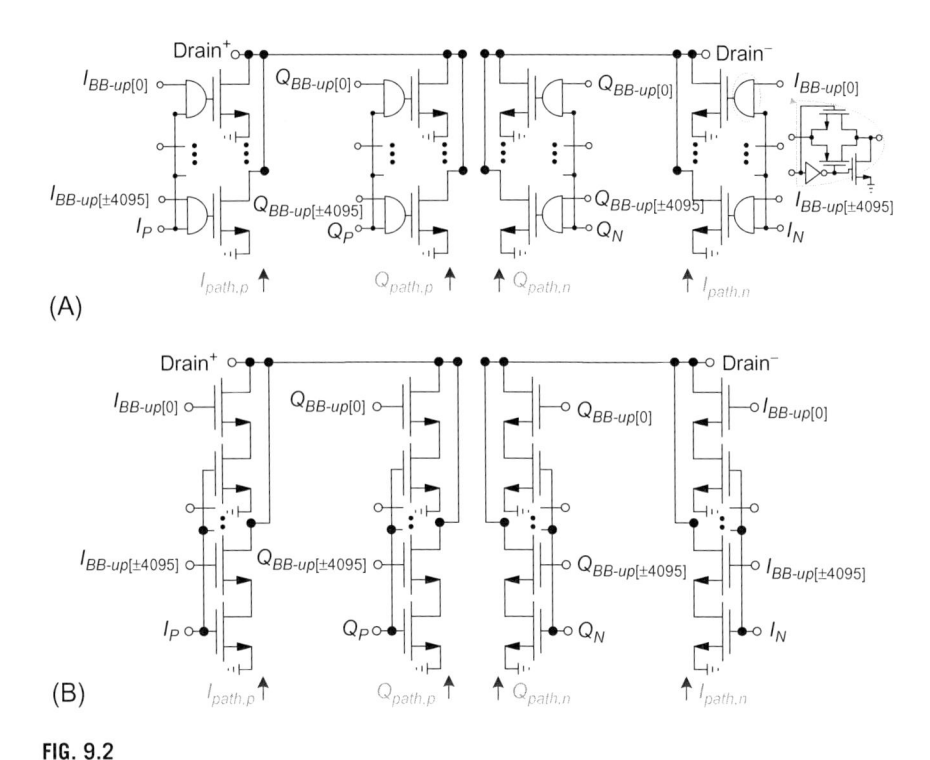

FIG. 9.2

Schematic diagram of the 2 × 13-bit differential I/Q power switch array : (A) simple DPA (Chapter 7); (B) switchable cascode DPA structure.

Fig. 9.1 exhibits an idealized power-combining network which could be connected to either of the following two DRAC structures:

1. According to Fig. 9.2A, the DRAC is a push-pull quadrature structure, in which each path comprises unit cell mixers (AND gates), and the composite DPA consists of the parallel combination of 4096 single transistor units. Hence, its resolution is 2 × 12 bits. This DRAC structure is incorporated into the 2 × 2-bit digital I/Q modulator of Chapter 7 (see Fig. 7.6) and is referred to as a single-switch DRAC structure.
2. Based on Fig. 9.2B, the DRAC is a push-pull quadrature structure, in which each path is composed of unit cell mixers (switchable cascode transistor), and the composite DPA consists of the parallel combination of 4096 switchable cascode transistor units. Thus, its resolution is also 2 × 12 bits. Note that it is, indeed, a 2 × 13-bit considering the sign bits. This structure is referred to as switchable cascode DRAC structure.

The peak voltage swing of $Drain^+$ and $Drain^-$ nodes could be more than 2.4 V which can cause a device breakdown if the switchable cascode DRAC structure is

FIG. 9.3

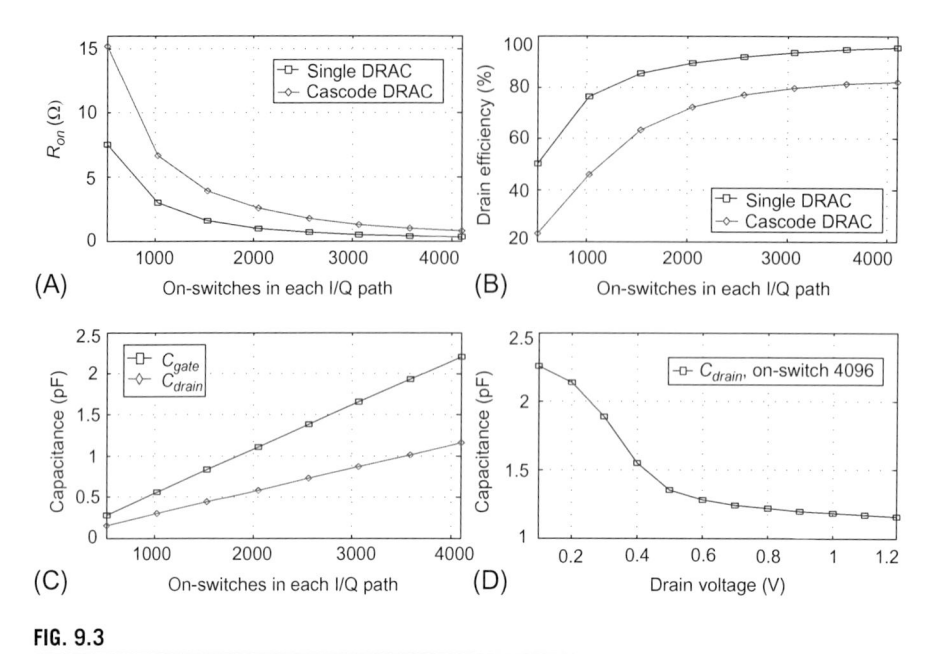

Simulations of each I/Q path for (A) R_{on}; (B) drain efficiency; (C) gate/drain capacitance versus on-switches; and (D) drain capacitance versus drain voltage.

not utilized. Using the switchable cascode as a unit cell, however, increases the on-resistance (R_{on}) of the unit cell switches (see Fig. 9.3A), which subsequently incites higher power loss as well as lowered drain efficiency. Note that all simulations of Fig. 9.3 are performed employing matching network of Fig. 9.1 with a single/cascode unit cell switch using a channel length of 60 nm while its gate width is 500 nm. As stated, the DRAC resolution is 12 bits which requires 4096 switch-array unit cells in each orthogonal path of $I_{path,p}$, $Q_{path,p}$, $I_{path,n}$, and $Q_{path,n}$. In this work, the targeted maximum RF output power (P_{max}) exceeds 22 dBm while maintaining V_{DD} at 1.2 V. Therefore, the maximum RF power of each orthogonal path should be 1/4 of P_{max}. According to simulations, using $W = 500$ nm minimum-length switches in 12-bit RFDAC configuration ensures that each orthogonal path provides more than 16 dBm.

Fig. 9.3B indicates that the drain efficiency of the cascode switch is lower than that of a simple switch due to its higher R_{on}. In this simulation, the power-combining network of Fig. 9.1 is lossless, which would result in 100% drain efficiency if R_{on} was, hypothetically, zero. Increasing the number of unit cell on-switches from 512 to 4096 improves the drain efficiency as a result of the lower R_{on} (less overall power loss due to increased turned-on switches). Note that the cascode switch not only mitigates the related breakdown issue, but it is also used as an upconverting unit cell mixer. Controlling each switchable cascode transistor unit based on its related baseband data (i.e., I_{bb-up}/Q_{BB-up}), the equivalent on-resistor of I_{path}/Q_{path} is changed. Therefore, this can modulate the amplitude and phase of the reconstructed RF output signal (see Eq. 6.21). Finally, perhaps the most significant advantage of this switchable cascode structure is to effectively isolate the I and Q paths, which results in improved EVM and linearity.

In addition to its on-resistance, the cascode MOS switch also has a considerable gate or drain capacitance, which is proportional to its channel width that was previously discussed in Eq. (6.13) and their capacitance simulation results are depicted in Fig. 9.3C. Selecting wider cascode switches in order to achieve higher efficiency, unfortunately, exacerbates the power consumption of the preceding RF clock buffers, which subsequently reduces the overall system efficiency. As a result, the selected channel width of 500 nm appears to be an effective compromise between the overall system efficiency and maximum RF power.[1] Note that the drain capacitance also depends on the drain voltage (see Eq. 6.14). Fig. 9.3D illustrates that the drain capacitance at $V_{DD} = 0.1$ V is almost double of that at $V_{DD} = 1.2$ V. Therefore, turning on the switches as well as varying the drain voltage changes the drain capacitance, which eventually results in AM-AM and AM-PM nonlinearities. As a result, the selected power-combining network must also manage the drain capacitors.

The power-combining network is an important part of the RFDAC as it determines its output power, efficiency, and quadrature accuracy. Its importance is verified using load-pull simulations and demonstrated in Fig. 9.4A. Note that, for simplicity, the load-pull simulation is only performed for the $Drain^+$ node, and its related drain efficiency, power, and modulation error contours are plotted. The modulation error is defined as a deviation of the modulated RF output signal from its ideal position. Load-pull simulation of Fig. 9.4A indicates that the orthogonality is degraded for loads corresponding to high efficiency and power contours. This reveals that employing upconverting clocks with $D = 25\%$ is a necessary, but not sufficient, condition for the orthogonal operation. Note that the modulation orthogonality can be easily substantiated with the assumption of the linear time-invariant model of the switched PA (Fig. 6.9), which is valid when the effective transistor impedance (total switch resistance, R_{on}) in Eq. (6.17) is much larger than the loading matching network impedance R_{load} in Fig. 9.3A. Equivalently, the RF output power in that region of operation is much less than its saturation level. Adding an incremental switch conductance will linearly increase the RF output envelope independent from the total instantaneous conductance level. Stated differently, at low RF power, the I and Q paths barely interact with each other. However, at higher RF power, R_{on} is lower and the drain capacitance is higher (lower capacitance reactance), therefore the I and Q paths begin interacting (loading) with each other.

---

[1] As indicated in Chapter 8, Eq. (8.1), in DPA-based TX, there is no such a concept as an analog/RF input power. The overall system efficiency is determined considering all related DC power losses including peripheral circuits.

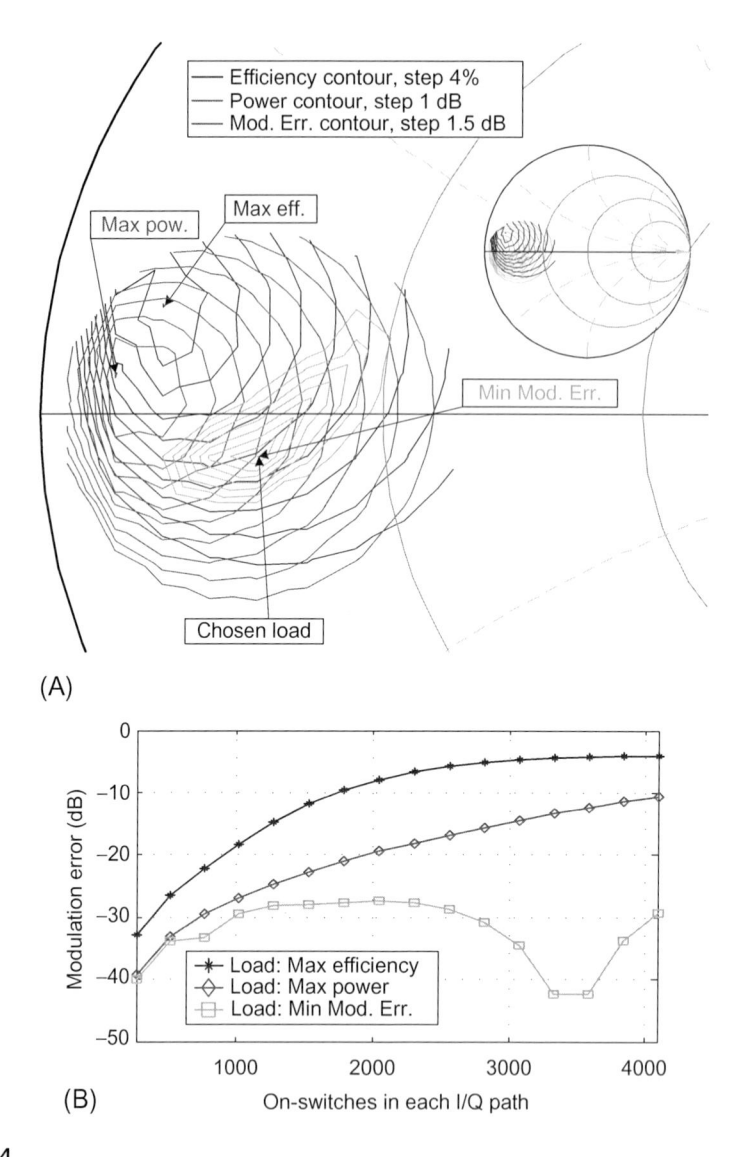

FIG. 9.4

(A) Load-pull simulations for *Drain+* node. (B) Modulation error of different load types (max efficiency, max power, and min modulation error) versus count of on-switches in each orthogonal path.

Note that, according to the simulated load-pull contours, three possible loads could be chosen: load based on the maximum efficiency, maximum power, and minimum modulation error. Fig. 9.4B exhibits the simulated modulation error versus the number of turned-on switches for the three mentioned load scenarios. This simulation confirms that the most appropriate choice for the modulation accuracy better than −28 dB is selecting the load based on a minimum modulation error, which is indicated in Fig. 9.4A. This load provides the best modulation accuracy and reasonable efficiency (better than 50%) as well as generating the desired RF output power. By doing so, I and Q interaction becomes less while the digital predistortion can be simpler. In other words, as will be discussed in Chapter 10, it is possible to use only 2×1D DPD instead of 2D counterpart due to uncorrelated I and Q paths. In conclusion, to maintain $I_{path,p}$, $Q_{path,p}$, $I_{path,n}$, and $Q_{path,n}$ orthogonal at all RF power levels, the circuit elements of power-combining network must also be incorporated in all quadrature paths. It should be pointed out that all circuit simulations are performed at 2.4 GHz. This signifies that in order to maintain the orthogonality for other carrier frequencies, the optimum loading condition, which is selected based on the best modulation accuracy, must be adjusted accordingly.

9.2 A DIFFERENTIAL I/Q CLASS-E-BASED POWER COMBINER

In order to achieve high efficiency at high RF power, and considering I_P, Q_P, I_N, and Q_N as being digital clock signals of rectangular pulse shape with a duty cycle of 25%, the class-E-type matching network [141–144] is adopted because it is well suited for pulse-shaped signals. Furthermore, the class-E matching can absorb the drain capacitance of the cascode switches. Based on Fig. 9.2B, the drain capacitance of the cascode switches is also considered in the eventual shunt capacitance, C_s in Fig 9.1. It should be mentioned that, due to the electrical summation of I_{path} and Q_{path}, the overall duty cycle at differential nodes of $Drain^+$ and $Drain^-$ in Fig. 9.2B is 50% at equal component power levels. In addition, in a class-E matching network, the loading condition for an RF signal with $D = 50\%$ is completely different than at $D = 25\%$ [143]. This explains why the efficiency/power contours of Fig. 9.4A significantly differ from the modulation error contours.

Based on the above considerations, the design of an orthogonal power-combining network is divided into four identical class-E-type matching networks, which are clearly depicted in Fig. 9.1. In this idealized power combiner, L_{BFI} provides the required DC current of DRAC; C_{AC} decouples the drain node from the output. There are three yet-to-be-defined components: R_{load}, C_s, and L_{add}, which will be addressed later in this section. The idealized power combiner of Fig. 9.1 is rather impractical. It should be modified such that it does not contain bulky components such as L_{BFI} and C_{AC}. Moreover, the eventual RF output must drive the single-ended load of 50 Ω. As a result, this power combiner should be modified.

Fig. 9.5A demonstrates the conventional class-E network [141–143] terminated with the transformed load. According to Figs. 9.1 and 9.2, the RFDAC comprises four separate sections, and the total RF output power is the summation of the power delivered by each individual section. Therefore, the output load for each section can be viewed as

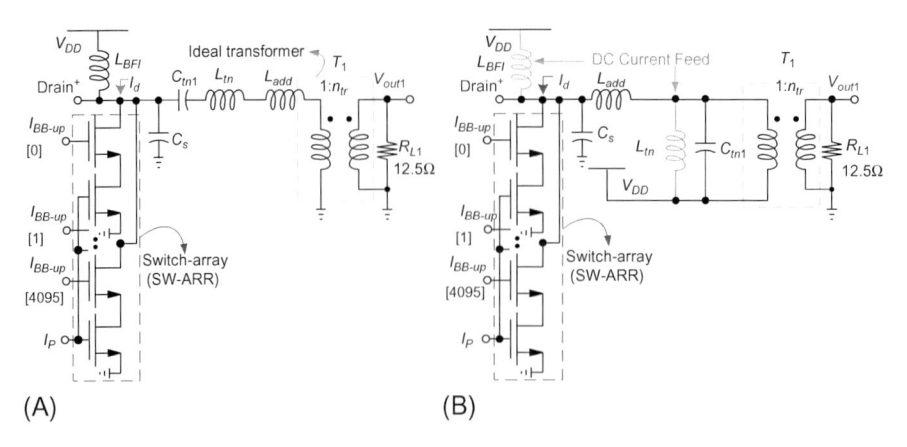

FIG. 9.5

Power-combining network: (A) first and (B) developing step.

$$R_{L1} = \frac{1}{4} \times R_L = \frac{50}{4} = 12.5 \, \Omega \tag{9.1}$$

However, Fig. 9.5A matching network is not an effective selection for monolithic implementations. The use of two inductors (L_{BFI}, ($L_{add} + L_{tn}$)) and one transformer (T_1) substantially increases the occupied area. The series tank filter of L_{tn} and C_{tn1} allows only the fundamental current component to pass. Equivalently, the series filter could be replaced by a parallel tank filter, as shown in Fig. 9.5B. Although L_{tn} is employed as a component of the fundamental-frequency selective circuitry, it also could be used as the DC current feed, thus eliminating L_{BFI}.

This modification is illustrated in Fig. 9.6A. Moreover, a practical transformer consists of magnetizing and leakage inductors [126]. Now, L_{BFI}, which is typically large, is fortunately eliminated. In addition, this circuit incorporates only one transformer and does not require the additional inductors. Indeed, L_{tn} and L_{add} are the magnetizing and leakage inductors of the transformer T_1 that is highlighted by the dotted gray box. C_{tn1} moves into the secondary side of the transformer and is now labeled as C_{tn}. Since the turns ratio of the transformer is $n_{tr} \geq 1$, then $C_{tn} \leq C_{tn1}$. Note that Fig. 9.6A schematic depicts only the in-phase (i.e., I) switch array and not the complete quadrature modulator. Therefore, for the complete quadrature configuration, another C_s and L_{add} should be added parallel to the circuit, which is exhibited in Fig. 9.6B. In contrast to the primary side of the transformer, the summation of I and Q signals in the secondary side is performed in series. Hence, R_{L2} is defined as:

$$R_{L2} = 2 \times R_{L1} = \frac{50}{2} = 25 \, \Omega \tag{9.2}$$

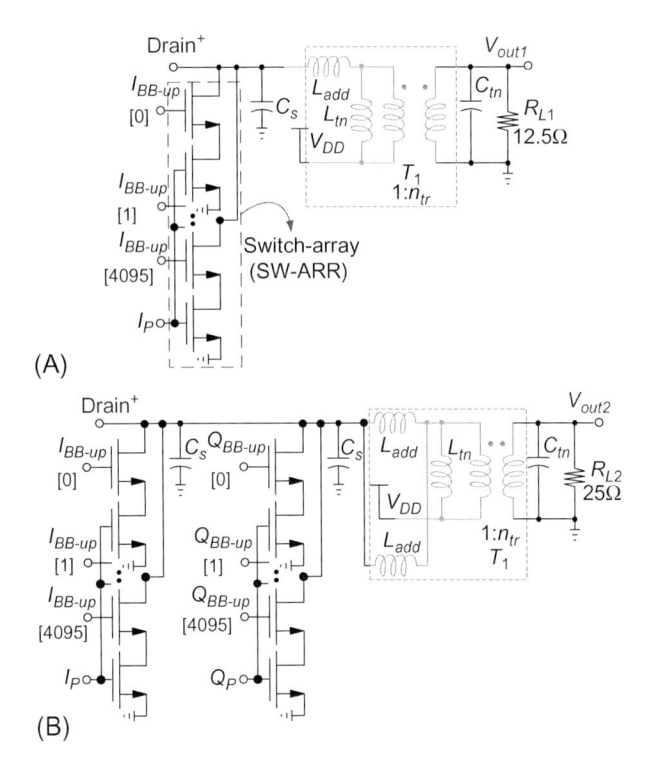

9.2 A differential I/Q class-E-based power combiner

FIG. 9.6

Power-combining network: (A) modified version; (B) I/Q single-ended.

As mentioned previously, the targeted power-combining network should be differential (see also Section 6.2, Section 8.1, and Fig. 9.7) and should also act as a balun converter. Accordingly, the transformer T_1 consists of leakage and magnetizing inductors of L_{leak} and L_{tn}, respectively, as well as an ideal transformer with N_1:N_2 turns ratio [126]. The signal summation on the secondary side of T_1 is performed in the voltage domain and, as a result, $R_L = 2 \times R_{L2} = 50 \, \Omega$.

$$R_L = 2 \times R_{L2} = 4 \times R_{L1} = 50 \, \Omega \tag{9.3}$$

For designing the power-combining network, the values of C_{shunt}, C_s, L_{add}, L_{tn}, C_{tn}, C_{out}, R_{in}, and n_{tr} should be derived. Based on [143], the values of C_s and L_{add} depend on D, f_0, R_{L1}, and n_{tr} (transformer turns ratio, that is, $N_2/N_1{}^2$) and could be expressed as follows:

---

[2] Considering coupling factor, the eventual transformer turn ratio is $n_{tr} = \frac{N_2}{N_1 \times k_m}$ (see [126]).

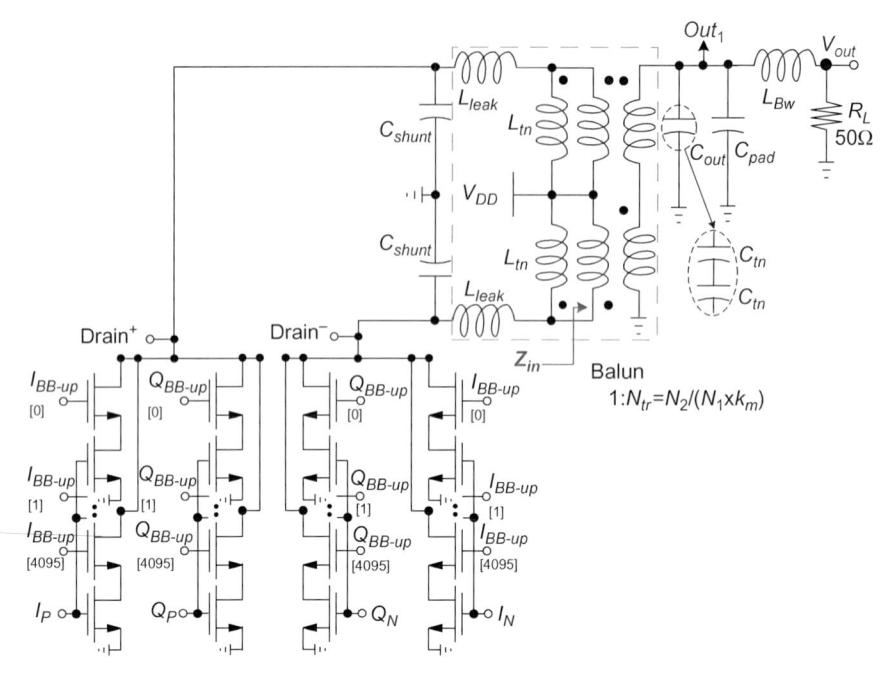

FIG. 9.7

Power-combining network: final differential I/Q RFDAC.

$$C_s(D, f_0, R_{L1}, n_{tr}) = \frac{f(D=25\%)}{2\pi f_0 \times \left(R_{L1}/n_{tr}^2\right)} = \frac{f(D=25\%)}{2\pi f_0 \times R_{load}} \quad (9.4)$$

$$L_{add}(D, f_0, R_{L1}, n_{tr}) = \frac{gt(D=25\%) \times \left(R_{L1}/n_{tr}^2\right)}{2\pi f_0} = \frac{gt(D=25\%) \times R_{load}}{2\pi f_0} \quad (9.5)$$

where $f(D)$ and $gt(D)$ represent two different functions, which depend on the duty cycle and are described in Appendix B (Eqs. B.6, B.8). Furthermore, based on Eqs. (B.10), (B.11) for a 25% duty cycle, C_s and L_{add} are as follows

$$C_s(f_0, R_{load}) \cong \frac{0.21322}{2\pi f_0 \times R_{load}} \quad (9.6)$$

$$L_{add}(f_0, R_{load}) \cong \frac{3.5619 \times R_{load}}{2\pi f_0} \quad (9.7)$$

Moreover, R_{load}, which is depicted in Fig. 9.1 is the equivalent input referred resistor of Fig. 9.6A, and defined as follows:

$$R_{load} = R_{L1} \times \left(\frac{N_1}{N_2}\right)^2 = \frac{1}{4} \times R_L \times \left(\frac{N_1}{N_2}\right)^2 = R_L \times \left(\frac{1}{2n_{tr}}\right)^2 = 12.5 \times \left(\frac{1}{n_{tr}}\right)^2 \quad (9.8)$$

According to Fig. 9.6B L_{tn} and C_{tn} resonate at f_0 and filter out higher-order harmonics. The value of L_{tn} affects the area, transformer coupling factor (k_m), quality factor (Q_F), and, consequently, the output power. The k_m of planar monolithic transformer is usually $0.6 \leq k_m \leq 0.9$. Therefore, for example, choosing $k_m = 0.71$, the value of L_{tn} (magnetizing inductance) is equal to the leakage inductance of each balun leg [126], which is:

$$L_{tn} = L_{leak} = L_{add}/2 \qquad (9.9)$$

C_{tn} resonates with L_{tn} which is:

$$C_{tn} = \frac{1}{(2\pi f_0)^2 \times L_{tn} \times n_{tr}^2} \qquad (9.10)$$

C_{Shunt} and C_{out} are the parallel and series combinations of C_s and C_{tn}, respectively, and based on Figs. 9.6B and 9.7, can be expressed as:

$$C_{shunt} = 2 \times C_s \qquad (9.11)$$

$$C_{out} = C_{tn}/2 \qquad (9.12)$$

Note that, based on the new arrangement of Fig. 9.7, L_{tn} should resonate out with the combination of C_{out} and C_{pad}:

$$C_{out} + C_{pad} = \frac{1}{(2\pi f_0)^2 \times (2L_{tn}) \times n_{tr}^2} \qquad (9.13)$$

where C_{pad} is the bond-pad capacitance. L_{BW} is a bond-wire inductor, which only slightly affects the power-combining network and has also been examined previously in Section 7.1.6 and was specifically depicted in Fig. 7.9C. Generally, the desired L_{leak} determines the size and structure of the selected transformer which subsequently determines the value of L_{tn} for a given value of the magnetic coupling factor k_m. To conclude, the balun performs the following:

1. It de-couples the drain DC condition from the load (elimination of C_{AC}).
2. It converts the differential signal to a single-ended output.
3. It provides a DC bias path for the DRAC transistor switches (elimination of L_{BFI}).
4. It transforms the 50 Ω load to the desired impedance at the drain nodes of DRAC.

Z_{in}, which is shown in Fig. 9.7, is indeed the single-ended matching network impedance of the modulator:

$$Z_{in} = R_{in} = \frac{R_L}{2 \times (n_{tr})^2} = 2 \times R_{load} \qquad (9.14)$$

The output power of Fig. 9.7 is expressed as:

$$P_{out} = \frac{(V_{out})^2}{2 \times R_L} = 4 \times P_{path} \tag{9.15}$$

where P_{path} is the power related to each orthogonal path of Figs. 9.1 and 9.2 or Fig. 9.7 and based on [143, eq. 3.29] expressed as

$$P_{path} = \frac{1}{2} \times \left(\frac{2 \times V_{DD}}{g(D)}\right)^2 \times \frac{1}{R_{load}} = \frac{2}{(g(D))^2} \times \frac{V_{DD}^2}{R_{load}} \tag{9.16}$$

where $g(D)$ is a unitless function and depends only on D, and thus, $g(D = 25\%) \approx 5.8$ (see Appendix B, Eqs. B.4–B.9). According to Eq. (B.10) the P_{path} is as follows:

$$P_{path} = \frac{2}{(5.79925)^2} \times \frac{V_{DD}^2}{R_{load}} \cong \frac{1}{16.8157} \times \frac{V_{DD}^2}{R_{load}} \tag{9.17}$$

As a result, the total output power can be expressed as:

$$P_{out} \cong \frac{4}{16.8157} \times \frac{V_{DD}^2}{R_{load}} \cong \frac{1}{4.2039} \times \frac{V_{DD}^2}{R_{load}} \tag{9.18}$$

In addition, according to Eq. (9.8)

$$P_{out} \cong \frac{(2n_{tr})^2}{4.2039} \times \frac{V_{DD}^2}{R_L} \cong \frac{(n_{tr})^2}{1.051} \times \frac{V_{DD}^2}{R_L} \tag{9.19}$$

Consequently, n_{tr} can be expressed as the output power:

$$n_{tr} \cong \frac{\sqrt{1.051 \times R_L \times P_{out}}}{V_{DD}} \tag{9.20}$$

If V_{DD} is kept constant and, since most of the time the off-chip load impedance (R_L) is fixed and equal to 50 Ω, according to Eqs. (9.19), (9.20), if n_{tr} increases, then the output power also increases, and the following statements can be concluded:

1. If the desired output power is considered to be in the range of $14.35 \leq P_{out\text{-}dBm} \leq 23.9$ while $V_{DD} = 1.2$ V, then $1 \leq n_{tr} \leq 3$.
2. If the targeted output power should be in the range of $20.75 \leq P_{out\text{-}dBm} \leq 30.3$, while $V_{DD} = 2.5$ V, then $1 \leq n_{tr} \leq 3$.

Note that, in the preceding conclusions, the passive components of the power-combining network are considered lossless. To validate the equations derived above, the RFDAC of Fig. 9.7 is simulated in Agilent ADS employing the following fixed design parameters: $V_{DD} = 1.2$ V, $f_0 = 2.4$ GHz, $L_{NMOS} = 60$ nm. All of the passive components are considered lossless. Table 9.1 reveals design components of the power-combining network for different n_{tr} values of the transformer turns

Table 9.1 Design Parameters of I/Q RFDAC Assuming a Lossless Transformer-Based Power Combiner: $V_{DD} = 1.2\,\text{V}$, $f_0 = 2.4\,\text{GHz}$, $L = 60\,\text{nm}$

n_{tr}	R_{in} (Ω)	P_{out} (dBm)	I_d (mA)	W (nm)	C_{shunt} (pF)	L_{leak} (pH)	N_{on-sw}
1	25	14.36	52	220	2.26	1500	1024
1.25	16	16.31	80	160	3.53	944	2048
1.5	11.11	18.29	126	120	5.09	656	4096
1.75	8.16	19.22	156	155	6.93	482	4096
2	6.25	20.40	200	210	9.05	369	4096
2.25	4.94	21.42	268	280	11.45	292	4096
2.5	4	22.31	317	345	14.14	236	4096
2.75	3.31	23.16	384	415	17.10	195	4096
3	2.77	23.89	451	485	20.36	164	4096

ratio.[3] As it is expected from Eqs. (9.19)–(9.20) and substantiated by simulation results, increasing n_{tr} results in higher output power. Moreover, for all cases, the drain efficiency is $\eta_{drain} \approx 45\%$.[4] As will be concluded later, the efficiency drops in comparison to class-E primarily due to an increase of shunt capacitance (Eq. 9.11), which, subsequently increases the modulator DC current.

By sequentially and simultaneously increasing the I and Q modulator codes, the summing and orthogonality can be examined in view of the simulation results exhibited in Fig. 9.8. In this context, the I/Q modulator of Fig. 9.7 is simulated again in Agilent ADS using the following fixed design parameters: $V_{DD} = 1.2\,\text{V}$, $f_0 = 2\,\text{GHz}$, $L_{NMOS} = 100\,\text{nm}$. Moreover, for simplicity, the resolution (N_b) is reduced to 10-bit. Note that, in the conventional equation of quadrature modulator, the resultant IQ signal can be expressed as:

$$IQ = I_{BB} \times \cos(\theta) - Q_{BB} \times \sin(\theta) \qquad (9.21)$$

If $I_{BB} = Q_{BB}$, then

$$IQ = \sqrt{2} \times I_{BB} \times \cos\left(\theta + \frac{\pi}{4}\right) \qquad (9.22)$$

As a result, the amplitude and phase of the IQ signal are

$$\rho_{IQ} = \sqrt{2} \times I_{BB} \qquad (9.23)$$

$$\theta_{IQ} = \frac{\pi}{4} \qquad (9.24)$$

---

[3] C_{shunt} in Table 9.1 is the value of the required shunt capacitor. This capacitor consists of the total drain capacitance of the related unit cell DPAs as well as additional explicit shunt capacitors.

[4] For all cases, their drain voltage swing is the same. Moreover, although employing a wider switch decreases its related R_{on}, yet its related DC current increases, consequently, $P_{loss} = V_{on} \times I_{DC} \approx cte$.

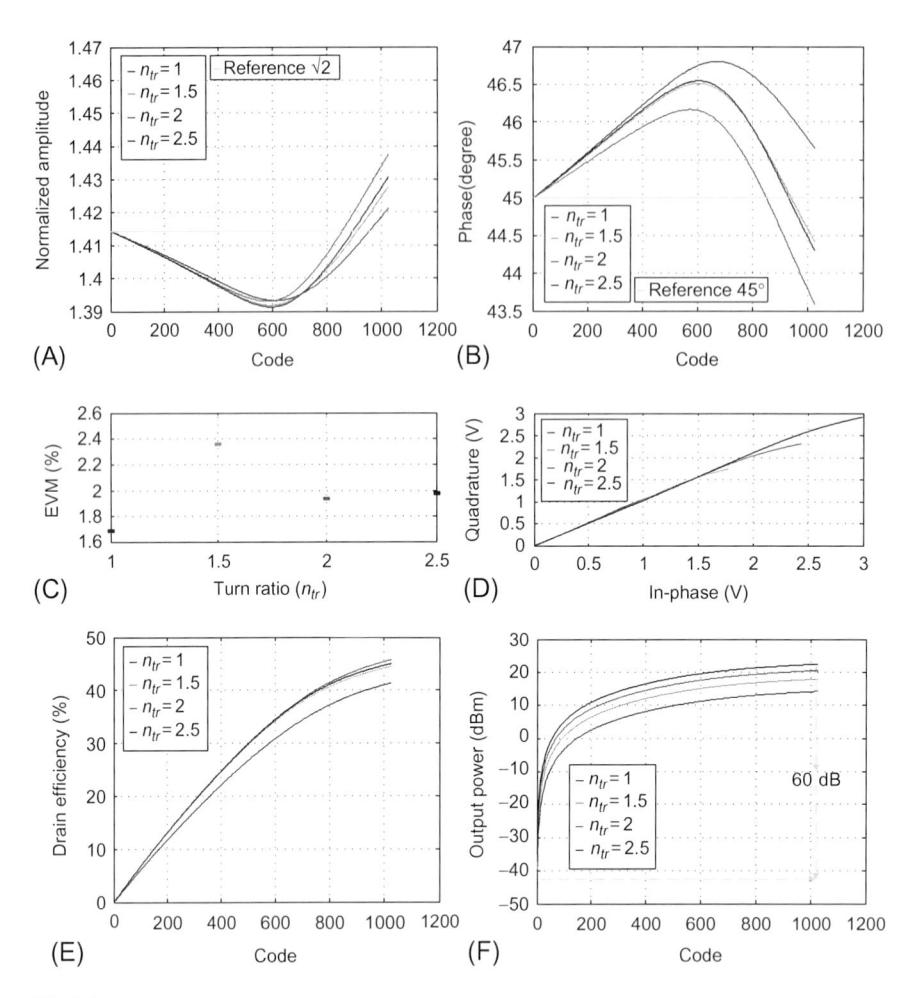

FIG. 9.8

Simulation results of power-combining network for equal I/Q codes (10 bits of resolution): (A) Normalized amplitude and (B) phase of RF output node; (C) corresponding EVM; (D) first quadrant constellation diagram; (E) drain efficiency; (F) output power.

Based on Eq. (9.23), the output power of Fig. 9.6B is:

$$P_{out2} = 2 \times P_{out1} = 2 \times P_{path} \tag{9.25}$$

where P_{out1} is the output power of Fig. 9.5A and B and Fig. 9.6A. Furthermore, Eq. (9.25) substantiates that each orthogonal path provides one quarter of P_{out}, which was previously indicated in Eq. (9.15). As derived in Eqs. (9.23), (9.24), the normalized amplitude and phase of the modulator should be $\sqrt{2}$ and $\pi/4$, respectively. Based on the simulation results, there are small deviations from these target numbers, which are illustrated in Fig. 9.8A and B. As is previously stated and also is seen in Fig. 9.4B, the reason for these amplitude and phase deviations are related to the fact that changing the G_{sw} would slightly change the phase and amplitude of the drain node and, consequently, the output node. Fig. 9.8C exhibits the corresponding EVM calculation based on Eq. (1.3). Simulation results reveal that the EVM for all of the simulation conditions is less than 2.5%, which confirms the orthogonal summing and power-combining operation. Fig. 9.8D shows that the linearity of the modulator is improved when the transformation ratio is increased which entails increasing of the peak output power. The reason is that, for higher output power, R_{in} is lower and, consequently, the drain voltage swing is lower. Fig. 9.8E and F shows that the drain efficiency and output power are selected to be maximum when the code is maximum. Moreover, based on Fig. 9.8E, the dynamic range of the modulator is approximately 60 dB, which could correspond to 10 bits of the RFDAC resolution.

9.3 EFFICIENCY OF I/Q RFDAC

Based on Table 9.1 and also as was illustrated in Fig. 9.8E, the maximum drain efficiency of the I/Q RFDAC is approximately half of the ideal class-E. The efficiency drop is investigated by considering the effect of the circuit component values on the drain current, voltage, and, subsequently, the resultant DC current and output power. In this context, the drain voltage of Fig. 9.6A, which is depicted in Fig. 9.9A, is considered periodic, thus, it can be represented using Fourier series. As a result, the periodic drain voltage could be expressed as:

$$V_{drain}(\theta) = V_{DD} + \sum_{n=1}^{\infty} a_n \cos(n\theta) + b_n \sin(n\theta) \qquad (9.26)$$

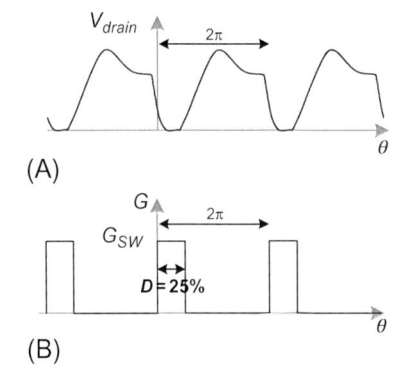

FIG. 9.9

Waveforms of (A) the drain voltage, $V_{drain}(\theta)$; (B) switchable conductance, $G(\theta)$.

in which a_n and b_n are in-phase and quadrature components of the periodic drain voltage in units of volts, respectively. In addition, the Fourier representation of a periodic pulse with $D = 25\%$ is:

$$\Pi_{D25}(\theta) = \frac{1}{4} + \left(\frac{2}{\pi}\right) \times \sum_{m=1}^{\infty} \left(\frac{1}{m}\right) \times \sin\left(\frac{m\pi}{4}\right) \times \cos(m\theta) \tag{9.27}$$

therefore, the switch conductance, which is illustrated in Fig. 9.9B, is expressed as:

$$G(\theta) = G_{sw} \times \Pi_{D25}(\theta) \tag{9.28}$$

where G_{sw} was defined in 6.11. It is worth reiterating that the DRAC switch conducts only 25% of the clock period. According to Ohm's law, the switch (transistor drain) current depends on V_{drain} and G:

$$I_d(\theta) = V_{drain}(\theta) \times G(\theta) \tag{9.29}$$

Applying Kirchoff's voltage and current law as well as equating all cosine and all sine terms, and considering only the first two harmonics of Eqs. (9.26)–(9.28), I_{DC}, a_1, and a_2 can approximately be estimated:

$$a_1 \cong \frac{V_{DD}}{\sqrt{2}\pi \times DET_1} \times H\left(\frac{C_s}{G_{sw}}, \frac{R_L}{L_{add}}\right) \tag{9.30}$$

$$a_2 \cong \frac{1}{DET_2} \times K\left(\frac{C_s}{G_{sw}}, \frac{R_L}{L_{add}}\right) \tag{9.31}$$

$$b_1 \cong \frac{V_{DD}}{\sqrt{2}\pi \times DET_1} \times T\left(\frac{C_s}{G_{sw}}, \frac{R_L}{L_{add}}\right) \tag{9.32}$$

$$I_{DC} \cong G_{sw} \times \left(\frac{V_{DD}}{4} + \frac{a_1}{\sqrt{2} \times \pi} + \frac{a_2}{2 \times \pi}\right) \tag{9.33}$$

where DET_2, DET_1, H, K, and T are defined in Appendix B (Eqs. B.12, B.16). Note that H, K, and T depend on the following variables:

$$B_{Cs} = 2\pi f_0 C_s \tag{9.34}$$

$$X_{Ladd} = 2\pi f_0 L_{add} \tag{9.35}$$

$$|Z_{RX}|^2 = R_L^2 + X_{Ladd}^2 \tag{9.36}$$

In fact, the phasor representations of drain voltage and output voltage at fundamental frequency of operation (f_0) of Fig. 9.6A are:

$$V_{P\text{-}drain} = a_1 + jb_1 \tag{9.37}$$

$$V_{P\text{-}out} = V_{P\text{-}drain} \times \left(\frac{R_L}{R_L + jX_{Ladd}}\right) \tag{9.38}$$

9.3 Efficiency of I/Q RFDAC

Table 9.2 Design Parameter of Class-E in Fig. 9.6A for $V_{DD} = 1V$, $f_0 = 2\,GHz$, $R_{on} = 0.4\,\Omega$, $C_{tn} = 10\,pF$, $L_{tn} = 633\,pH$, $n_{tr} = 1$

Type	Figure	C_s (pF)	L_{add} (pH)	R_{L1} (Ω)
Case I	Fig. 9.6A	16.97	284	1
Case II	Fig. 9.6B	33.94	142	2

Table 9.3 Comparison Between Derived Equations and Circuit Simulation

Type	$V_{P\text{-}drain}$ (V)	$V_{P\text{-}out}$ (V)	P_{out1} (dBm)	I_{DC} (mA)	η_{drain1} (%)
Equations Case I	−1.100−j0.317	−0.163+j0.263	16.805	54	89.061
Simulation Case I	−1.082−j0.431	−0.191+j0.250	16.950	55	89.500
Equations Case II	−0.742+j0.149	−0.340+j0.451	19.019	137	58.065
Simulation Case II	−0.726+j0.210	−0.300+j0.478	19.007	142	56.148

Consequently, based on the voltage and current equations of Eqs. (9.26)–(9.38), the output power, DC power, and drain efficiency are:

$$P_{out1} = \frac{1}{2} \times \Re\left(V_{P\text{-}out} \times \text{conj}\left(\frac{V_{P\text{-}out}}{R_L}\right)\right) \tag{9.39}$$

$$P_{DC1} = V_{DD} \times I_{DC} \tag{9.40}$$

$$\eta_{drain1} = \frac{P_{out1}}{P_{DC1}} \tag{9.41}$$

For validating these equations, two different cases are considered. The first case (case I) is related to the class-E circuit (Fig. 9.6A) with $R_{on} = 0.4\,\Omega$. The second case (case II) is the circuit of Fig. 9.6B in which the quadrature switch arrays are off. The design parameters of the two cases are tabulated in Table 9.2. Note that for the sake of simplicity, in this context, the R_{L1} is considered $1/2\,\Omega$. As a result, Table 9.3 compares the derived equations and circuit simulation results of these two cases. Table 9.3 reveals certain significant conclusions:

1. There is a positive agreement between the derived equations and circuit simulation results. The moderate differences between them arise because of neglecting the higher harmonics of Eq. (9.26).
2. Since $C_{s\text{-}caseII} = 2 \times C_{s\text{-}caseI}$, $L_{add\text{-}caseI} = 2 \times L_{add\text{-}caseII}$, and $R_{L1\text{-}caseII} = 2 \times R_{L1\text{-}caseI}$, according to Eqs. (9.30)–(9.36), these entail an increase of a_1, a_2, I_{DC}, and reduction in quality factor. Consequently, the drain efficiency of the modulator decreases compared to a class-E implementation.
3. The phase and amplitude of the drain node and, consequently, the output node depend only on the value of G_{sw} when the other passive components are fixed. Therefore, by sequentially turning on the switches, the phase of output would be

changed (see eq. 6.21). In other words, turning on and off the switches modulate the RF output signal and can address the desired constellation diagram points.
4. The output power of the second case ($P_{out2\text{-}2}$) is higher than case one (P_{out1}):

$$P_{out2\text{-}2} = \alpha_p \times P_{out1} \tag{9.42}$$

where α_p is an incremental factor of power. Simulation results indicate that this value primarily depends on R_{on}, and α_p varies between $1 \leq \alpha_p \leq 2$. As was previously mentioned, the second case is a special case of Fig. 9.6B in which the Q path is completely deactivated (off). Thus if the Q path turns on completely, then the output power will be doubled.

9.4 EFFECT OF RISE/FALL TIME AND DUTY CYCLE

As thoroughly explained in this book, in order to isolate between the I_{path} and Q_{path}, the differential quadrature upconverting clocks with the duty cycle of 25% are utilized. Thus, due to their shorter pulse width, the related rise/fall time of the upconverting clocks should be fast enough to properly charge and discharge the drain capacitances in a short period of time in order to perform reasonable isolation between I_{path} and Q_{path}. Consequently, the I/Q RFDAC is implemented in 65 nm CMOS which affords moderate rise/fall time. According to the post layout simulation results of the I/Q RFDAC, the rise/fall time at the gate of digital power amplifier switches is approximately 45 ps. Employing a 2.4 GHz upconverting clock while its related duty cycle is 25%, the rise/fall time must be less than 104.16 ps in order to avoid nonorthogonal operation. To validate the effect of rise/fall times, they are swept from 30 to 100 ps. According to the simulation results of Fig. 9.10A–C, the related modulation error of the I/Q RFDAC will be less than -25 dB as long as the corresponding rise/fall time of the upconverting clocks are less than 80 ps. Moreover, as depicted in Fig. 9.10A–C, the drain efficiency of the RFDAC along with its corresponding output RF power will be diminished for the slower clock edges. Nonetheless, implementing the proposed RFDAC in a more advanced nanoscale CMOS process, such as 28 nm, will improve the performance.

Note that, in order to preserve orthogonality as stated throughout the book, the duty cycle of the upconverting differential, quadrature clocks is selected at 25%. As explained previously, this selection, based on the fact that the timing overlap of the differential quadrature clocks, must be as minimal as possible. As a result, any duty cycle less than 25% is also acceptable. Nonetheless, there would be three primary disadvantages in not selecting the 25%. First, generating differential quadrature upconverting clocks with a 25% duty cycle is less complicated. Second, the corresponding clock generation circuitry consumes less power. Third, as depicted in Fig. 9.10D, the related RF output power for a 25% duty cycle is much higher due to the fact that the fundamental frequency of the RF current is proportional to the clock pulse width which makes the RF output power proportional to the pulse width of the

FIG. 9.10

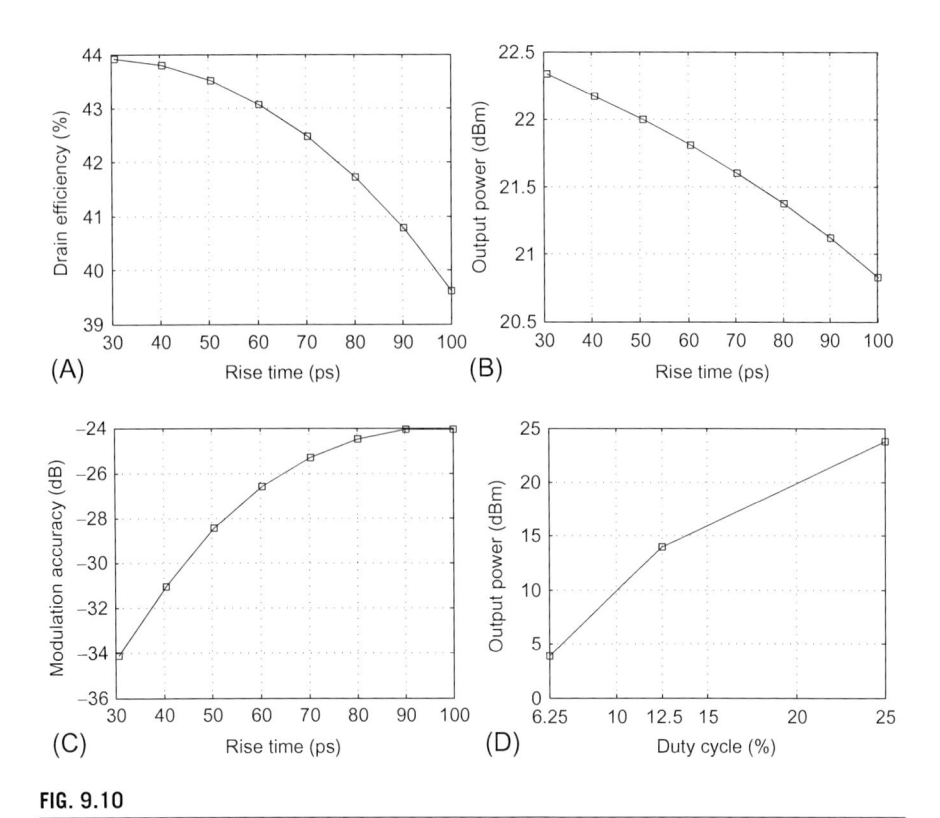

Simulation results based on sweeping the rise/fall time of the upconverting clock in order to illustrate their effects on (A) drain efficiency; (B) output power; (C) modulation accuracy. (D) Output power versus the duty cycle of the upconverting quadrature clocks.

differential, quadrature upconverting clock. It should be pointed out that the related drain efficiency and modulation accuracy of them, that is, upconverting clocks with $D \leq 25\%$, are more or less identical.

9.5 EFFICIENCY AND NOISE AT BACK-OFF LEVELS

Although the power-combining network, at full power, has been designed based on a class-E matching network, yet at back-off power, the RFDAC performs more like a class-B power amplifier. The reason is related to the fact that, at full power, the entire transistor arrays perform equivalently as a switch due to the low drain-source, on-resistance (R_{on}) of the equivalent switch. According to the simulation result in Fig. 9.3A, the R_{on} at full power, that is, the baseband code of 4096, is less than 2 Ω. Otherwise stated, the drain-source voltage of the aforementioned switchable cascode transistors is approximately 0.25 V indicating that the equivalent switch operates in the triode region. On the other hand, at back-off power levels, the number of on-switches decreases which subsequently increases the R_{on} of the equivalent switch. As a result, the drain-source voltage of the equivalent switch increases. This causes the equivalent switch (switchable cascode transistor) to operate in the saturation region. Thus, it performs as a current source rather than a switch, which indicates that the RFDAC operates as a class-B power amplifier. In order to confirm the foregoing discussion, the performance of the RFDAC efficiency is simulated while the input baseband code is swept from code 1 up to 4096. Simultaneously, a class-B power amplifier, which employs an idealized matching network, is also simulated. The two simulation results are depicted in Fig. 9.11A. According to that, the RFDAC at back-off power levels more or less performs the same as a class-B power amplifier. Additionally, it indicates that the drain efficiency diminishes approximately 26% at 6 dB back-off power.[5] Note that, if the RFDACs were also to operate in class-E mode at the power back-off as illustrated in Fig. 9.11A, it must maintain its drain efficiency as in the full RF power even at the back-off power levels.

The RFDAC noise performance is also simulated and depicted in Fig. 9.11B. Two scenarios are exhibited which are the noise performance of the RFDAC when the baseband input code is 1 and 4096, respectively. As expected, the RFDAC noise performance will improve provided it operates at higher power levels in which the number of on-switches are increased. The noise floor is better than -194 dBc/Hz

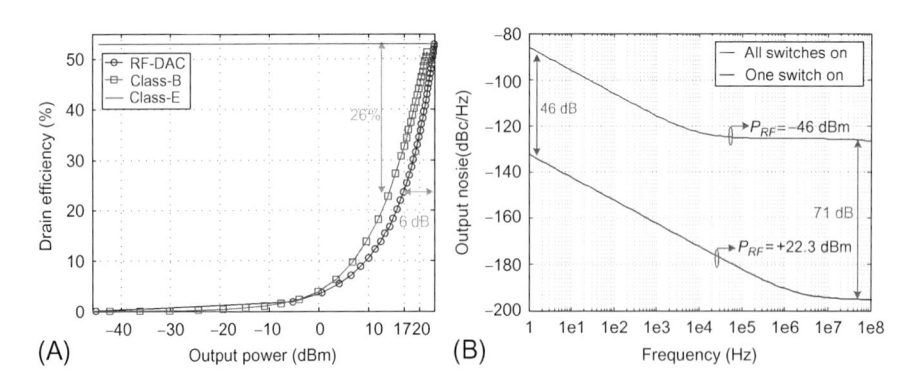

FIG. 9.11

(A) Comparison of back-off power profiles between the proposed RFDAC and a class-AB power amplifier. (B) Noise performance of the RFDAC at low and full power mode of operation.

---

[5]Note that at 6 dB back-off power levels, the power is one-fourth of its saturated RF power. In this context, the drain's fundamental as well as its DC current component is one-half of their maximum value. Nonetheless, its related DC voltage is fixed, that is, $V_{DD} = 1.2$ V, while its fundamental drain voltage becomes half. Consequently, the efficiency drops to half of its maximum value.

at full power which indicates that the noise of the switches is not a limiting factor in the total noise floor of the RFDAC. Indeed, as will be presented in Chapter 10, Section 10.4, Fig. 10.4D, the noise of the clock generator sources are the dominant noise sources. Note that, based on the simulation results of Fig. 9.11B, at low frequencies, the noise of the switches are almost dominant ($1/f$ region), however, at higher frequencies, the noise sources related to the power-combining network overtakes the noise of the switches.

9.6 DESIGN AN EFFICIENT BALUN FOR POWER COMBINER

As stated previously, the power-combining network consists of input/output capacitor tuners and the balun. The transformer, however, should be properly designed as it affects efficiency, output power, and modulation accuracy. In the previous sections, the balun role and its impact on the eventual RF output power as well as the modulation accuracy of the proposed I/Q RFDAC have been thoroughly discussed. Nonetheless, this passive component introduces certain power losses and should be carefully designed.

Fig. 9.12 exhibits six different balun transformers (T_1). The turns ratio (N_2/N_1) of these baluns are 1:2, in which the top row of Fig. 9.12 is a conventional 1:2. According to Table 9.1, for producing ≥ 22 dBm RF output power, the balun must manage high currents from 317 mA up to 451 mA while $n_{tr} \geq 2.25$. To support this substantial current, the balun could employ three parallel traces in the primary winding (see Fig. 9.14) that are interdigitated with the secondary winding to satisfy electromagnetic rules of the technology [19–48]. The bottom row of Fig. 9.12 utilizes the interdigitated transformer structure with 1:2 turns ratio. Furthermore, according to Fig. 9.12, the size of the outer diameter varies between 300 and 450 µm. The transformers use metal layers 6 and 7 as well as an aluminum layer for decreasing losses due to the series resistance. Note that the thickness of metal 6, 7, as well as the aluminum layer are 0.9, 3.4, and 1.45 µm, respectively. Thus, metal 7 is the thickest layer. The transformer traces are 12 µm wide except in Fig. 9.12D that uses 6 µm traces for the primary side. The traces are separated with 3 µm gaps between them in order to satisfy the metal density rule of the process technology. The 2×13-bit I/Q RFDAC is simulated together with these baluns in the power-combining network. The simulations have been performed using the following conditions: $f_0 = 2.4$ GHz, $V_{DD} = 1.2$ V, $W = 500$ nm, $L = 60$ nm, and $R_L = 50\,\Omega$. Note that, according to simulation results based on Table 9.1, for the lossless power-combining network, the drain efficiency would be in the range of 45%. Moreover, based on the simulation results of Table 9.3, the drain efficiency for a lossless power-combining network can exceed 55%. The difference between these two results is related to the selected width of their corresponding switches. Consequently, the efficiency of Table 9.1 would increase using wider switches. As stated earlier, selecting unit cell switches with 500 nm width is an effective compromise between the overall system efficiency, output power, and modulation accuracy. Table 9.4 summarizes the simulation results.

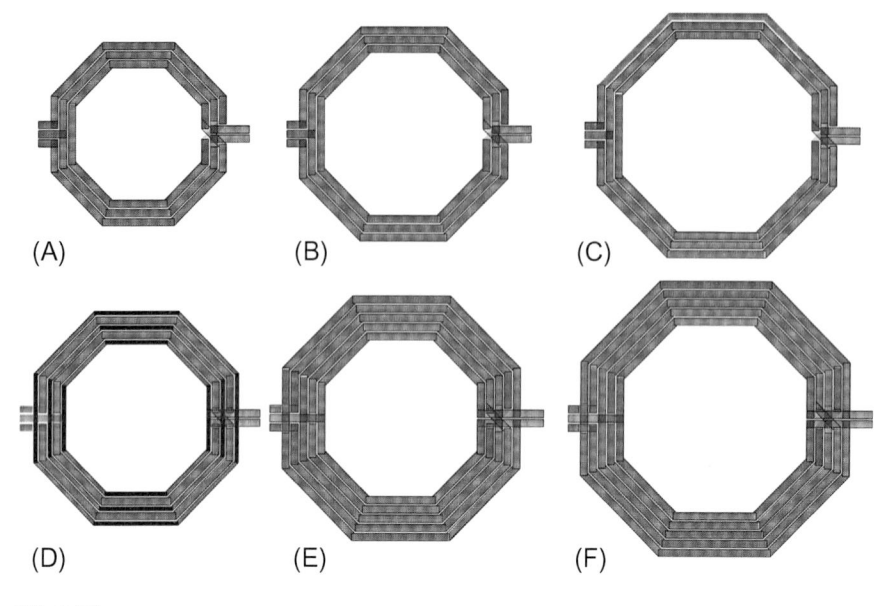

FIG. 9.12

Six different baluns (T_1) design example: Conventional 1:2, (A) 300 × 300 μm²; (B) 350 × 350 μm²; (C) 400 × 400 μm². Interdigitated 1:2, (D) 350 × 350 μm²; (E) 400 × 400 μm²; (F) 450 × 450 μm².

Table 9.4 Simulation Results of RFDAC at $f_0 = 2.4$ GHz Using the Baluns of Fig. 9.12

T_1	L_p (nH)	C_{iw} (fF)	k_m	Q_p	Q_s	P_{loss} (dB)	P_{out} (dBm)	EVM (%)	η_{drain} (%)	η (%)
(a)	0.6	23	0.74	7.5	10.5	1.3	21.3	3.9	51	38
(b)	0.76	31	0.77	8.1	10.9	1.1	21.3	4.3	51	39
(c)	0.93	39	0.79	8.3	10.7	1.1	21.3	5.2	51	40
(d)	0.52	52	0.82	10.7	9.6	1	22.1	2.5	53	42
(e)	0.53	64	0.82	15.9	8.4	0.9	22.4	3	53	43
(f)	0.67	76	0.84	16.6	8.5	0.8	22.4	3.5	54	44

Note that Q_p and Q_s are the primary and secondary quality factors of the balun, respectively, and the following statements can be inferred:

1. The conventional balun manifests moderate Q_p, output power, and efficiency. In addition, its insertion loss (P_{loss}) is also not better than 1.1 dB;

2. The corresponding differential primary inductance of conventional transformer, L_p, is higher. Note that L_p of Fig. 9.7 is defined as follows:

$$L_p = 2 \times (L_{leak} + L_{tn}) \tag{9.43}$$

The L_p of conventional transformer is higher than the interdigitated one due to smaller effective width of the primary winding. Consequently, the current through the conventional transformer is more constant than through the interdigitated one.[6]

3. The coupling factor (k_m) of an interdigitated balun is higher than the conventional one. This is due to fact that the primary winding sandwiches between the secondary winding trances. Moreover, since the effective width of primary winding is larger, the power losses are subsequently smaller and Q_p is higher.

4. The interwinding capacitance of interdigitated transformer, C_{iw},[7] which is the capacitance between the primary and secondary winding is higher due to five interdigitated metal traces. These distributed capacitors electrically couple the primary to secondary winding that provide paths between the single-ended output node to the differential primary inputs. As a result, due to lower impedance of the interwinding capacitance at higher frequencies, the transmission of the even and odd harmonic contents of the differential drain nodes to the output node is easier. Thus, both the even and the odd harmonic contents of the interdigitated balun are higher than the conventional one. Stated differently, the balanced-to-unbalanced performance of the interdigitated transformer is inferior than the conventional structure.

As noted previously, the RFDAC must produce more than 22 dBm RF output power. The width of the digital NMOS switches is selected 500 nm to provide the necessary R_{on} for generating that amount of targeted RF power. Based on the simulation results of Table 9.4, the conventional balun structures of Fig. 9.12 could not transform the optimum impedance to the drain nodes in Fig. 9.7. Therefore, the desired 22 dBm RF output power could not be generated. This also affects the efficiency and modulation accuracy of the RFDAC. On the other hand, the interdigitated transformer balun can produce the expected output power level due to its higher coupling factor and lower primary inductance. Note that, according to Aoki et al. the transformer efficiency can be expressed as follows [8]

$$\eta_{Balun}(X_p) = \frac{R_L/n_{tr}^2}{\left(\frac{X_p/Q_s + R_L/n_{tr}^2}{k_m X_p}\right) \cdot \frac{X_p}{Q_p} + \frac{X_p}{Q_s} + R_L/n_{tr}^2} \tag{9.44}$$

---

[6] Note that the drain DC power of DPA switches is fed using L_{tn}. This inductance should be large enough to provide only DC path from V_{DD}.
[7] C_{iw} is represented by C_{o1} and C_{o2} in Fig. 7.8.

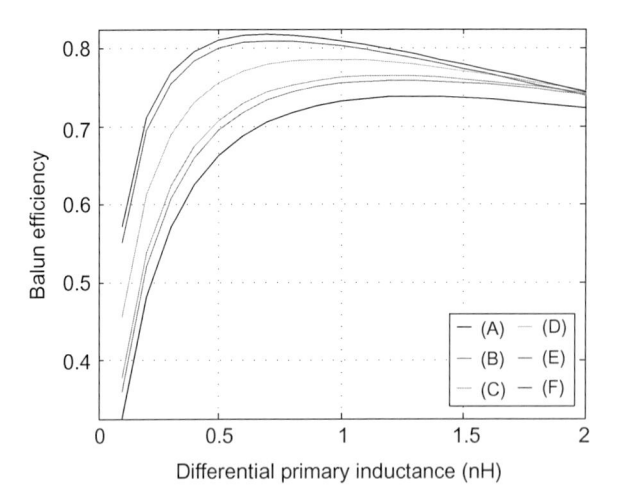

FIG. 9.13

The balun efficiency comparison of all transformers of Fig. 9.12.

where $X_p = \omega_0 L_p$ while $\omega_0 = 2\pi f_0$. Based on this Eq. (9.44), to obtain adequate efficiency, the balun must be designed in such a way that its Q_p, Q_s, and k_m are as high as possible. Using Eq. (9.44), the balun efficiency of all transformers of Fig. 9.12 is compared in Fig. 9.13 versus L_p. According to Fig. 9.13, the highest efficiency (82%) is obtained by employing the interdigitated transformer "f" in Fig. 9.12. Note that the maximum balun efficiency occurs when $L_p = 0.7$ nH. This is an interesting result since, based on Table 9.3, the L_p of the balun (f) is approximately 0.67 nH. Based on the foregoing explanations, the balun of Fig. 9.14A is selected (same as Fig. 9.12F), which is the interdigitated $450 \times 450\,\mu m^2$ with 1:2 turns ratio. In summary, the interdigitated balun structure introduces two primary advantages [19].

1. Due to effective wider primary traces, the DC ohmic series losses are reduced.
2. Due to interdigitated traces, the AC resistance caused by proximity effect[8] is dramatically decreased.

Fig. 9.14B–E illustrates the ADS Momentum simulation results of the selected balun structure. According to Fig. 9.14B, the insertion loss is better than 1 dB in the frequency range of 1.6–3.4 GHz. Furthermore, the Q_p is higher than 14.5 within the frequency of 1.5–3.3 GHz (Fig. 9.14C). In addition, the related k_m of up to 6 GHz is 0.84 and the equivalent primary loss resistance is less than 1 Ω while the frequency is less than 3.2 GHz. It is worth mentioning that, based on the simulation results in Table 9.4, drain efficiency is 54%. Moreover, based on Eq. (9.44), the balun efficiency

---

[8] At RF frequencies, the opposite currents of two parallel traces tend to be concentrated along adjacent edges (nonuniform current distribution), which causes higher insertion loss [19].

FIG. 9.14

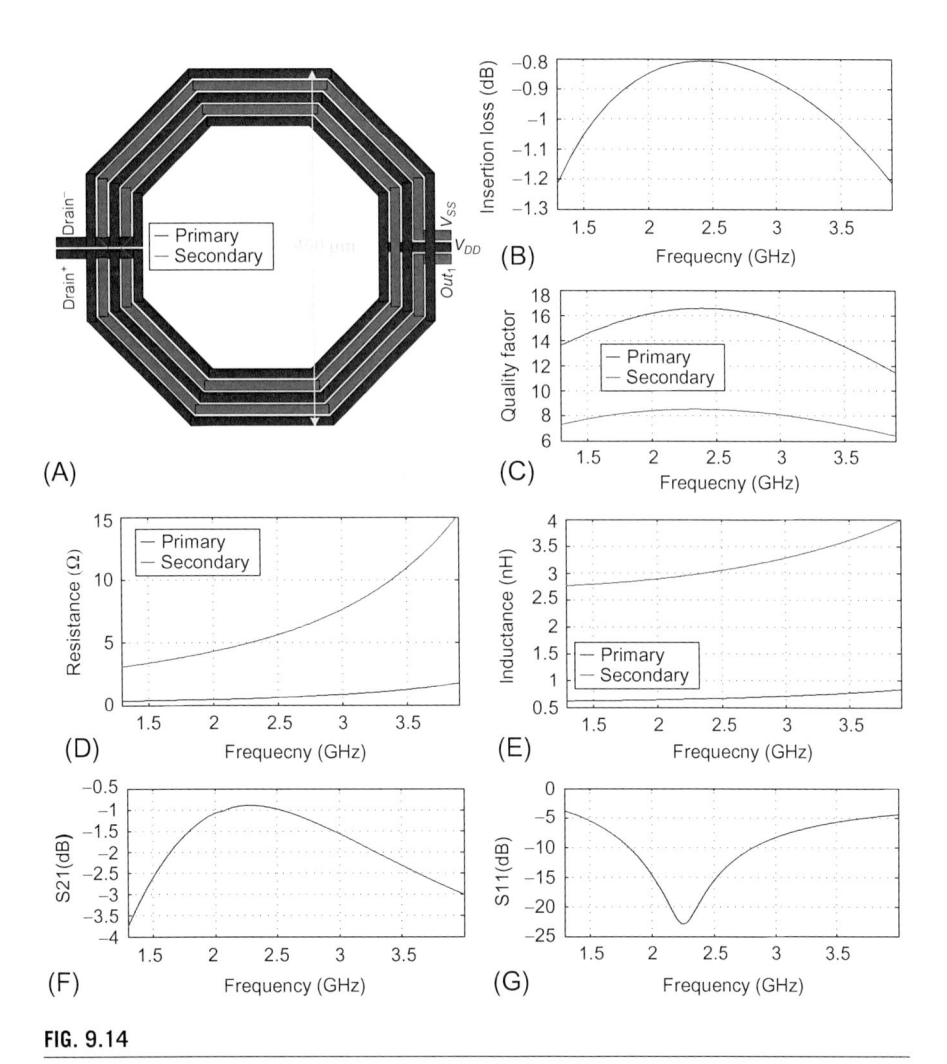

(A) Chosen balun; (B) insertion loss; (C) quality factor; (D) loss series resistance; (E) inductance versus frequency. Loaded (F) S21, and (G) S11 versus frequency.

is 82%. As a result, the eventual efficiency is $0.54 \times 0.82 = 0.44$, which agrees to the simulated total efficiency that is reported in Table 9.4.

Note that, although the power-combining network has been designed based on the class-E matching network topology, its operational "carrier" bandwidth (see Fig. 9.14F and G) is much higher than with the traditional narrowband Class-E PA. The reason lies in the fact that, in this context, the balun transformer is employed instead of an inductor. In general, the transformer's bandwidth is much larger than inductor's due to its tolerance to lower quality factor requirements as well as a larger

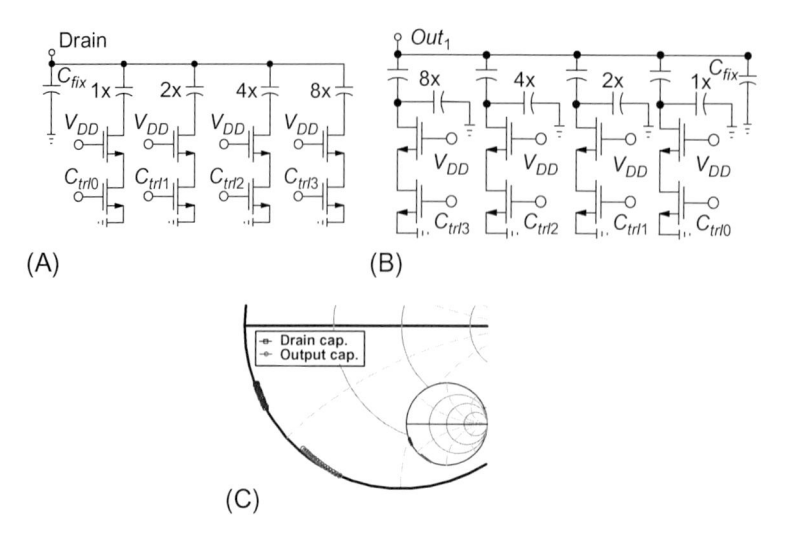

FIG. 9.15

(A) The primary input shunt tuning switched-capacitors; (B) the output tuning switched-capacitors; (C) The simulation results of the related load reflection coefficients.

number of transfer function poles. It should be pointed out that, as stated previously, the subsequent power-combining network is not a pure class-E matching network. Based on simulation results of Fig. 9.14F and G, the designed power combiner can manage more than 500 MHz RF bandwidth.

The shunt input and output capacitors of the transformer balun are utilized to fine tune the amplitude and phase relationships of the I/Q modulator for the desired frequency. For this purpose, two 4-bit binary-weighted capacitor banks are added at the primary and secondary sides of Fig. 9.7 and are illustrated in Fig. 9.15A and B. Since the entire design is accomplished using 1.2 V standard thin-oxide transistors, the voltage swings at the transformer connections are too high to be managed by a single transistor. Consequently, cascode switches are used. Moreover, the voltage swing at the secondary side can be as high as 4 V. Therefore, series-capacitors are incorporated to reduce the cascode drain node swing to, at most, 2 V (Fig. 9.15B) [145]. The load reflection coefficient of primary/secondary capacitor tuners is shown in the Smith chart of Fig. 9.15C. Based on the simulations, the primary capacitance varies between 4.8 and 7.8 pF while the secondary capacitance changes between 1.9 and 2.7 pF. Note that, although in both input and output capacitor tuners, many lossy switches are employed, yet their quality factors ($\geq$32) are still significantly higher than the designated balun (see Fig. 9.14C).

In addition, the reliability of RFDAC is evaluated with the assistance of Fig. 9.16A and B. Based on that, the peak drain voltage of nodes $Drain^+$/$Drain^-$

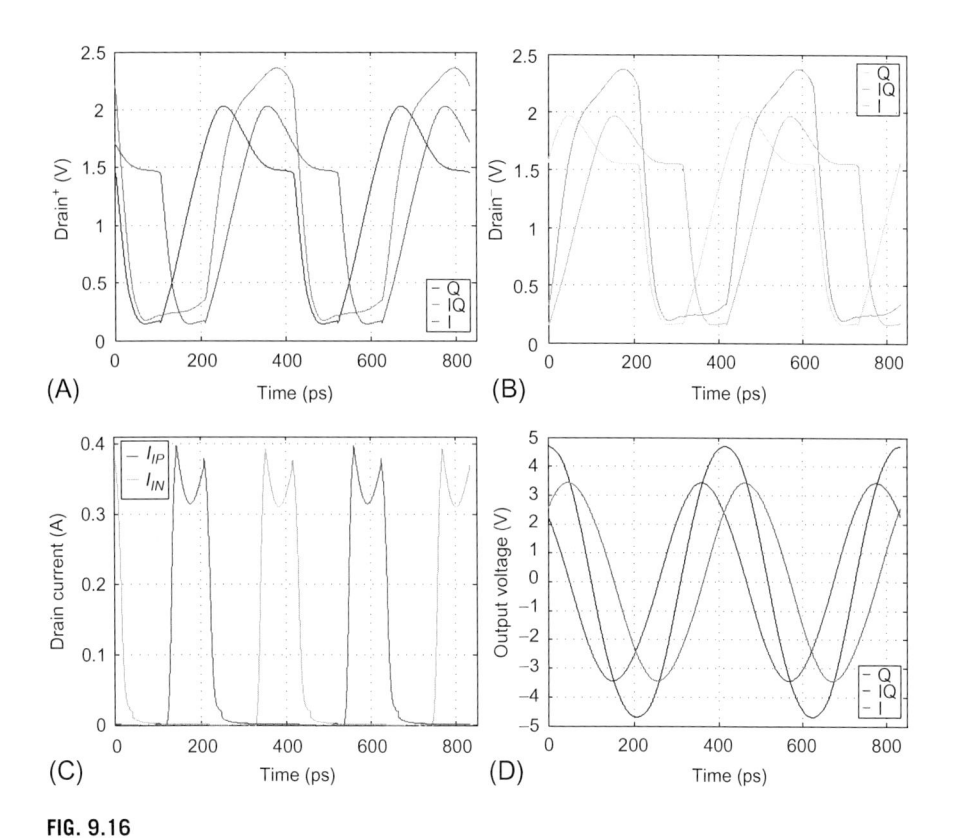

9.6 Design an efficient balun for power combiner

FIG. 9.16

I/Q RFDAC voltage/current waveforms of three simulation results of Fig. 9.7 in which Q path is on, on and off, respectively. Moreover, I path is off, on and on, respectively; (A) $Drain^+$; (B) $Drain^-$; (C) In-phase drain currents; (D) RF output voltage.

is less than 2.4 V, which indicates that the breakdown will not occur.[9] Moreover, according to this simulation, the minimum drain voltage is approximately 0.25 V, which results in an appropriate drain efficiency. Fig. 9.16C depicts the simulated 25% drain current of $I_{path,p}$ and $I_{path,n}$. To achieve the highest possible RF power and drain efficiency, as could be seen, the NMOS power switches must carry almost 400 mA while they enter in the triode region. Indeed, these NMOS switches operate

---

[9] Note that due to the fact that the DPA unit cell is, indeed, a switchable cascode structure, the peak gate-drain voltage could be as high as 2.2 V provided that one or more cascode switches are off (i.e., the gate of cascode switch is shorted to ground). Based on the study in Ref. [146], the gate-drain breakdown voltage in 65-nm CMOS could be as high as 2.1 V. As a result, the breakdown will not occur. This assumption has also been verified in the lab: The RFDAC test-chip has been measured over long time and no performance degradation has been observed. This indicates that the gate-drain breakdown does not occur.

in a "voltage limited" regime [147]. Within this region of operation, the drain-source voltage of the switches becomes minimal, which subsequently causes the drop of the drain current. The dimple at the peak of the drain current in Fig. 9.16C indicates that the switches operate in a "voltage-limited" regime. Note that, in the lower power mode of operation in which the turn-on power switches are fewer, the DPAs operate in a "current limited" regime that the NMOS power switches work at in the saturation region. The eventual peak RF output power of the I/Q RFDAC exceeds 22.4 dBm while the drain efficiency including power-combining network losses (η) is better than 44%. Also, the appropriate modulation accuracy of I/Q RFDAC could be quickly ascertained from Fig. 9.16A, B, and D. Based on these simulations, the IQ signal is the result of orthogonally summing of I and Q signals ($IQ = I\angle(\pi/4)$).

9.7 CONCLUSION

In this chapter, the theory and the design procedure of an innovative, differential, orthogonal power-combining network, which is employed in the proposed all-digital modulator, is thoroughly explained. It is demonstrated that, in order to maintain an orthogonal operation between the in-phase and quadrature-phase paths, the effect of the power combiner on the in-phase and quadrature-phase paths must be considered, otherwise, the linear summation will not occur. As a result, the EVM and linearity performance will diminish. The power combiner consists of a transformer balun as well as its related programmable primary and secondary shunt capacitors. In order to achieve high efficiency at full power of operation, a class-E-type matching network is adopted and subsequently modified in order to obtain a minimum modulation error. A switchable cascode structure is exploited to mitigate a reliability issue as well as to perform a mixer operation. Moreover, utilizing a switchable cascode structure also improves the isolation between quadrature paths. Furthermore, it is explained that the power combiner efficiency is primarily related to the transformer balun efficiency. A procedure is introduced in order to design an efficient, compact balun transformer. Also, it is explained that the RFDAC operates as a class-B power amplifier at the power back-off levels. As a result, its performance in the power back-off region is lowered.

CHAPTER 10

A wideband 2 × 13-bit all-digital I/Q RFDAC*

CHAPTER OUTLINE

10.1 Clock Input Transformer	202
10.2 High-Speed Rail-to-Rail Differential Dividers	203
10.3 Complementary Quadrature Sign Bit	204
10.4 Differential Quadrature 25% Duty Cycle Generator	205
10.5 Floorplanning of 2 × 13-Bit DRAC	207
10.6 Thermometer Encoders of 3-to-7 and 4-to-15	209
10.7 DRAC Unit Cell: MSB and LSB	210
10.8 MSB/LSB Selection Choices	214
10.9 Digital I/Q Calibration and DPD Techniques	216
10.9.1 IQ Image and Leakage Suppression	217
10.9.2 DPD Based on AM-AM and AM-PM Profiles	219
10.9.3 DPD Based on I/Q Code Mapping	220
10.9.4 DPD Required Memory and Time	224
10.9.5 DPD Effectiveness Against the Temperature and Aging	225
10.9.6 Verification of DPD I/Q Code Mapping	226
10.10 Conclusion	228

This chapter comprehensively explains the circuit-level design of the 2 × 13-bit all-digital I/Q RFDAC [119, 120]. The measurement result of this chip will be subsequently revealed in the following chapter, that is, Chapter 11. Fig. 10.1 exhibits a block diagram of the implemented 2 × 13-bit RFDAC transmitter. In the remainder of this section, its building blocks will be sequentially disclosed and their circuit design techniques described. Section 10.1 discusses the input transformer which converts a 4× single-ended clock into a differential one. Section 10.2 reveals high-speed low-noise divide-by-2 circuits which are employed as dividers throughout the chip. Section 10.3 reveals the high-speed rail-to-rail complementary quadrature sign bit, while Section 10.4 introduces the utilized 25% duty cycle generator. The

---

*The authors acknowledge substantial contributions from Prof. Dr. Leo de Vreede (TU Delft).

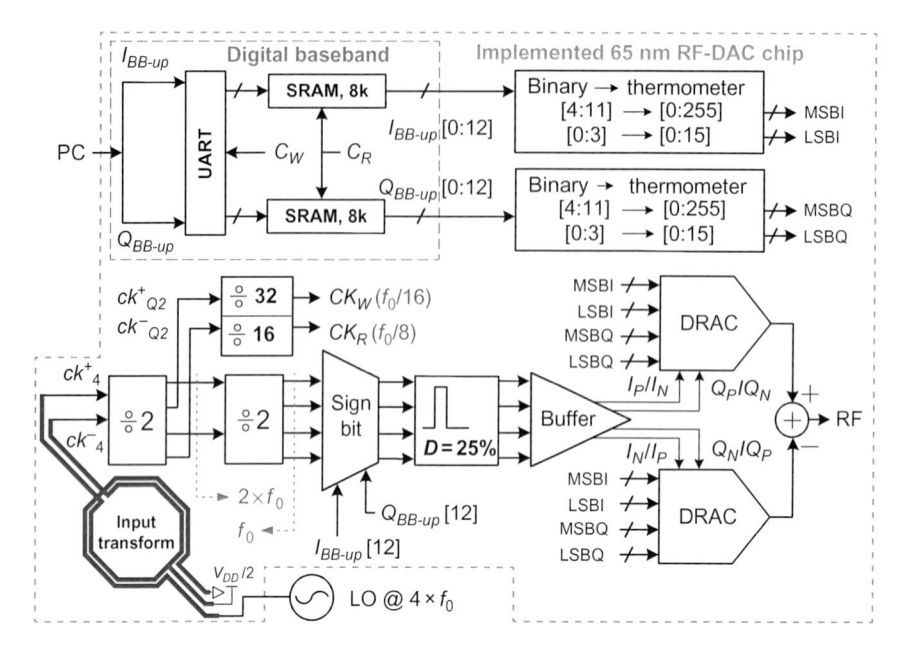

FIG. 10.1

Block diagram of the implemented transmitter based on 2 × 13-bit all-digital I/Q RFDAC.

floorplan of the DRAC is proposed in Section 10.5, which requires its related binary-to-thermometer encoder addressed in Section 10.6. The design of MSB and LSB DRAC unit cell is discussed in Section 10.7, while the philosophy supporting the selected segmentation configuration is revealed in Section 10.8. The digital calibration techniques as well as DPD procedures are addressed in Section 10.9. Finally, Section 10.10 concludes the chapter.

10.1 CLOCK INPUT TRANSFORMER

An off-chip single-ended clock at $4 \times f_0$ frequency is applied to an on-chip transformer to convert it to differential clock signals (ck_4^+/ck_4^-). The transformer size is selected at $150 \times 150\,\mu m^2$ with 1:1 turns ratio, while the center tap is located at its secondary winding and connected to a common mode voltage of $V_{DD}/2$. The windings are $6\,\mu m$ wide with $3\,\mu m$ gaps in-between. Momentum simulation demonstrates that the coupling factor within a frequency range of 7–13 GHz is $k_m \approx 0.625$. Note that the simulated k_m is related to the coupling between the input and output turns of the transformer. Based on that, the circuit simulations indicate that the transformer converts a 3.84 V_{p-p} single-ended clock to $2 \times 1.2\,V_{p-p}$ differential signals that swing around $V_{DD}/2$.

Due to nonidentical differential layout traces that introduce differing parasitic capacitance, the differential signals could arrive at the following ÷2 divider misaligned and that might corrupt its operation. Therefore, the phases of ck_4^+/ck_4^- clocks are aligned employing back-to-back inverters.

10.2 HIGH-SPEED RAIL-TO-RAIL DIFFERENTIAL DIVIDERS

The differential $4 \times f_0$ clock, ck_4^+/ck_4^-, is applied to two cascaded $\div 2$ dividers to generate the desired carrier LO at f_0 (see Fig. 10.2A and B). The $\div 2$ divider is implemented as a flip-flop-based frequency divider, which consists of four C^2MOS latches [123] arranged in a loop (Fig. 10.2C). This topology produces four differential quadrature clock signals (ck_{I2}^+, ck_{Q2}^+, ck_{I2}^-, and ck_{Q2}^- in Fig. 10.2A) that operate at $2f_0$. The back-to-back inverters of Fig. 10.2C ensure that no illegal states will occur. They also align the differential clock phases (ck_{I2}^+/ck_{I2}^- and ck_{Q2}^+/ck_{Q2}^-). The input and output nodes of C^2MOS latches experience rail-to-rail voltage swing.

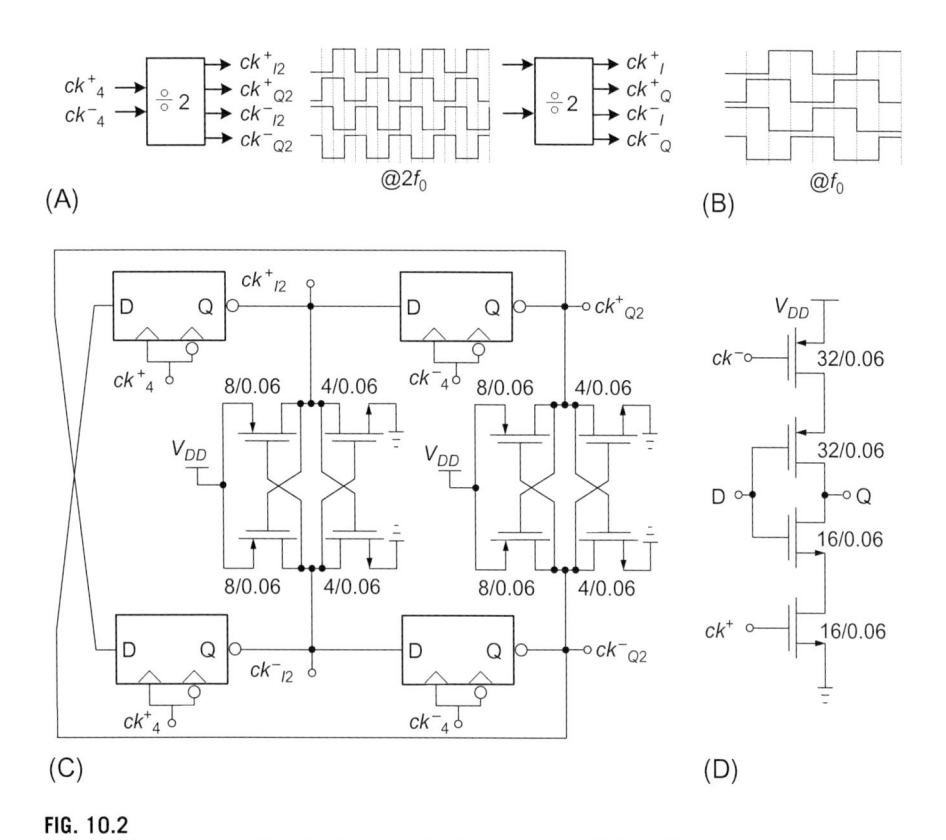

FIG. 10.2

Clock $\div 2$ dividers with corresponding waveforms: (A) first; and (B) second; (C) schematic; (D) C^2MOS latch with swapped data/clock inputs.

Consequently, they exhibit a superior noise performance over the low-swing CML latches. On the other hand, due to their high current bias and lower voltage swing, the operational frequency of the CML latches can be much higher than that of C^2MOS. Since the noise performance and power consumption are crucial design considerations and recognizing that the speed advantages of the CML latches are not always needed, the C^2MOS latches are thus adopted in this instance. The $2f_0$ clock signals, however, could be as high as 7 GHz and the divider should be operational for all PVT conditions, which might be difficult to achieve. Dissipating more current (e.g., by employing wider transistors) could improve the speed of C^2MOS latches as the supply voltage is fixed. Hence, their power consumption increases, which would decrease the overall system efficiency of the transmitter.

In this work, however, in lieu of increasing the power, the data and clock inputs of C^2MOS are swapped (Fig. 10.2D). By doing so, the D-to-Q delay of the latch and, subsequently, the overall loop time period of the divider, decreases. Based on simulation and measurement results, the RFDAC frequency of operation can be as high as 3.5 GHz at $V_{DD} \leq 1.3$ V. Note that all other ÷2 divider circuits also utilize the same structure. The transistor sizing, however, is adjusted based on its operational frequency. For instance, the width of all transistors in the next ÷2 divider in both the main RF clock path (÷2) as well as the baseband clock path (÷16/32) of Fig. 10.1 are reduced by a factor of two. Furthermore, every other differential output clock of the first divider (ck_{2I}^+/ck_{2I}^- and ck_{2Q}^+/ck_{2Q}^-) is applied to the next divide-by-two circuits. By doing so, all C^2MOS latches experience identical loading conditions. Thus, their fan-outs are equal.

Note that all clocks in the digital baseband circuitry (CK_W and CK_R) as well as the final RF fundamental clocks, I_P, Q_P, I_N, and Q_N, are synchronized. The amplitude and phase imbalances of the I and Q paths would deteriorate the IQ image and leakage performance of the transmitter, thus they should be calibrated. The baseband and RF phase synchronization makes the IQ calibration much simpler. Furthermore, employing two cascaded ÷2 dividers (i.e., divide-by-4 circuit) will ameliorate the quadrature accuracy of the fundamental clocks since all the phases of the fundamental f_0 clock are derived from the same rising edge of the $4 \times f_0$ master clock even in face of a non-50% duty cycle.

10.3 COMPLEMENTARY QUADRATURE SIGN BIT

As depicted in Fig. 10.1, the second ÷2 divider is followed by a sign bit circuitry (see Fig. 10.3A). It is implemented as two pseudo-differential (i.e., complementary) NAND-gate-based multiplexers with input selection control signals $I_{BB-up}[12]$ and $Q_{BB-up}[12]$. Based on the 2-bit (i.e., 4-state) selection control, the differential clock pairs of ck_I^+/ck_I^- or ck_Q^+/ck_Q^- can be swapped and thus the entire four-quadrant constellation diagram can be addressed (see Fig. 10.3B). Contradictory to Section 7.1.4 [116, 118], the sign bit is located between the second divider and the 25% duty-cycle

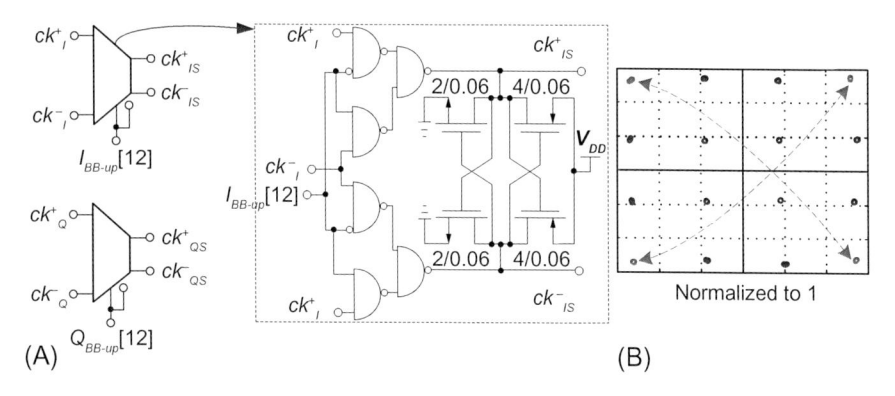

FIG. 10.3

(A) Complementary NAND-gate-based sign-bit multiplexer symbol and its related schematic; (B) 16-point constellation diagram.

generator. In this new arrangement, the sign bit circuitry manages the 50% duty cycle clock instead of 25%, which reduces power consumption. Moreover, a simple back-to-back inverter pair (see Fig. 10.3A) is employed for further phase alignment which was not feasible in Section 7.1.4 [116, 118].

As a result, by exploiting smaller devices, faster rise/fall times are achievable. Moreover, compared to the transmission-gate-based multiplexer used in Section 7.1.4 [116, 118], the NAND-based multiplexer produces faster rise/fall times. This is because, in the transmission gate, the charging/discharging of the MOS channel is decelerated (see Fig. 7.6A).

10.4 DIFFERENTIAL QUADRATURE 25% DUTY CYCLE GENERATOR

The sign bit outputs of ck_{IS}^+, ck_{QS}^+, ck_{IS}^-, and ck_{QS}^- are applied to a 25% duty cycle generator (see Fig. 10.4A). As stated previously, the orthogonal summing of the I and Q paths is achieved by employing the differential quadrature clocks with a 25% duty cycle. As a result, the 25% duty cycle generator is one of the most important building blocks of the clock generator chain.

The circuit utilized in Section 7.1.3 [116, 118] provides unmatched narrow/wide clock pulses. It signifies that the duty cycle for one pulse might be 31% while 27% for the others. In this work, however, the 25% duty cycle circuit generator of [60] is adopted. It is conceptually illustrated in Fig. 10.4A. Based on this approach, the 25% clocks at f_0 (ck_{ID}^+, ck_{QD}^+, ck_{ID}^-, and ck_{QD}^-) are generated by an AND operation between clocks of (ck_{I2}^+/ck_{I2}^-) and (ck_{IS}^+/ck_{IS}^-, ck_{QS}^+/ck_{QS}^-) where they operate at $2f_0$ and f_0, respectively. Thus, the 50% duty cycle clocks of ck_{I2}^+/ck_{I2}^- are used as a reference

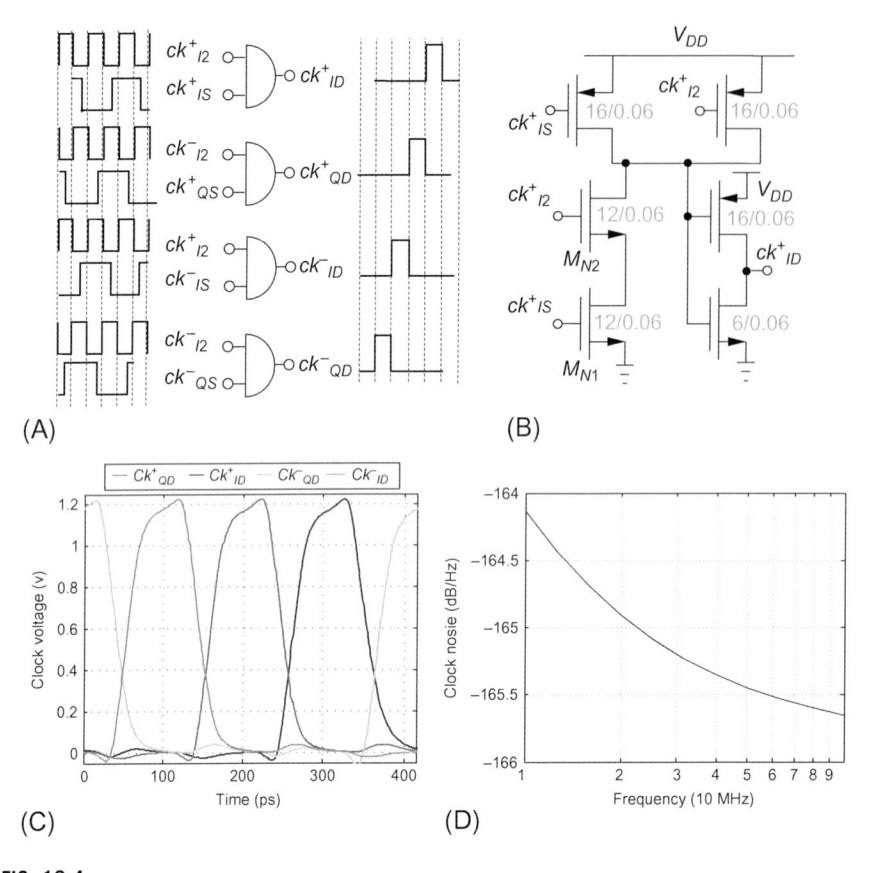

FIG. 10.4

(A) 25% duty cycle generator schematic [60]; (B) its AND logic circuit. Postlayout simulation results of the clock generation part: (C) 25% clock waveforms; (D) their corresponding clock phase noise.

pulse width for generating ck_{ID}^+, ck_{QD}^+, ck_{ID}^-, and ck_{QD}^-. Namely, their pulse width is the same as ck_{I2}^+/ck_{I2}^- while operating at f_0. Hence, the circuit creates clocks with a precise 25% duty cycle.

The AND operation of the 25% duty cycle generator as well as the sign bit are accomplished utilizing Fig. 10.4B. This is an asymmetric circuit with respect to the gates of M_{N1} and M_{N2}. The gate capacitance of M_{N2} is smaller than that of M_{N1} due to the cascode configuration of M_{N1}/M_{N2}. Otherwise stated, C_{gs} of M_{N2} is in series with the combination of drain-bulk capacitance of M_{N1} and source-bulk capacitance of M_{N2}, while C_{gs} of M_{N1} is directly connected to the ground. Therefore, ck_{I2}^+ and ck_{IS}^+ are applied to the M_{N2} and M_{N1} gates, respectively. Thus, the AND gate consumes less power. Note that the desired 25% duty cycle clocks could also be generated using an AND operation of every two adjacent clocks of ck_{IS}^+, ck_{QS}^+, ck_{IS}^-, and ck_{QS}^-. The disadvantages would be the asymmetric AND inputs that create unmatched wide/narrow pulses. Thus, the circuit illustrated in Fig. 10.4A is the preferred approach.

The postlayout simulation results of the differential quadrature clocks of ck_{ID}^+, ck_{QD}^+, ck_{ID}^-, and ck_{QD}^- are depicted in Fig. 10.4C. All clock signals comprise a 25% duty cycle and, most importantly, are matched to each other. Note that their corresponding rise time and fall time will be improved later by utilizing a number of appropriate clock buffers. Furthermore, the postlayout clock phase noise is simulated and depicted in Fig. 10.4D for the frequency offset range of 10–100 MHz. According to this simulation result, the clock noise floor is better than −164 dBc/Hz, which is an adequate number. Thus, the RFDAC noise floor is primarily limited to the clock generation circuitry, and it will be verified by measurement results.

10.5 FLOORPLANNING OF 2 × 13-BIT DRAC

As mentioned previously, the targeted transmitter is an all-digital RFDAC with 2 × 13-bit (including sign bit) resolution. $I_{BB\text{-}up}$ and $Q_{BB\text{-}up}$ represent binary digital codes which must be converted to thermometer codes in order to avoid nonmonotonic behavior and mid-code transition glitches [129, 148–150].[1] The use of the pure thermometer code, however, increases the complexity of the encoders, the chip area, interconnect parasitics, and power consumption. Thus, a segmented approach is adopted in this aspect [47]. The segmentation is selected such that 8 bits are used for the MSB and 4 bits for the LSB of the binary input. The philosophical explanation of the selected MSB and LSB in this segmentation arrangement will be disclosed later in Section 10.8. Therefore, the DRAC implementation requires 256 MSB and 16 LSB units.

The design of such a complex RFDAC requires several iterations between the schematic and layout. The 256 MSB units further split into two sections while the clock generator circuits are situated in the middle (see Fig. 10.5A). Moreover, the 128 MSB units of each part are arranged such that they comprise 8 rows and 16 columns (8 × 16). Subsequently, the I/Q segmented thermometer code requires two types of in-phase and quadrature-phase baseband row and column thermometer codes, which are referred to as Row_I/Row_Q as well as Col_I/Col_Q and are generated by row and column encoders. The right MSB unit bank covers the low thermometer code values (i.e., 0–127) while the remaining (i.e., 128–255) are managed by the left bank. Furthermore, the LSB unit comprises 16 small DRAC unit cells, which

---

[1] Glitches occur when the switching time of different bits in a binary weighted DAC is unmatched, and they add a signal-dependent error to the output signal. Moreover, the glitches can be produced by the dynamic behavior of the switches such as charge-injection and clock feedthrough. The generated power of glitches should be less than quantization noise so as to not degrade the dynamic range of RFDAC.

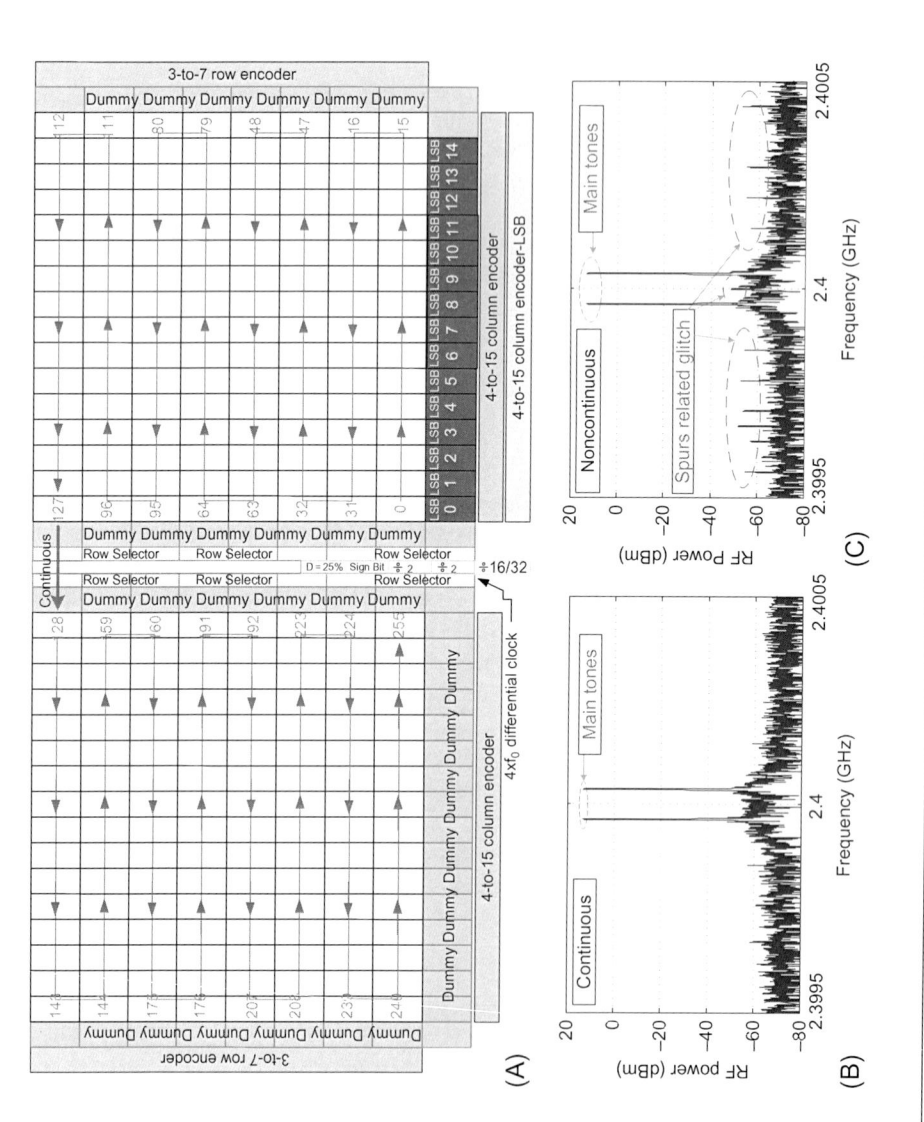

208 CHAPTER 10 A wideband 2 × 13-bit all-digital I/Q RFDAC

FIG. 10.5

(A) 2 × 13-bit DRAC floor plan; (B) continuous traversal: glitch-free spectrum; (C) non-continuous traversal: spectrum with glitches.

occupies only one row (1 × 16) at the bottom of the right MSB DRAC unit bank. The MSB DRAC units in each row must be situated in close proximity to each other. Moreover, the dummy DRAC cells are placed at the beginning and end of each row which globally improves the matching of the DRAC unit cells with respect to each other.

In addition, odd rows begin from the left side while the even rows begin from the right side. This "snake" traverse movement is indicated with arrow lines in Fig. 10.5A. By doing so, the MSB thermometer units are continuously traversed from an odd to an even row and vice versa. As a result, the DNL of RFDAC as well as the glitch related to the dynamic switching of DRAC units are retained below one LSB. Note that the clock trees (clock generating blocks) force DRAC to split into two sections, thus, there would be unavoidable discontinuity between the left and right DRAC banks which would possibly introduce considerable glitches. In order to cross over from the right bank to the left one, the DRAC must adopt the nearest possible path, which is the direct path between cell 127 and 128. This movement is, indeed, referred to as continuous traverse and is indicated in Fig. 10.5A. On the other hand, if the DRAC might intentionally follow the faraway probable, hypothetical path, which is the path between cell 127 and 255, consequently, a noncontinuous traverse would have occurred. To further justify it, Fig. 10.5B and C compares the two aforementioned movement scenarios from the right bank to the left one, that is, continuous and intentionally noncontinuous traverse. Fig. 10.5C illustrates that noncontinuous movement generates a significant number of spurs and should thus be avoided. Therefore, as exhibited in Fig. 10.5A, the movement from the right bank to the left must be performed gently. In conclusion, the continuous traverse, prudent layout as well as employing dummy cells would almost entirely eliminate the dynamic glitch problem.

10.6 THERMOMETER ENCODERS OF 3-TO-7 AND 4-TO-15

Based on the above-segmented arrangement, two 3-to-7 and three 4-to-15 (including the LSB encoder) binary-to-thermometer encoders are employed (five in total) and placed at the left, right, and bottom sides of the DRAC (see Fig. 10.5A). The encoders are implemented based on a 2-to-3 binary-to-thermometer encoder depicted in Fig. 10.6A. In this approach, the LSB (BB_0) and MSB (BB_2) of the thermometer code are produced by OR and AND operations of the two input binary bits (A_0 and A_1), respectively. Moreover, the middle bit of the thermometer code (BB_1) is equal to the input MSB (A_1). The 3-to-7 encoder, however, is implemented in two increments. First, the intermediate 3-bit thermometer codes of Fig. 10.6A are created. Using these codes, B_0, B_1, B_2, B_4, B_5, and B_6 bits of the eventual seven-bit thermometer code are generated by OR and AND operations of BB_0, BB_1, BB_2 by A_2, respectively. Moreover, B_3 is also equal to A_2 (Fig. 10.6B). Similarly, the 4-to-15 encoder (see Fig. 10.6C) is created in two increments employing intermediate 3-to-7 thermometer bits and again applying OR/AND logic operations of the intermediate bits with A_3.

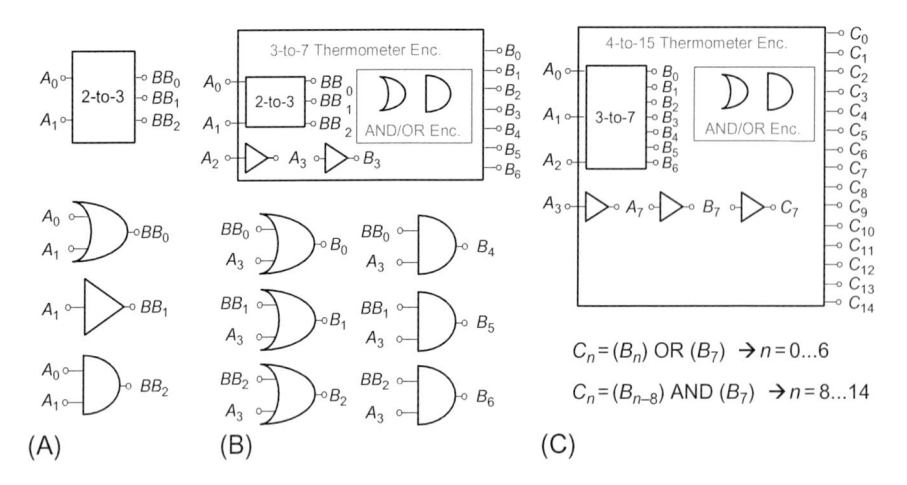

FIG. 10.6

(A) 2-to-3; (B) 3-to-7; and (C) 4-to-15 thermometer encoders.

$C_n = (B_n)$ OR (B_7) → $n = 0...6$

$C_n = (B_{n-8})$ AND (B_7) → $n = 8...14$

10.7 DRAC UNIT CELL: MSB AND LSB

The DRAC design was fully described in Section 9.1. In this section, the DRAC unit cell is explained in more detail. The MSB DRAC unit is illustrated in Fig. 10.7A. This unit consists of four equal and well-matched subsections (sub DRAC) each comprising its own data and clock inputs. The quadrature input clocks are I_P, Q_P, I_N, and Q_N and, based on these signals, the sub DRACs are referred to as SD_{IP}, SD_{QP}, SD_{IN}, and SD_{QN}, respectively. Moreover, as mentioned earlier, the related input data thermometer bits are Row_I, Col_I, Row_Q, and Col_Q along with two extra control bits of Row_{I+1} and Row_{Q+1} in which they guarantee that all DRAC unit cells of the previous rows are activated. The sub DRAC section comprises two parts; a pure digital (logic) and a digital-to-RF conversion part.

The logic part consists of a decoding logic (AND-OR) and a time synchronizer flip-flop. Based on logic condition of its inputs, the AND-OR decoder (see Fig. 10.7B) determines whether or not the sub DRAC cell should be activated. The master/slave edge-triggered flip-flop is employed for synchronizing all DRAC unit cells to its input clock, namely CI_P, CQ_P, CI_N, and CQ_N, in order to reduce undesirable harmonic distortion related to early or late arrival of input data of each DRAC unit cell. Additionally, this flip-flop also behaves as a ZOH interpolator. It comprises two cascaded multiplexer-based latches [123], as shown in Fig. 10.7C. In the sense mode of operation, the input clocks CK_P/CK_N are low/high and, consequently, the input data (D_I) passes through the "lower" pass-gate logic of M_1/M_2 and is subsequently buffered by the cascaded inverters of M_3/M_4 and M_5/M_6. It signifies that the path between D_I and Q_O is transparent. In the store mode, on the other hand, CK_P/CK_N are high/low and, as a result, the "top" pass-gate logic of

FIG. 10.7

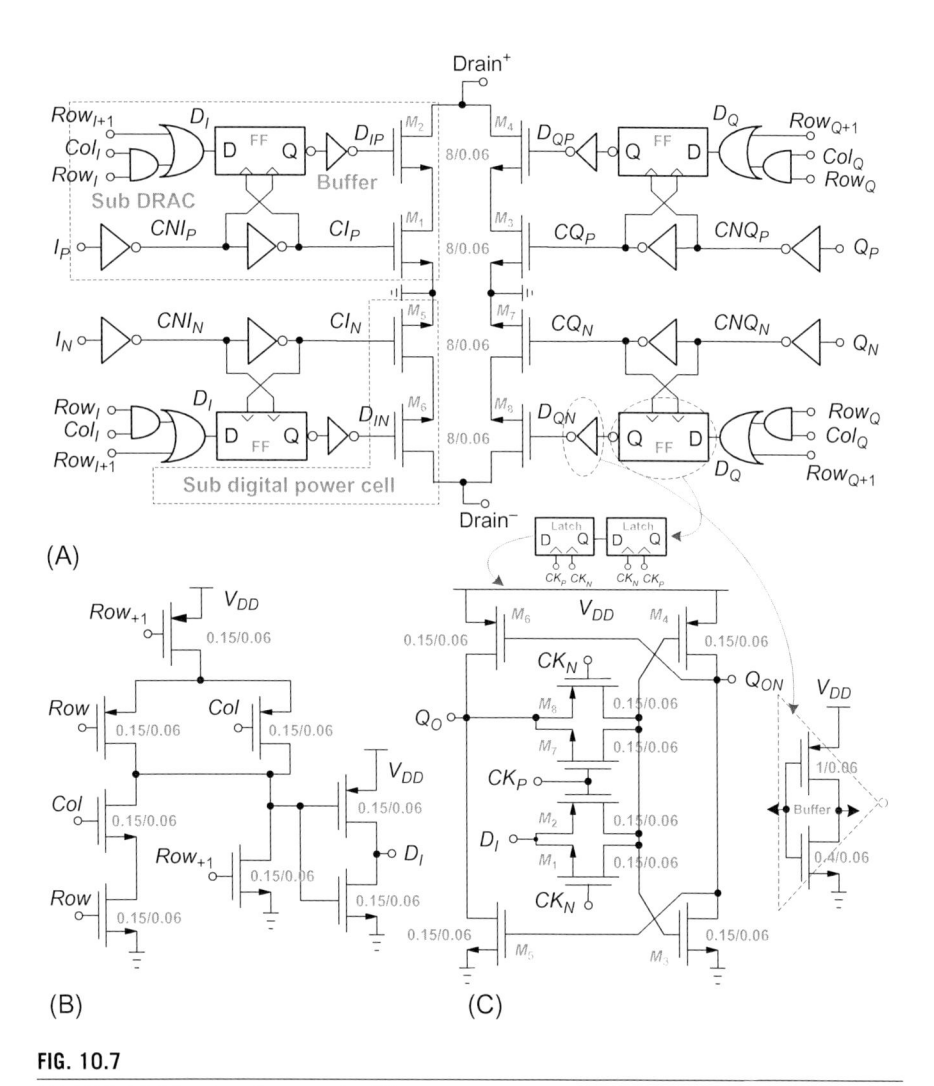

(A) MSB DRAC unit cells: pseudo differential quadrature digital power mixer. Schematics of (B) AND-OR decoder; (C) multiplexer-based latch including last data buffer.

M_7/M_8 is transparent, and the "lower" one is opaque. Therefore, the two inverters of M_3/M_4 and M_5/M_6 are cross-coupled with each other and latch the digital input signal. All transistors of both the AND-OR decoder logic and flip-flop circuit are implemented with the most minimal possible aspect ratio in 65 nm CMOS, that is, $W/L = 0.15\,\mu\text{m}/0.06\,\mu\text{m}$ to minimize area and power consumption.

As depicted in Fig. 10.7A, the flip-flop output of the sub DRAC cell is buffered and subsequently connected to the cascode transistor (M_2, M_4, M_6, or M_8) to handle the input gate capacitance and, consequently, to improve the rise/fall time performance. As stated previously, the gate capacitance of the cascode transistor with an aspect ratio of $W/L = 8\,\mu m/0.06\,\mu m$ is much lower than the input capacitance of M_1 with the same transistor sizing. Therefore, using a moderated buffer size is sufficient enough to satisfy the required data transition conditions. The buffer sizing is indicated in Fig. 10.7C.

Note that, in lieu of employing these digital components at every sub DRAC cells, it is feasible to situate them at the on-set of each row and column and share them with the entire row and column. The disadvantages would be inferior timing synchronization as well as greater power consumption due to the fact that they should feed a number of sub DRAC cells along with enormous routing metal lines. Moreover, the flip-flop requires as a synchronizing signal either the high-speed upconverting clock or CK_R upsampling clock. If the digital part is not situated in the DRAC cell and is placed at the on-set of each row/column, consequently, the aforementioned synchronizer clock should also be provided at the beginning of each row/column which results in a more complex layout/floorplanning as well as higher power consumption. As a result, the proposed DRAC cell appears the best viable configuration.

The digital-to-RF conversion part consists of a gated cascode switch (M_1/M_2, M_3/M_4, M_5/M_6, and M_7/M_8) that yields the upconverting 1-bit mixer operation. Furthermore, it is perceived as a sub digital power cell. The cascode transistor (M_2, M_4, M_6, and M_8) alleviates the reliability issue related to the high-voltage swing that appears on the output nodes ($Drain^+$, $Drain^-$). Moreover, the cascode configuration also increases the output impedance, which results in improved isolation between the I and Q paths that assists with improved orthogonal combination. The overall quadrature mixer digital power unit cell is formed by electrically combining the outputs of two individual quadrature upconverter mixers (the upside M_1–M_4 and downside M_5–M_8 of Fig. 10.7A) that are driven by quadrature input clocks (which also act as four sub digital power cells). Consequently, the entire RFDAC is now created by simply connecting the corresponding drain nodes of each 256 MSB along with 16 LSB DRAC unit cells together.

As stated, each DRAC unit cell consists of SD_{IP}, SD_{QP}, SD_{IN}, and SD_{QN} unit cells, and their layout arrangement affects the performance of the entire RFDAC. Fig. 10.8A shows one possible solution in which each quadrature sub DRAC pair, that is, SD_{IP}/SD_{QP} and SD_{IN}/SD_{QN}, is juxtaposed in two different subrows which indicates that the DRAC unit cell is expanded horizontally. In this arrangement, the high-frequency 25% duty cycle quadrature clock pairs of I_P/Q_P and I_N/Q_N are laid out alongside each other. This, subsequently, increases the parasitic coupling capacitance of these clock lines and, as a result, deteriorates the clock rise/fall times. Moreover, since the position of Q_P/Q_N clock lines is different than I_P/I_N, their line capacitances also vary. Thus, Q_P/Q_N and I_P/I_N clock pulses are narrower and wider, respectively. Postlayout circuit simulations of Fig. 10.8A demonstrate the rise/fall time as well as narrow/wide pulse problems related to the horizontal layout.

FIG. 10.8

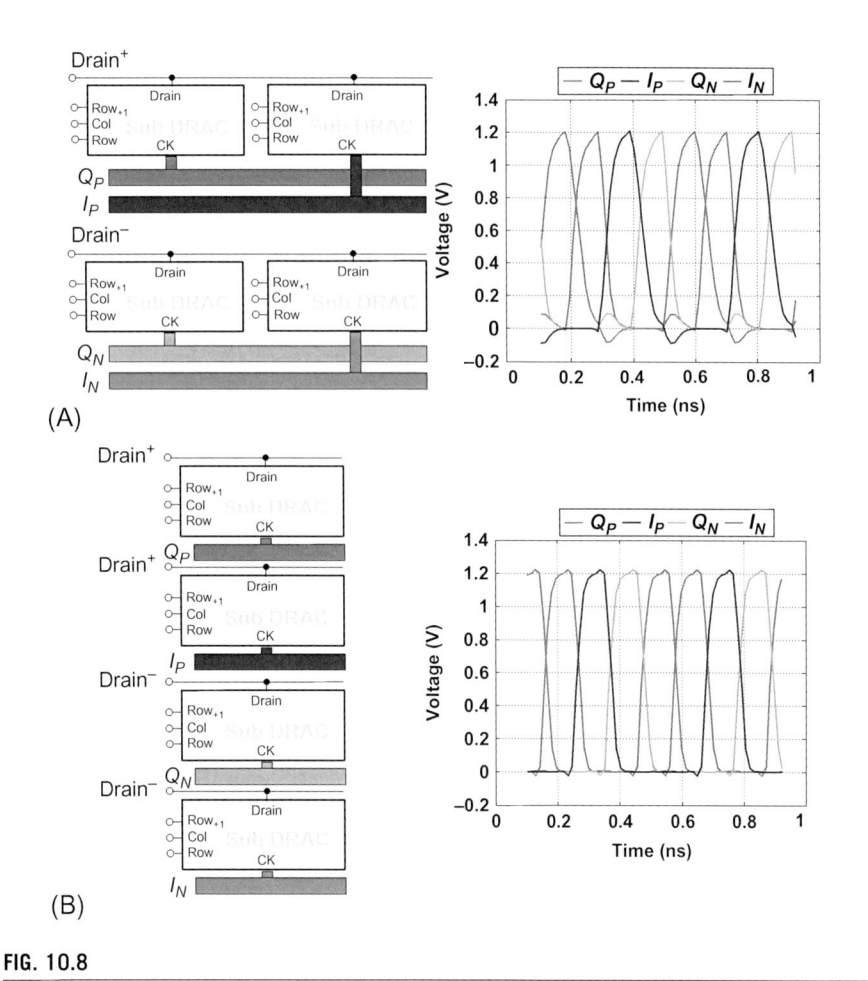

DRAC unit sub-cells layout: (A) Horizontal; and (B) vertical with their related differential quadrature clock simulations.

The better solution, however, is to expand the DRAC unit cell vertically and place SD_{IP}, SD_{QP}, SD_{IN}, and SD_{QN} sub DRAC unit cells in four subrows, as illustrated in Fig. 10.8B. In this arrangement, the parasitic coupling capacitance between the clock lines are almost negligible. The clock lines are also situated in the same positions and are sandwiched between the same sub DRAC cells. Hence, their related rise/fall time and pulse width are matched with one another. Postlayout simulations in Fig. 10.8B substantiate that the vertical expansion is the most appropriate selection. To compensate for the extra vertical area related to the vertical expansion of DRAC unit cell, the entire 256-MSB-cell, as stated previously, comprise 8 rows and 16 columns. Thus, left/right MSB DRAC banks

FIG. 10.9

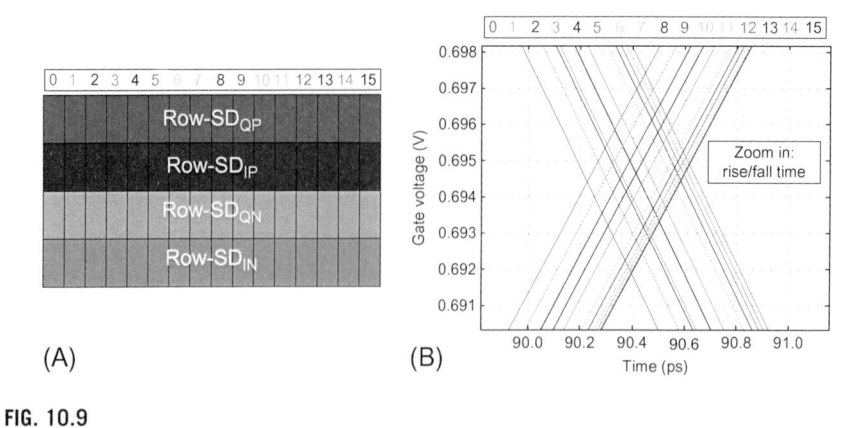

(A) DRAC row layout of Fig. 10.5; (A) Its corresponding gate voltage simulations.

become "squarish," which is beneficial for improved area efficiency and shorter clock distribution (i.e., less power dissipation). Fig. 10.9A exhibits the eventual DRAC row employing vertical layout solution of Fig. 10.8B. Note that each DRAC unit cell should be clocked almost simultaneously, otherwise, the AM-AM and AM-PM nonlinearity emerge. Based on the postlayout simulation results of multiple rows of Fig. 10.5, the time difference between the first (0-cell) and the last column (15-cell) is less than 500 fs, which is a very minimal delay and will not affect the performance of I/Q RFDAC. Furthermore, according to postlayout simulation results, the rise/fall time at all temperature and process corners does not exceed 50 ps. Note that, based on the discussion in Chapter 9, the proposed I/Q RFDAC can tolerate more than 80 ps rise/fall time while $f_0 = 2.4$ GHz.

10.8 MSB/LSB SELECTION CHOICES

As stated, each left/right DRAC banks should be arranged such that they comprise X row and Y column in which $Y = 2 \times X$ to maintain their squarish layout form. As mentioned, the DRAC requires a large number of binary-to-thermometer encoders due to employing the thermometer code. As disclosed, each $(\log_2 Y)$-to-$(Y - 1)$ binary-to-thermometer encoder is constructed employing a sub-$(\log_2 X)$-to-$(X - 1)$ binary-to-thermometer one. Nonetheless, employing the pure 100% thermometer code increases complexity of the subsequent encoders, the chip area, interconnect parasitic, and, most importantly, power consumption. Thus, the segmentation approach is adapted which consists of an MSB and an LSB part. The minimum number of LSB is 2 bits. Thus, in that case, each left/right MSB DRAC banks comprise 9 bits while, in each bank, $X = 16$ and $Y = 32$. As a result, they require two 4-to-15 as well as two 5-to-31 encoders. Employing the aforementioned structure in 65 nm CMOS will still not be an adequate or reasonable approach as it occupies more area, adds design complexity, and, on top of that, dramatically increases the power consumption. Consequently, the following arrangement is finally adopted: $X = 8$ and $Y = 16$. It should be mentioned that, in the segmentation approach, the selected LSB number of bits should be as small as possible in order to diminish the corresponding DNL, glitches and, in turn, its related distortion. Note that increasing LSB/MSB weight ratio will increase the distortion, hence the optimum ratio must be discovered [148, 149]. For that reason, MSB and LSB are selected as 8 and 4 bits of binary coding, respectively. It should be pointed out that, although, in this work, DPD is employed, it is better to construct a DAC in such a way that it does not require the DPD process to repair the glitches. Note that the DPD also could not manage a substantial amount of distortions, which are related to dynamic glitches. Note that the traverse operation in DRAC is an independent operation and does not rely on DPD. In this work, the DPD process is not assigned to remove the gradient mismatch error.[2] and its corresponding glitches related to traversing MSB unit cells of the DRAC. As a result, the DRAC arrangement strongly affects the RFDAC static and dynamic performance. It is worth noting that, as stated, the LSB part of the RFDAC is also implemented fully segmented in order to affect the following:

1. to improve the matching between LSB cells as well as boost DNL performance; and
2. to decrease the glitch related to the dynamic switching of DRAC units.

As previously stated, the segmented LSB approach only requires an extra 4-to-15 binary-to-thermometer encoder which introduces negligible power and area penalty in 65-nm CMOS technology. Fig. 10.10 illustrates the measured DNL of the RFDAC for which the input code is statically swept from 1 to 40. It should be pointed out that the measured DNL is calculated as follows [127–129]:

$$DNL(k) = \frac{(V_{out}(k+1) - V_{out}(k)) - \Delta}{\Delta} \qquad (10.1)$$

where $V_{out}(k)$ is the RF output voltage of the kth digital baseband code and the step size Δ is equal to the analog value of LSB. According to the measurement result of Fig. 10.10, the DNL is bounded to a ± 0.2 LSB which indicates an adequate matching between the consecutive codes. As a result, employing a fully segmented LSB part leads to an acceptable matching and, in turn, a reasonable DNL. Note that the jump points at the DNL profile of Fig. 10.10 are related to the transition between the MSB and LSB parts of the RFDAC.

---

[2]There are three different types of mismatches: systematic mismatch, random mismatch, and gradient mismatch. The gradient mismatch is related to first- or second-order fluctuations over longer length across the chip. It could be minimized by employing similar size, orientation, location, supplies, and also temperature. Moreover, utilizing layout techniques such as common-centroid as well as interdigitation is beneficial.

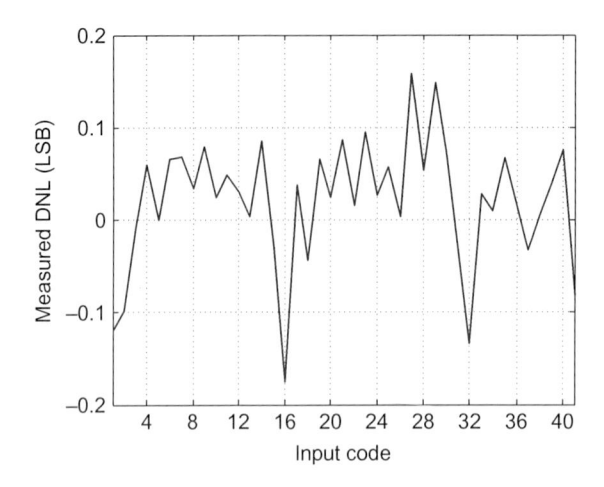

FIG. 10.10

The measured differential nonlinearity of the RFDAC.

10.9 DIGITAL I/Q CALIBRATION AND DPD TECHNIQUES

The proposed digital I/Q RFDAC-based transmitter, just as with a typical I/Q transmitter [6, 24], requires an I/Q calibration to balance the I path with respect to the Q path in order to mitigate LO leakage and I/Q image issues. Moreover, as stated above, the I/Q RFDAC comprises the efficient DPA arrays which produce more than 22 dBm of saturated RF power. Operating at the aforementioned power level leads to compression that causes AM-AM nonlinearity. Otherwise stated, as depicted in Fig. 10.11A, the $G_{on} = 1/R_{on}$ of the turn-on switches changes nonlinearly with respect to the input code and thus creates the AM-AM nonlinearity. Namely, the AM-AM nonlinearity is the result of the code-dependent conductance of the drain node [20]. Furthermore, as stated previously, turning on the switches as well as varying the drain voltage changes the drain-bulk capacitance of the digital power switches (see Fig. 10.11B). These varying capacitances in combination with the code-dependent conductance of switches cause a large impedance shift at their related drain nodes, which, subsequently, leads to the AM-PM nonlinearity. Fig. 10.11C illustrates the shifting of the load reflection coefficient ($\Gamma_{switch-on}$) of the related DRAC's drain node while sweeping the turn-on switches. Note that both I_{path} and Q_{path} contribute to the AM-AM and AM-PM nonlinearities. In addition, as elaborated above, due to the fact that the passive power-combining network affects the RFDAC's orthogonality, the imperfect orthogonal summing of the I and Q quadrature paths, as a result of inaccurate components of the passive combining network, leads to spectral regrowth [24]. Consequently, the RFDAC must be digitally predistorted to meet the spectral mask of a communication standard. To address these issues, techniques to address

10.9 Digital I/Q calibration and DPD techniques

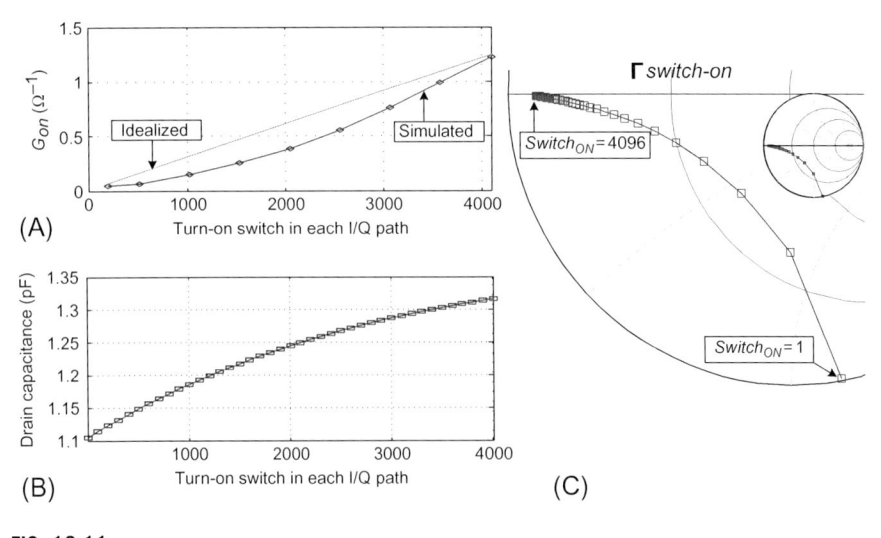

FIG. 10.11

Sweeping the turn-on switches (A) G_{on}; (B) drain capacitance; (C) $\Gamma_{switch\text{-}on}$.

these nonidealities are presented here. It should be pointed out that the following calibration as well as DPD techniques are frequency dependent. In other words, all reported calibration and DPD processes of this IC chip are performed employing carrier clock at 2.4 GHz. Consequently, they should be updated for other carrier frequencies.

10.9.1 IQ IMAGE AND LEAKAGE SUPPRESSION

To improve the LO leakage and IQ image suppression, the I/Q RFDAC should be calibrated. First, Eq. (6.4) is rewritten according to clock pulses of I_P, Q_P, I_N, and Q_N.

$$IQ_{Ideal}(t) = I_{path} + Q_{path}$$
$$= \cos(2\pi f_{BB}t) \times \left\{ A_{ip}\Pi\left(\frac{t}{T_0/2}\right) - A_{in}\Pi\left(\frac{t}{T_0/2} - 2\right) \right\}$$
$$+ \sin(2\pi f_{BB}t) \times \left\{ A_{qp}\Pi\left(\frac{t}{T_0/2} + 1\right) - A_{qn}\Pi\left(\frac{t}{T_0/2} - 1\right) \right\} \quad (10.2)$$

where f_{BB} is the baseband frequency, T_0 is upconverting clock period, and $\Pi\left(\frac{t}{T_0/2}\right)$ represents a 25% duty cycle rectangular pulse clocked at f_0. In other words, the corresponding Fourier series of Q_P, I_P, Q_N, and I_N can be expressed as follows:

$$Q_P = \Pi\left(\frac{2t}{T_0} + 1\right) = \frac{1}{4} - \frac{\sqrt{2}}{\pi}\sin(\omega_0 t) - \frac{1}{\pi}\cos(2\omega_0 t) + \frac{\sqrt{2}}{3\pi}\sin(3\omega_0 t) + \ldots \quad (10.3)$$

$$I_P = \Pi\left(\frac{2t}{T_0}\right) = \frac{1}{4} + \frac{\sqrt{2}}{\pi}\cos(\omega_0 t) + \frac{1}{\pi}\cos(2\omega_0 t) + \frac{\sqrt{2}}{3\pi}\cos(3\omega_0 t) + \ldots \quad (10.4)$$

$$Q_N = \Pi\left(\frac{2t}{T_0} - 1\right) = \frac{1}{4} + \frac{\sqrt{2}}{\pi}\sin(\omega_0 t) - \frac{1}{\pi}\cos(2\omega_0 t) - \frac{\sqrt{2}}{3\pi}\sin(3\omega_0 t) + \ldots \quad (10.5)$$

$$I_N = \Pi\left(\frac{2t}{T_0} - 2\right) = \frac{1}{4} - \frac{\sqrt{2}}{\pi}\cos(\omega_0 t) + \frac{1}{\pi}\cos(2\omega_0 t) - \frac{\sqrt{2}}{3\pi}\cos(3\omega_0 t) + \ldots \quad (10.6)$$

Moreover, A_{ip}, A_{qp}, A_{in}, and A_{qn} are amplitudes of $I_{path,p}$, $Q_{path,p}$, $I_{path,n}$, and $Q_{path,n}$, respectively. In ideal conditions, their amplitudes are identical and equal to one. As a result, after some iterations, Eq. (10.2) is rewritten as

$$IQ_{Ideal} = \frac{2\sqrt{2}}{\pi}\cos(2\pi(f_0 + f_{BB})t) \quad (10.7)$$

Note that, as stated in Section 10.2, due to the phase synchronization between the RF and baseband paths as well as the precise quadrature clock generation utilizing divide-by-four circuitry, the phase imbalance between I_{path} and Q_{path} is zero. This is one of the significant advantages of the proposed I/Q RFDAC. In reality, however, because of mismatches between A_{ip}, A_{qp}, A_{in}, and A_{qn}, after some iterations and simplifications, Eq. (10.7) changes to the following equation:

$$IQ_{nonideal} = IQ_{Ideal} + C_{image}\cos(2\pi(f_0 - f_{BB})t) + C_{Leakage} \quad (10.8)$$

in which C_{image} and $C_{Leakage}$ are the carrier image and leakage, respectively. To cancel $C_{Leakage}$, a proper DC value (i.e., $-C_{Leakage}$) is added to the original complex-valued baseband signal. Moreover, exploiting a very simple algorithm, the amplitudes of I_{path} and Q_{path} (A_{ip}, A_{qp}, A_{in}, and A_{qn}) change such that C_{image} decreases. As a result, the calibration algorithm should improve the LO leakage and I/Q image. It is worth mentioning again that A_{ip}, A_{qp}, A_{in}, and A_{qn} are the baseband amplitude codes and can easily be set to any value between $(-4095\ldots 4095)/4096$.[3]

To prove effectiveness of a simple IQ calibration algorithm, a 2.234 MHz I/Q baseband signal is applied to the chip. Fig. 10.12 illustrates that the simple calibration algorithm can significantly improve the LO leakage and IQ image suppression. In this scenario, $f_0 = 2.4$ GHz while the output power is 19.62 dBm. Based on this measurement, the I/Q image exceeds -58 dBc after five iterations while the LO leakage converges to more than -80 dBc.

Furthermore, to improve the transfer function linearity of the RFDAC, eight IC chips have been measured and two well-known DPD algorithms have been employed.

---

[3] The I and Q DC codes, in LO leakage calibration, are selected between $(-10\ldots 10)/4096$ while their corresponding I and Q codes for IQ image calibration are selected between $(-4095\ldots -4000)/4096$ and $(+4000\ldots +4095)/4096$.

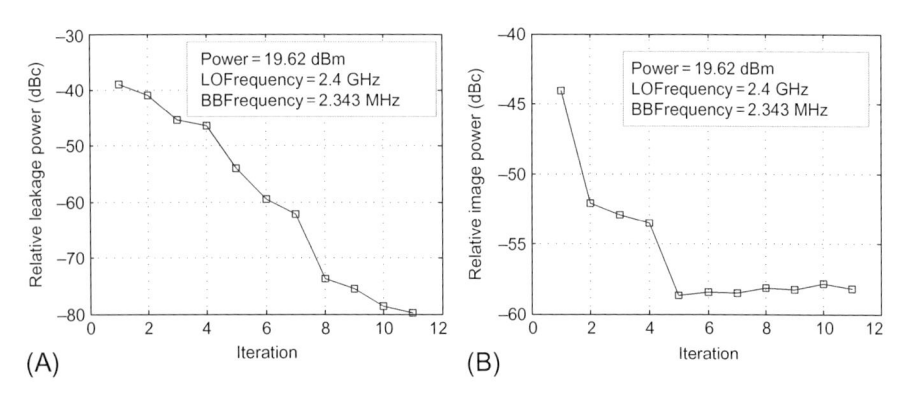

FIG. 10.12

Measurements at 2.4 GHz of (A) carrier leakage; (B) image suppression.

10.9.2 DPD BASED ON AM-AM AND AM-PM PROFILES

In this approach, a two-tone sinusoidal signal with a frequency of f_m is applied at the baseband input. The AM-AM and AM-PM profiles of the I/Q RFDAC are then evaluated [151]. First, the LO leakage and IQ image are calibrated (as discussed in Section 10.9.1) and the down-converted envelope and phase of the probed RF output are subsequently collected. After rearranging the measured envelope and phase signals based on the signed 12-bit baseband code range, that is, −4095 to +4095, the AM-AM and AM-PM characteristics are obtained and are depicted in Fig. 10.13. Note that, even though the RFDAC's input signals are digital codes, they could be normalized to 1 V and represented as a continuous voltage input. Assuming the RFDAC is a memoryless nonlinear system, its input-output voltage relationship can be approximated as:

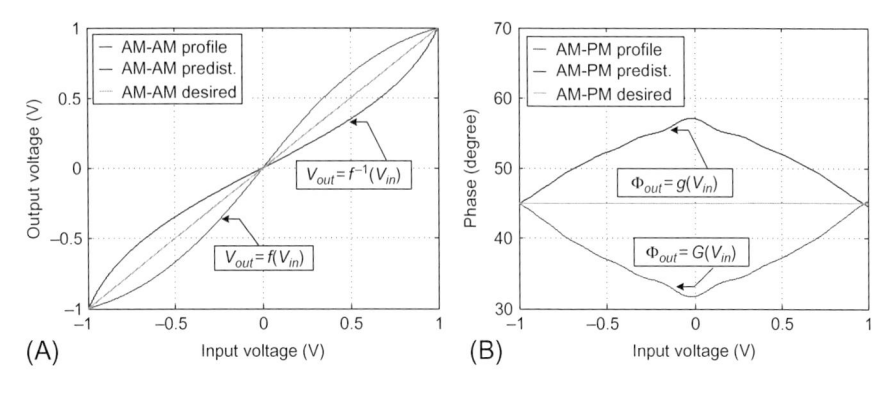

FIG. 10.13

Two-tone test envelope and phase profiles: (A) AM-AM (B) AM-PM.

$$V_{out} = f(V_{in})$$
$$= a_1 V_{in} + a_2 V_{in}^2 + a_3 V_{in}^3 + a_4 V_{in}^4 + \ldots \quad (10.9)$$
$$\cong 1.45 V_{in} - 0.0006 V_{in}^2 - 0.46 V_{in}^3 + 0.0005 V_{in}^4$$

using MATLAB's curve-fitting toolbox, the $V_{out} = f(V_{in})$ profile can be fitted and thus, the $a_1 \ldots a_4$ are evaluated. Moreover, the output-phase/input-voltage relationship (AM-PM profile) can also be modeled by a polynomial representation as:

$$\Phi_{out}|_{deg} = G(V_{in})$$
$$= \phi_0 + \phi_1 V_{in} + \phi_2 V_{in}^2 + \phi_3 V_{in}^3 + \phi_4 V_{in}^4 + \ldots \quad (10.10)$$
$$\cong 34 + 0.34 V_{in} + 16.86 V_{in}^2 - 0.21 V_{in}^3 - 4.7 V_{in}^4$$

According to Fig. 10.13 and Eqs. (10.9)–(10.9), the inverse functions of the envelope are as follows:

$$V_{out} = f^{-1}(V_{in}) = f^{-1}(a_1 V_{in} + a_2 V_{in}^2 + a_3 V_{in}^3 + a_4 V_{in}^4 + \ldots)$$
$$= \alpha_1 V_{in} + \alpha_2 V_{in}^2 + \alpha_3 V_{in}^3 + \alpha_4 V_{in}^4 + \ldots \quad (10.11)$$
$$\cong 0.572 V_{in} + 0.0002 V_{in}^2 + 0.403 V_{in}^3 + 0.001 V_{in}^4$$

Moreover, the required predistorted phase profile of Fig. 10.13B could be expressed as:

$$\Phi_{out}|_{deg} = g(V_{in})$$
$$= -G(V_{in}) + 90$$
$$= \psi_0 + \psi_1 V_{in} + \psi_2 V_{in}^2 + \psi_3 V_{in}^3 + \psi_4 V_{in}^4 + \ldots$$
$$\cong 56 - 0.34 V_{in} - 16.86 V_{in}^2 + 0.21 V_{in}^3 + 4.7 V_{in}^4 \quad (10.12)$$

in which Eqs. (10.11), (10.12) are applied to the baseband code. Based on Fig. 10.13, applying the AM-AM predistorted profile makes the desired AM-AM transfer function as a straight line, that is, $V_{out} = V_{in}$. Moreover, the desired AM-PM characteristic is a constant line, that is, $\phi_{out} = cte = 45$. It should be mentioned that the AM-AM and AM-PM nonlinearities are added vectorially and, therefore, they cannot cancel each other (see Appendix B.3). Thus, each profile should be corrected accordingly. The measurement results based on these profiles will be presented in Chapter 11, Section 11.3.1. It should be reiterated that the AM-AM and AM-PM profiles are carrier frequency dependent due to the fact that the power combiner circuit elements change their impedance at other frequencies.

10.9.3 DPD BASED ON I/Q CODE MAPPING

The second predistortion approach, which is utilized for this IC chip, is performed using a constellation-mapping-based DPD algorithm [152–154]. This work, however, proposes a very simple, modified constellation-mapping DPD, which is based on 1D mapping of $I_{BB\text{-}up}$ and $Q_{BB\text{-}up}$, that is, $2\times 1D$. A complex modulated baseband data is defined as:

$$IQ_{BB}(I_{BB\text{-}up}, Q_{BB\text{-}up}) = I_{BB\text{-}up} + j \times Q_{BB\text{-}up}$$
$$= A_{IQ}(I_{BB\text{-}up}, Q_{BB\text{-}up}) \angle \phi_{IQ}(I_{BB\text{-}up}, Q_{BB\text{-}up}) \quad (10.13)$$

$I_{BB\text{-}up}$ and $Q_{BB\text{-}up}$ are demonstrated in Fig. 10.14A. Moreover, A_{IQ} and ϕ_{IQ} are envelope and phase information of the corresponding baseband data, respectively. Thus, ideally, the modulated RF output of the RFDAC is expressed as:

$$V_{IQ}(I_{BB\text{-}up}, Q_{BB\text{-}up}) = IQ_{BB}(I_{BB\text{-}up}, Q_{BB\text{-}up}) \times \exp(j\omega_0 t)$$
$$= A_{IQ} \times \exp(j(\omega_0 t + \phi_{IQ})) \quad (10.14)$$

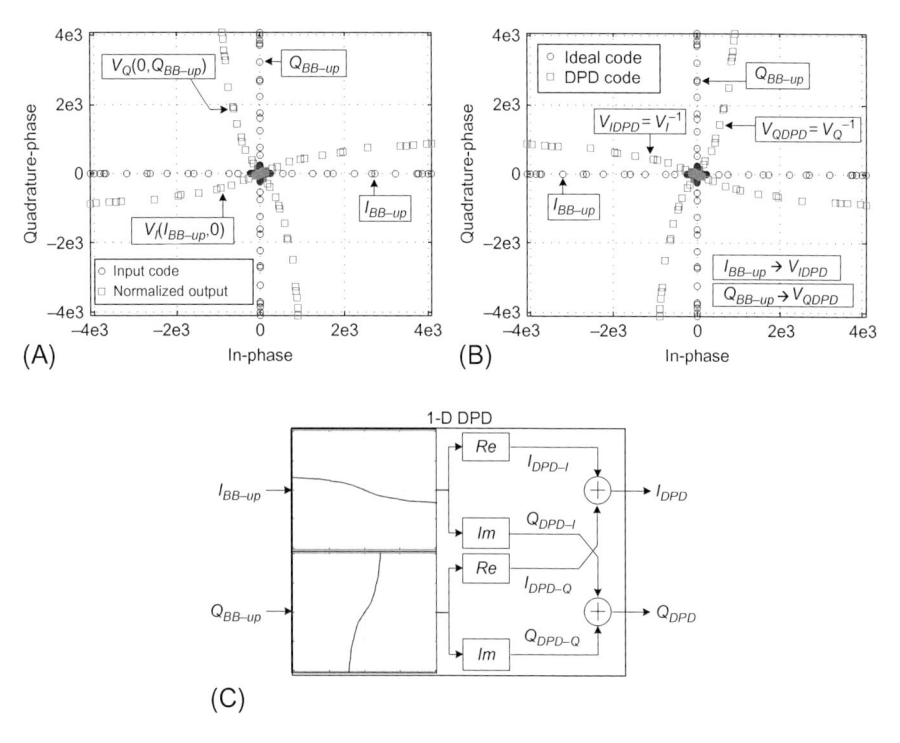

FIG. 10.14

(A) Input codes along with their corresponding nonlinear output voltages; (B) DPD in-phase and quadrature-phase input code mapping diagram; (C) Illustration of the open-loop, $2\times 1D$ DPD.

Nonetheless, due to the fact that RFDAC is a nonlinear transmitter, as a result, the RF output of the RFDAC becomes:

$$V_{IQ}(I_{BB\text{-}up}, Q_{BB\text{-}up}) = (V_I(I_{BB\text{-}up}, 0) + j \times V_Q(0, Q_{BB\text{-}up})) \times \exp(j\omega_0 t) \quad (10.15)$$

where $V_I(I_{BB\text{-}up}, 0)$ and $V_Q(0, Q_{BB\text{-}up})$ are the corresponding nonlinear complex profiles of $I_{BB\text{-}up}$ and $Q_{BB\text{-}up}$ in which they are normalized to their related input codes. These profiles are indicated in Fig. 10.14A. In practice, $V_I(I_{BB\text{-}up}, 0)$ and $V_Q(0, Q_{BB\text{-}up})$ are acquired as follows: First, due to orthogonal operation of RFDAC, I_{BB} and Q_{BB} are individually swept from -4095 to $+4095$. The subsequent RF output is down-converted, and the related baseband complex signals, that is, $V_I(I_{BB\text{-}up}, 0)$ and $V_Q(0, Q_{BB\text{-}up})$, are obtained. Next, the inverse function of $V_I(I_{BB\text{-}up}, 0)$ and $V_Q(0, Q_{BB\text{-}up})$ are evaluated and depicted in Fig. 10.14B. The in-phase and quadrature-phase DPD profiles are as follows:

$$V_{IDPD}(I, Q) = V_I^{-1}(I_{BB\text{-}up}, 0)$$
$$= I_{DPD\text{-}I} + j \times Q_{DPD\text{-}I} \quad (10.16)$$

$$V_{QDPD}(I, Q) = V_Q^{-1}(0, Q_{BB-up})$$
$$= I_{DPD\text{-}Q} + j \times Q_{DPD\text{-}Q} \quad (10.17)$$

Otherwise stated, the following relationships are established between $I_{BB\text{-}up}$ and $V_{IDPD}(I, Q)$ as well as $Q_{BB\text{-}up}$ and $V_{QDPD}(I, Q)$.

$$I_{BB\text{-}up} = V_I\left(V_I^{-1}(I_{BB\text{-}up}, 0)\right)$$
$$= V_I(V_{IDPD}(I, Q)) \quad (10.18)$$
$$= V_I(I_{DPD\text{-}I}, Q_{DPD\text{-}I})$$

$$Q_{BB\text{-}up} = V_Q\left(V_Q^{-1}(0, Q_{BB\text{-}up})\right)$$
$$= V_Q(V_{QDPD}(I, Q)) \quad (10.19)$$
$$= V_Q(I_{DPD\text{-}Q}, Q_{DPD\text{-}Q})$$

Therefore, in this DPD process, $I_{BB\text{-}up}$ and $Q_{BB\text{-}up}$ are individually mapped to $V_{IDPD}(I, Q)$ and $V_{QDPD}(I, Q)$, respectively.

$$I_{BB\text{-}up} \to V_{IDPD}(I, Q) \quad (10.20)$$

$$Q_{BB\text{-}up} \to V_{QDPD}(I, Q) \quad (10.21)$$

Specifically, this DPD process can be inferred as 1D mapping of two individual signals of $I_{BB\text{-}up}$ and $Q_{BB\text{-}up}$, that is, $2\times 1D$. In particular, since I_{path} and Q_{path} are orthogonal, the DPD does not require a 2D exhaustive search of the entire constellation diagram which is utilized in Ref. [66]. Consequently, due to orthogonality, the subsequent I_{DPD} and Q_{DPD} are obtained as follows

$$I_{DPD}\left(I_{BB\text{-}up}, Q_{BB\text{-}up}\right) = I_{DPD\text{-}I} + I_{DPD\text{-}Q} \tag{10.22}$$

$$Q_{DPD}\left(I_{BB\text{-}up}, Q_{BB\text{-}up}\right) = Q_{DPD\text{-}I} + Q_{DPD\text{-}Q} \tag{10.23}$$

Fig. 10.14C illustrates the open loop, 2×1D DPD. Note that the DPD profiles of V_{IDPD} and V_{QDPD} are obtained only at the beginning of the measurement operation, and they will remain unchanged afterward.[4] It should be again pointed out that DPD profiles in Fig. 10.14 are carrier frequency dependent due to the power combining network.

Fig. 10.15 depicts the constellation mapping measurement setup structure. Using MATLAB, in-phase and quadrature-phase randomized symbols (I_{symb} and Q_{symb})

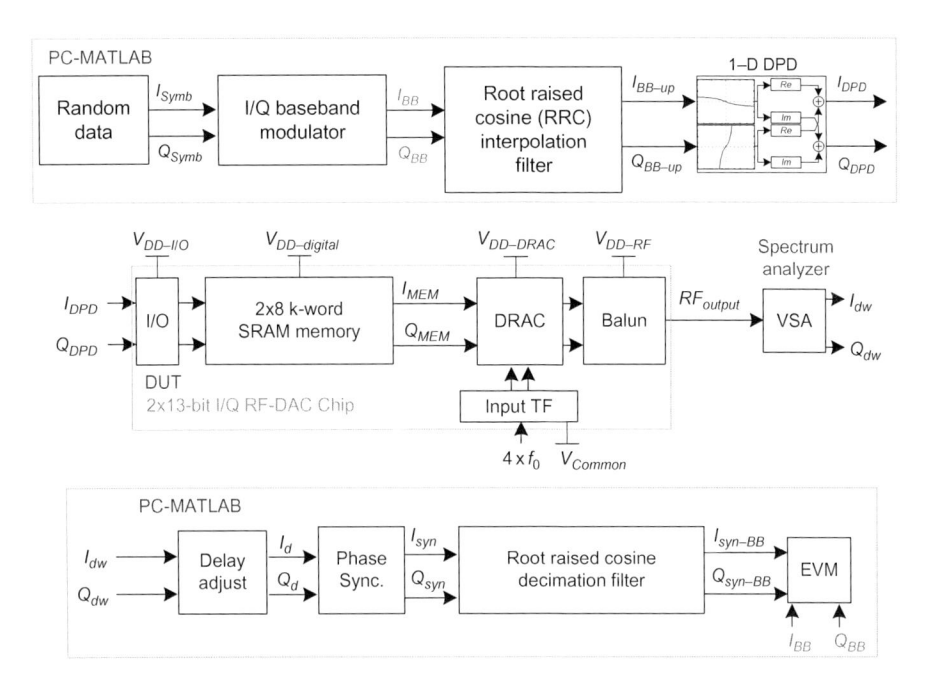

FIG. 10.15

DPD measurements constellation mapping flow.

---

[4] As stated previously, due to inaccurate components of the passive power combiner, the RFDAC test-chip exhibits imperfect orthogonal summation. As a result, it requires only simple 2×1D DPD. Although employing 2D instead of 1D results in a better linearity performance, it is extremely difficult (and very expensive in practice) to obtain 2D DPD profiles.

are generated and supplied to the I/Q baseband modulator. This block creates QAM signals of I_{BB} and Q_{BB}. Then, to confine the modulation bandwidth, I_{BB} and Q_{BB} get pulsed-shaped using an RRC interpolation filter and upsampled to as high as the CK_R rate, which is $f_0/8$ (see also Fig. 10.1). Afterward, $I_{BB\text{-}up}$ and $Q_{BB\text{-}up}$ are mapped utilizing Eqs. (10.20)–(10.23) and Fig. 10.14B. Next, the predistorted signals (I_{DPD} and Q_{DPD}) are uploaded into two designated on-chip SRAMs. Thereafter, the upconverted RF signal is down-converted using a VSA, and the subsequent down-converted digital in-phase (I_{dw}) and quadrature-phase (Q_{dw}) signals are fed back to MATLAB. Three important steps should be followed. First, the measurement time-delay should be calibrated. Then, the subsequent complex signal phase, that is, $\phi_d = \angle(I_d + jQ_d)$, should be rotated such that the eventual phase, that is, $\phi_{syn} = \angle(I_{syn} + jQ_{syn})$, is the same as the original complex phase, that is, $\phi_{BB\text{-}up} = \angle(I_{BB\text{-}up} + jQ_{BB\text{-}up})$. Finally, I_{syn} and Q_{syn} are down-sampled utilizing an RRC decimation filter to recover the original I/Q baseband modulated signals, that is, $I_{syn\text{-}BB}$ and $Q_{syn\text{-}BB}$. Comparing the measured $I_{syn\text{-}BB}/Q_{syn\text{-}BB}$ with the original I_{BB}/Q_{BB}, the EVM based on Eq. (1.3) or [118, eqs. 24] is calculated.

10.9.4 DPD REQUIRED MEMORY AND TIME

As stated, the DPD is performed at the initiation of the measurement process and is not a background operation. Namely, the DPD profiles, V_{IDPD} and V_{QDPD}, are generated at the initiation and they will subsequently be frozen for the remainder of the measurement process. It is an open loop procedure and, in this work, it is not automated. It is worth mentioning that the two designated on-chip SRAMs are employed in order to upload the upsampled complex baseband signals, that is, $I_{BB\text{-}up}$ and $Q_{BB\text{-}up}$, into them. Nonetheless, due to utilizing the DPD, the predistorted baseband signals, that is, I_{DPD} and Q_{DPD} are eventually uploaded into them. As stated previously, the memory length is fixed, that is, $l_{mem} = 8$ k-word. Thus, this work is severely hindered by this limitation. As explained earlier, $I_{BB\text{-}up}$ and $Q_{BB\text{-}up}$ should be individually swept from −4095 to +4095 in order to obtain the V_{IDPD} and V_{QDPD} DPD profiles, respectively. These two individual processes are illustrated in Fig. 10.16. Note that, as explained later in this chapter, to preserve the continuity in order to alleviate the unwanted spectral jump, the first data point and the last one should be identical. Thus, the total number of sweeping points should be doubled. In other words, as depicted in Fig. 10.16, this operation comprises two specific paths, from -4095 to +4095 and vice versa, which requires $2 \times 8\,\text{k}=16\,\text{k}$ points. Consequently, due to the fact that $l_{mem} = 8$ k-word, every other baseband code value from −4095 to +4095 can only be swept. Hence, the in-phase sweeping time is

$$\begin{aligned}
T_{I\text{-}sweep} &= 8192 \times \frac{1}{f_{CKR}} \\
&= \frac{8192}{300e6} \\
&\cong 27.31\ \mu\text{s}
\end{aligned} \quad (10.24)$$

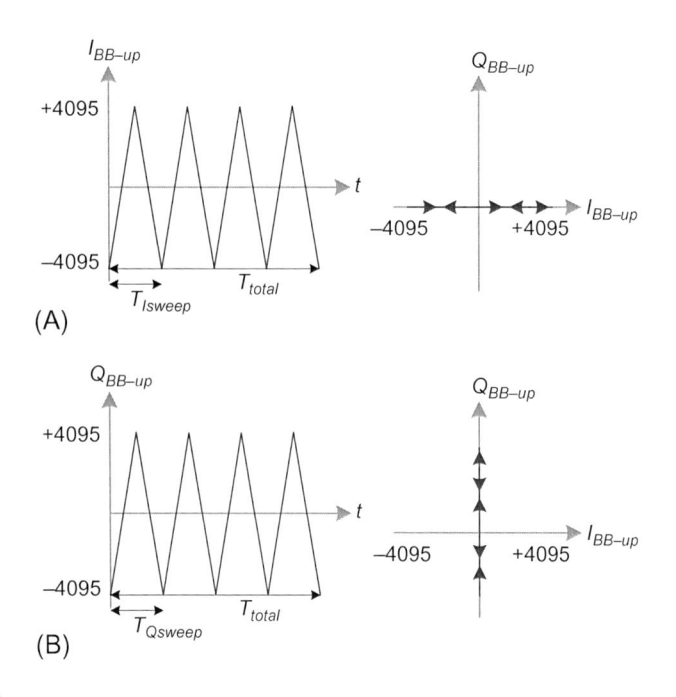

FIG. 10.16

DPD sweeping time of (A) I_{path}; (B) Q_{path}.

Also considering the quadrature-phase sweeping time (see Fig. 10.16B), the total sweeping time is doubled, that is, 54.62 μs. Nonetheless, to improve the accuracy of the DPD profiles, as depicted in Fig. 10.16, the sweeping time can be increased to four cycles whereby DPD profiles can be extracted while averaging these four cycles. Thus, the total sweeping time is increased to 218.48 μs. Note that, since this time is inversely proportional to f_{CKR}, increasing the upsampling rate not only improves the out-of-band spectral performance but also decreases the DPD calibration time.

10.9.5 DPD EFFECTIVENESS AGAINST THE TEMPERATURE AND AGING

DPD depends on the temperature. In order to validate that, a simple simulation is performed in which the corresponding circuit temperature is swept between −50 and 130°C. The simulation results are demonstrated in Fig. 10.17A–C. Moreover, the I/Q baseband input codes of the RFDAC are swept, and its corresponding drain efficiency, output power, and output voltage are simulated for the temperature of −50, 25, and 50°C, which are depicted in Fig. 10.17D–F. According to the simulation results in Fig. 10.17, the I/Q RFDAC performance depends on the temperature and, thus, its related DPD profiles must be updated for various temperature conditions. Moreover,

FIG. 10.17

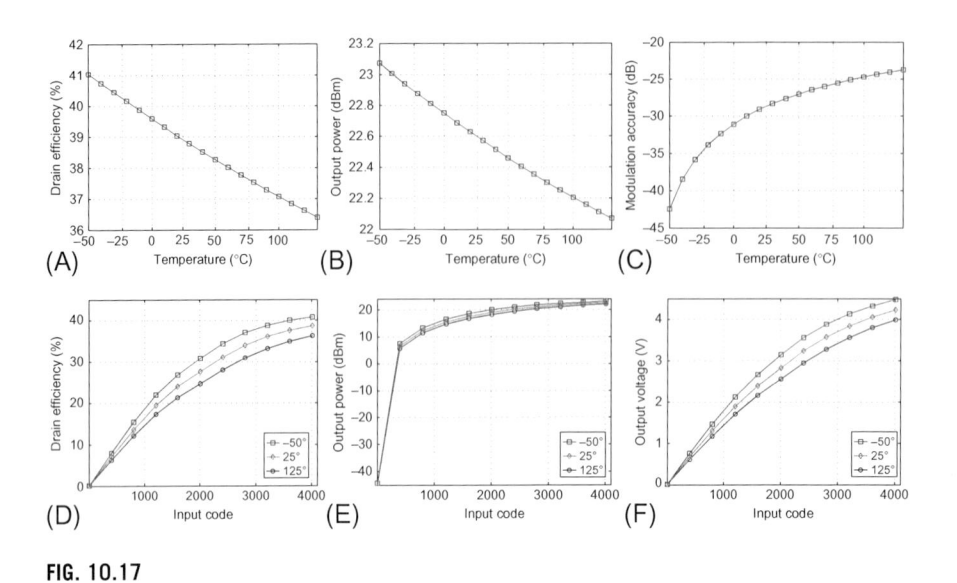

Effect of temperature sweeping on (A) drain efficiency; (B) output power; (C) modulation accuracy. Effect of baseband code sweeping on (D) drain efficiency; (E) output power; (F) output voltage for −50, 25, and 125°C.

the aging would not affect the DPD profiles. This argument is confirmed during the measurement process. A number of chips in the period of three to four months are measured, and their corresponding DPD performances were always within 1 or 2 dB of each other which can be tolerated. Therefore, at least, it could be expressed that, for brief periods of time, the effectiveness of the DPD will not be degraded. Checking with the TSMC manual confirms that a significant effect due to aging would not be expected.

10.9.6 VERIFICATION OF DPD I/Q CODE MAPPING

Examining this approach, a 256-symbol modulation is created. Based on the concept depicted in Fig. 10.18A, the constellation diagram is continuously swept from the top-left to top-right in a "snake"-like manner and traversed back again to its original point in order to preserve continuity. Note that, for simplicity, Fig 10.18A only illustrates a 16-symbol constellation diagram as well as their time domain representations, which is exhibited in Fig 10.18B. These signals are then upsampled and interpolated using an RRC interpolation filter to produce $I_{BB\text{-}up}$ and $Q_{BB\text{-}up}$ (see their I/Q trajectories in Fig. 10.18C). Next, the resultant signals are predistorted (I_{DPD} and Q_{DPD}) using the lookup table of Fig. 10.18B and loaded into two on-chip SRAMs. Fig. 10.18D shows the effect of the I/Q DPD mapping on the I/Q trajectories of the original modulated signals. The RF output signal is down-converted, and its

10.9 Digital I/Q calibration and DPD techniques

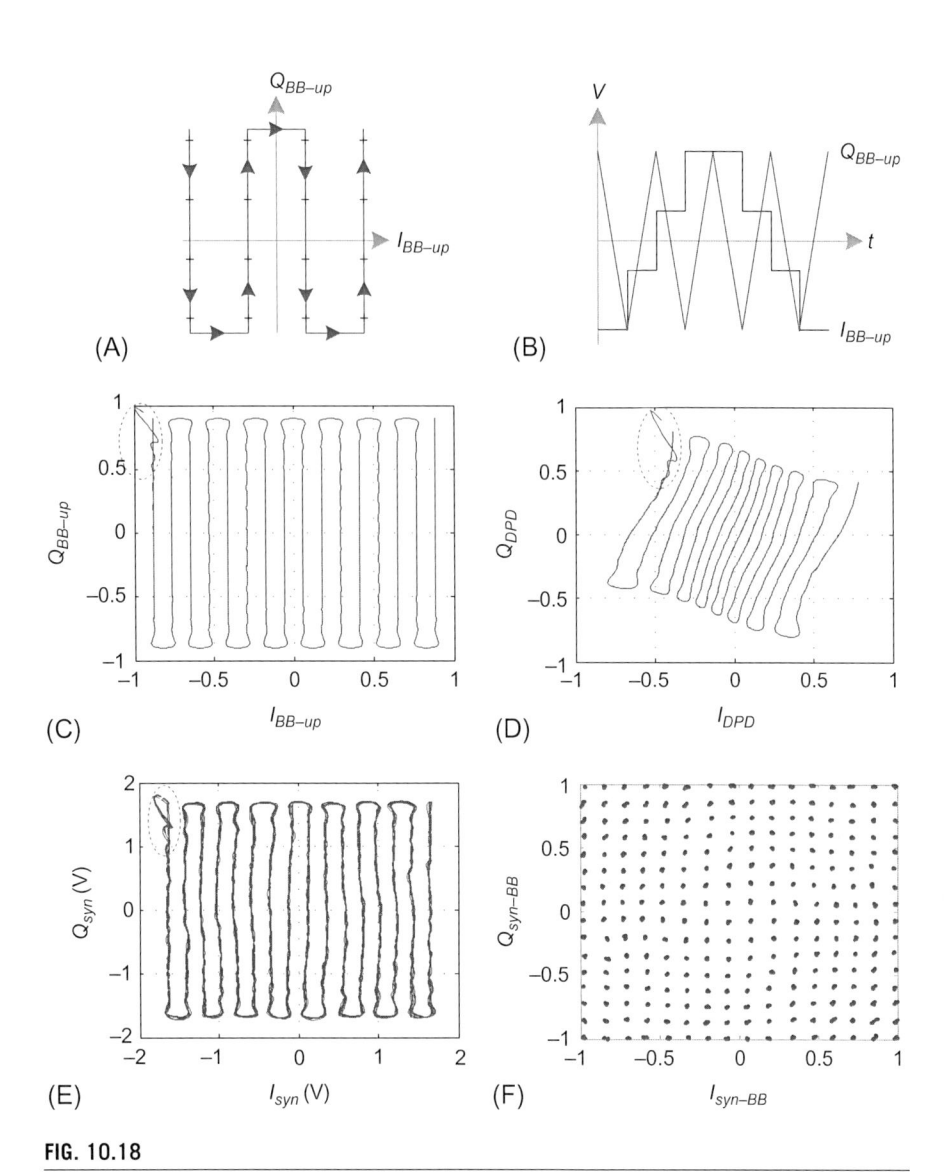

FIG. 10.18

DPD measurements: (A) Simplified diagram of 16-symbol I_{BB-up}/Q_{BB-up}; (B) Trajectories of their related time domain waveforms. DPD 256-symbol constellation mapping plots: (C) I_{BB-up}/Q_{BB-up} trajectories; (D) predistorted I_{DPD}/Q_{DPD} trajectories; (E) measured I_{syn}/Q_{syn} trajectories; (F) measured 256-point constellation.

corresponding I/Q trajectories are exhibited in Fig. 10.18E, which demonstrates a good agreement with the original I/Q trajectories of Fig. 10.15C. I_{syn} and Q_{syn} are then down-sampled and decimated to create the measured constellation diagram (Fig. 10.18F). Note that its related EVM, RF power, and drain efficiency are −32 dB,

16.1 dBm, and 19%, respectively. It should be mentioned that, due to the limited data length of I_{DPD}/Q_{DPD} (i.e., 8192), which are repeatedly fed to the DRAC circuit from the first data point to the last, any discontinuity between the first data point and the last one creates an undesirable spectral jump. To alleviate this issue and to preserve the continuity, the data length of I_{BB} and Q_{BB} are doubled and applied to the RRC interpolation filter and then only the half of the data length of the subsequent $I_{BB\text{-}up}$ and $Q_{BB\text{-}up}$ are exploited and applied to the DPD lookup table. This technique is referred to as wrap-around process. As a result, the beginning points of the I/Q trajectories of Fig. 10.18C–E, indicated with circles, have been shaped in such a way as to ensure the continuity of the I/Q signals.

10.10 CONCLUSION

In this chapter, the implemented wideband, 2 × 13-bit I/Q RFDAC-based all-digital transmitter realized in 65-nm CMOS is presented. Employing the orthogonal I/Q combining approach, which is proposed in Chapter 6, guarantees the isolation between I_{path} and Q_{path}. Nonetheless, due to inaccurate components of the passive power combiner, the I_{path} and Q_{path} interact with one another. The $4 \times f_0$ off-chip single-ended clock is converted to a differential version employing an on-chip transformer. The wide swing, low phase noise, high-speed dividers are incorporated to translate the $4 \times f_0$ differential clock to the fundamental frequency of f_0. In the meantime, the complementary quadrature sign bit is used to address four quadrants of the related constellation diagram. The 25% differential quadrature clocks are generated using logic-AND operation between $2 \times f_0$ differential clock and f_0 differential quadrature clocks. The 12-bit DRAC is implemented employing a segmentation approach, which consists of 256 MSB and 16 LSB thermometer unit cells. The layout arrangement of the DRAC unit cell proves to be very crucial. It was concluded that the vertical layout would be the most appropriate selection. The LO leakage and I/Q image rejection technique as well as two DPD memoryless techniques of AM-AM/AM-PM and constellation mapping are introduced, which will be extensively utilized in the measurement segment. The DPD techniques are utilized to alleviate the imperfect orthogonal summing of the I_{path} and Q_{path}.

CHAPTER 11

Measurement results of the 2 × 13-bit I/Q RFDAC*

CHAPTER OUTLINE

11.1 Measurement Setup 230
11.2 Static Measurement Results 231
11.3 Dynamic Measurement Results 234
    11.3.1 LO Leakage and IQ Image Suppression of I/Q RFDAC 234
    11.3.2 The RFDAC's Linearity Using AM-AM/AM-PM Profiles 236
    11.3.3 The RFDAC's Linearity Using Constellation Mapping 238
11.4 Conclusion 243

This chapter addresses the related measurement results of the proposed 2 × 13-bit chip presented in Chapter 10. Section 11.1 thoroughly explains the measurement setup structure. Section 11.2 demonstrates the performance of the chip under continuous-wave conditions. Section 11.3 exhibits its behavior when applying the complex-modulated baseband signal. Finally, Section 11.4 concludes the chapter. The proposed 2 × 13-bit all-digital I/Q RFDAC is implemented in a TSMC 65 nm LP CMOS process. Fig. 11.1A exhibits the chip micrograph. The chip occupies $1.27 \times 2 \, mm^2$ with an active area of $0.58 \times 1.03 \, mm^2$. Moreover, the designated SRAMs occupy an area of $1.27 \times 1 \, mm^2$ while the remainder is occupied by decoupling capacitors and I/O pads. Note that the DRAC in combination with its corresponding binary-to-thermometer encoders as well as decoupling capacitors occupies only $0.41 \times 0.41 \, mm^2$. The RFDAC employs only standard "Vt" transistors. All pads, including the single-ended RF input clock and RF output, are wire-bonded directly to the FR4 board. The RFDAC ground plane is improved utilizing the following approach. First, all ground pads are wire-bonded using flat bond wire, which decreases the equivalent inductance of the bond wire by approximately four times. Second, the chip is situated in a 300 μm deep hole. This makes the bond wires shorter and, thus, the interconnecting inductance is smaller.

---

*The authors acknowledge substantial contributions from Prof. Dr. Leo de Vreede (TU Delft).

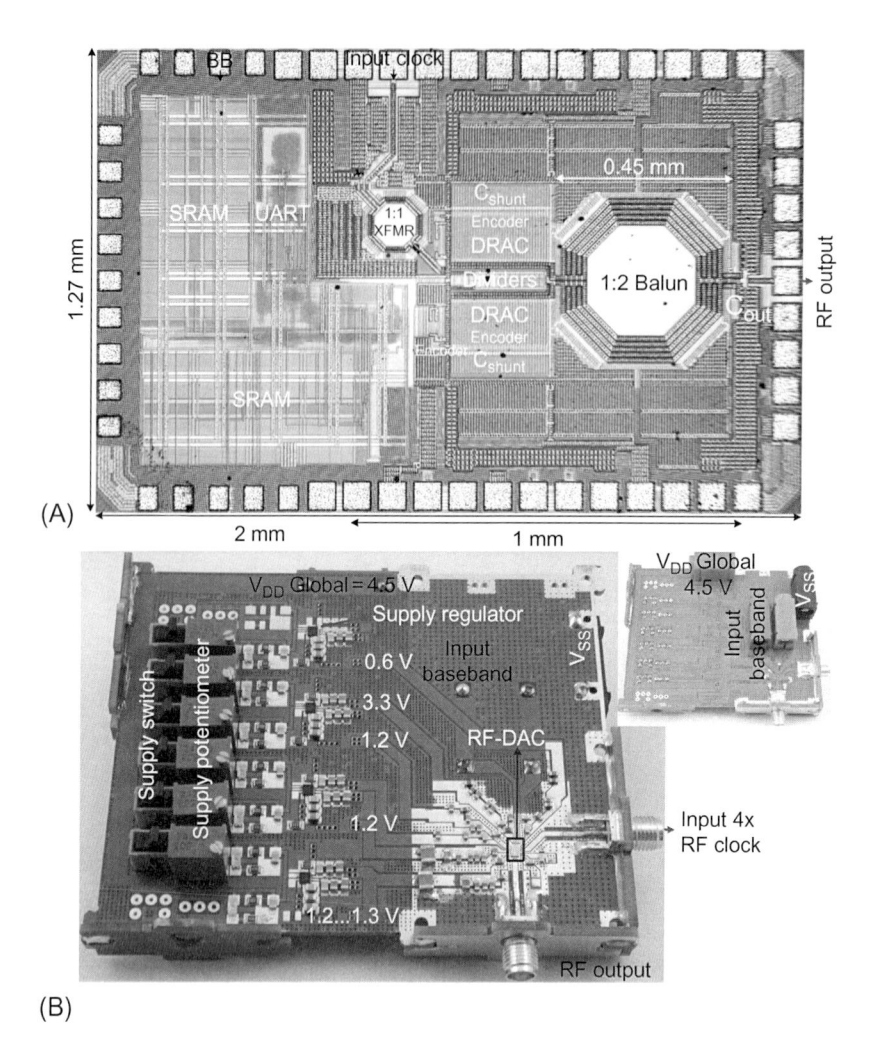

CHAPTER 11 Measurement results of the 2 × 13-bit I/Q RFDAC

FIG. 11.1

(A) Micrograph of the 2 × 13-bit all-digital I/Q RFDAC transmitter, (B) front and back view of its related PCB.

11.1 MEASUREMENT SETUP

For measurements, as depicted in Fig. 11.1B, the chip requires five different supply voltages namely, $V_{\text{DD-RF}} = 1.2\ldots1.3\,\text{V}$ for the balun center-tap node, $V_{\text{DD-DRAC}} = 1.2\,\text{V}$ for the RFDAC core, $V_{\text{Common}} = 0.6\,\text{V}$ for the input transformer center-tap node, $V_{\text{DD-digital}} = 1.2\,\text{V}$ for the SRAMs and UART interface,[1] and finally

---

[1] Courtesy of George Voicu from Professor Cotofana's group in TU-Delft, who designed and synthesized the UART interface.

$V_{DD\text{-}I/O} = 3.3$ V for I/O supply voltages. They are generated employing on-board regulators, ADP225ACPZ-R7 from analog devices, which use a common input supply voltage of 4.5 V. Note that on-board potentiometers are employed to set the required voltages of the on-board regulators in order to adjust the supply voltages. In addition, in order to turn on and off the supply voltages, on-board switches are employed.[2] This configuration allows the entire I/Q RFDAC chip to be tested using only one battery or supply voltage. Moreover, due to employing the on-chip input transformer, the input $4\times$ RF clock is a single-ended signal. In addition, as described in Chapter 10, Section 10.2, all required clock signals, including the baseband upsampling clock and the upconverting RF carriers, are generated via the on-chip frequency dividers. Thus, the I/Q RFDAC only requires one external clock generator, which results in a very straightforward board design and the test setup.

To verify the design through measurements, as it was fully explained in Chapter 8 (Fig. 8.1) and also in Chapter 10 (Fig. 10.15), first, the I_{BB} and Q_{BB} baseband signals are upsampled and interpolated in software (PC-MATLAB). These upsampled signals, $I_{BB\text{-}up}$ and $Q_{BB\text{-}up}$, are subsequently loaded via UART (see Appendix B, Section B.1) into two on-chip SRAMs that are clocked with CK_R of Fig. 10.1. Note that, as stated previously, the clock frequency is $f_{CKR} = (1/8) \times f_0$.

11.2 STATIC MEASUREMENT RESULTS

Earlier simulations demonstrate that the achievable maximum drain efficiency of the I/Q RFDAC output stage should be well above 44%. Due to the low power arrangement of the forgoing clocking and predriver circuitry, the overall system efficiency of the realized monolithic transmitter, based on Eqs. (1.2), (8.1), should be able to reach 37% at 2.4 GHz for a peak output power level of 22.6 dBm at 1.2 V. Experimental verification shows that, without using any correction for the PCB and SMA connector losses, the peak overall system efficiency occurs at 2.1 GHz and reaches 31.5% with a related peak output power of 22.3 dBm at 1.2 V.

Although the transmitter was verified to work properly from 60 MHz to 3.5 GHz, the most superior performance is achieved in the range of 1.36 to 2.51 GHz, where measurements illustrate an output power and overall system efficiency of more than 21 dBm and 21%, respectively (see Fig. 11.2). For this measurement, the carrier frequency is swept from 1.35 to 2.63 GHz in steps of 2 MHz. The supply voltage is also swept from 0.6 to 1.3 V. Fig. 11.2A and B only indicates the measurement results for 1.2–1.3 V. Based on these results, the peak output power is 22.8 dBm while its related drain efficiency and system efficiency are 42% and 34%, respectively. These results emphasize the wide-band operation of the realized on-chip output balun. Since the resolution of RFDAC is 2×12 bits, the input baseband codes are swept from -4095 to $+4095$, and the output power with its related voltage and phase are measured. The measurement results are depicted in Fig. 11.2C and D. Based

---

[2]Courtesy of my group colleagues Amir Ahmadi Mehr, Massoud Tohidian, and Masoud Babaie, who have designed the on-board regulator circuits.

FIG. 11.2

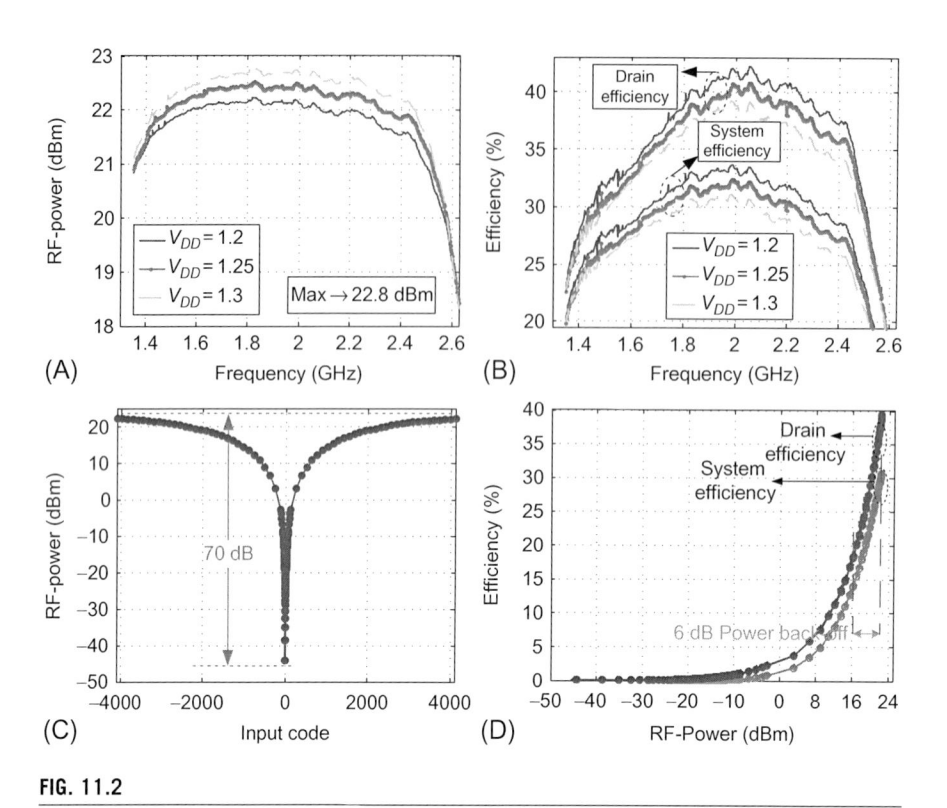

RF measurements: (A) RF output power; (B) efficiency of modulator versus frequency; (C) RF output power versus input code; (D) efficiency versus RF output power.

on Fig. 11.2C, the static carrier leakage level is more than 70 dB lower than the achievable maximum power. Fig. 11.2D exhibits the RFDAC efficiency versus RF output power. According to Fig. 11.2D, the drain and system efficiencies at the 6 dB back-off are 19% and 14%, respectively.

The static AM-AM nonlinearity of the digital I/Q transmitter is illustrated in Fig. 11.3A. As expected, at lower absolute codes (center of the curve), the output voltage changes linearly with respect to the input code. In contrast, at higher codes, the curve begins to saturate. Moreover, Fig. 11.3B and C indicates the static AM-PM nonlinearity profiles. Based on the measurement results of Fig. 11.3B, the maximum phase deviation of individual I and Q codes from lower to higher codes is less than 10°. Fig. 11.3C indicates that, by changing only the $I_{BB\text{-}up}$ or $Q_{BB\text{-}up}$ not only changes the amplitude but also changes the output phase that reveals the AM-PM distortion of the RFDAC. By applying the lookup table of Fig. 10.14B, the static I/Q constellation for a 256-symbol case is measured and depicted in Fig. 11.3D. Its related EVM is better than −30 dB while the maximum RF power is more than 22 dBm.

11.2 Static measurement results   233

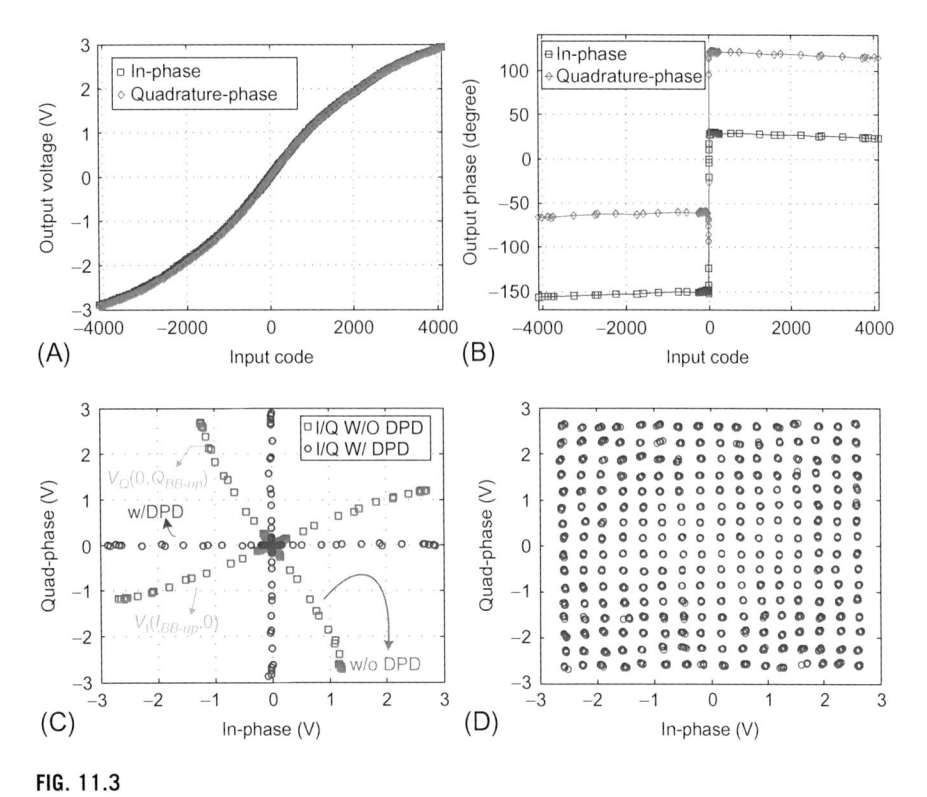

FIG. 11.3

Static measurement results: (A) output voltage versus input code; (B) output phase versus input code; (C) simple I/Q constellation diagram with and without DPD; (D) 256-point constellation diagram with DPD.

It should be pointed out that, as seen in Fig. 11.3D, the RFDAC exhibits memory effects which might be related to heating up the test-chip. Note that, as stated in Chapter 7, Section 7.2, the measurement results of Fig. 11.3B–D are obtained as follows: Time domain RF output signals are captured and saved. The FFT of these signals is subsequently calculated, and the amplitudes and phases are plotted to obtain the static constellation diagram of Fig. 11.3D.

The static phase noise of RFDAC is measured for various carrier frequencies between 1.5 and 2.5 GHz, and the noise floor is ascertained to be better than −160 dBc/Hz. Fig. 11.4A shows the RFDAC phase noise at 2.4 GHz. The maximum baseband code for I_{BB} and Q_{BB} is 4095, which produces 21.54 dBm of RF power. It should be noted that, at 200 MHz frequency offset, the phase noise is −160 dBc/Hz. The figure also indicates two "spurs" at 300 and 600 MHz, which are actually the spectral replicas discussed previously. In this aspect, the ZOH filter operation ensures

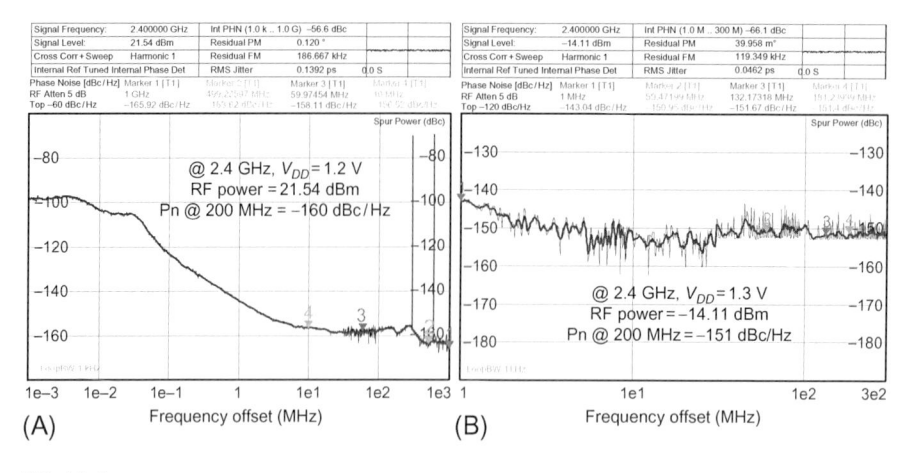

FIG. 11.4

Phase noise at 2.4 GHz: (A) full power; (B) low power.

that these replica levels are below -70 dBc/Hz.[3] Moreover, the RFDAC phase noise performance is reexamined for lower codes (e.g., 32). Based on Fig. 11.4B, its related RF power and noise floor reduces to -14 dBm and -165 dBm/Hz, respectively. Note that the shape of the noise floor is primarily related to the external $4\times$ clock generator source. Thus, the limiting factor in the noise floor is related to the noise of the input clock source as well as the corresponding noise of the divider circuitry.

11.3 DYNAMIC MEASUREMENT RESULTS

Dynamic measurements based on employing a modulated baseband signal have also been extensively performed. As stated in Chapter 10, the LO leakage and IQ image suppression performance of I/Q RFDAC ought to be examined. Moreover, the transmitter's linearity will be explored.

11.3.1 LO LEAKAGE AND IQ IMAGE SUPPRESSION OF I/Q RFDAC

First, LO leakage and IQ image suppression are examined. For this experiment, the LO frequency is set to 2.1 GHz, and the baseband frequency of $I_{BB\text{-}up}$ and $Q_{BB\text{-}up}$ signals are approximately 2.05 MHz. Fig. 11.5A demonstrates that, even without applying any I/Q calibration, the LO leakage and image levels are -62 dBc and

---

[3]The RFDAC works in the static mode of operation, hence the ZOH does not create any spectral replicas at multiples of the sampling frequency f_{CKR} away from the f_0 carrier. What is seen is 300/600 MHz clock leakage.

11.3 Dynamic measurement results

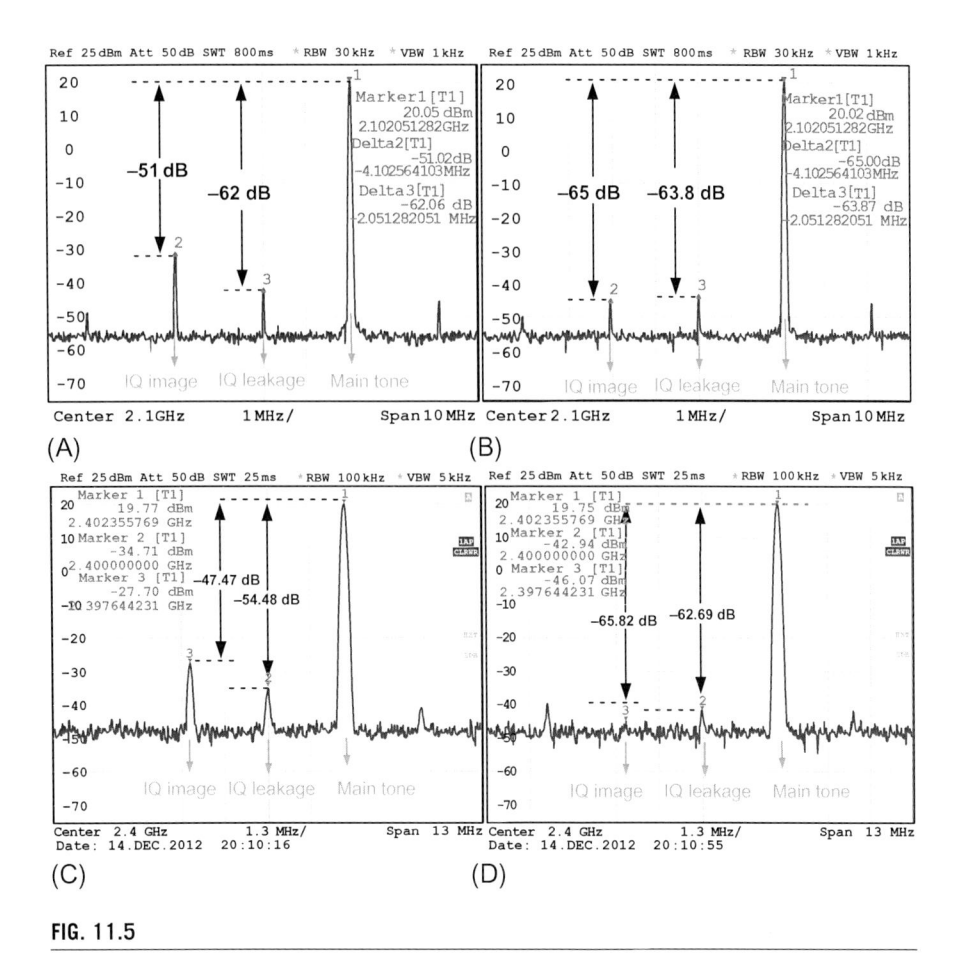

FIG. 11.5

Leakage and image suppression: at 2.1 GHz (A) without calibration; (B) with calibration; at 2.4 GHz (C) without calibration; (D) with calibration.

−51 dBc, respectively, at an output power of 20.03 dBm. As such, these numbers are sufficient to meet the specifications of most communication standards. The low image level indicates the superior matching of I and Q paths. Moreover, the use of a divide-by-4 circuit instead of a divide-by-2 approach also proves to be beneficial in improving the quadrature operation. By applying the I/Q calibration technique of Chapter 10 in Section 10.9.1, the image signal is further reduced by 14 dB (Fig. 11.5B). Moreover, the LO leakage and IQ image suppression performance of the RFDAC are also examined at other frequencies. For example, utilizing a 2.4 GHz carrier signal while applying 2.4 MHz I_{BB-up} and Q_{BB-up} single tone baseband signals, as depicted in Fig. 11.5C and D, the LO leakage and IQ image are approximately the same as in the 2.1 GHz scenario. Note that, based on the static measurement results in Fig. 11.2A and B, RFDAC performance is optimum at 2.1 GHz frequency instead of the targeted frequency of 2.4 GHz due to an imperfect estimation of the parasitic capacitances. This fact is also obvious by comparing Fig. 11.5A with Fig. 11.5C. According to these dynamic measurement results, without applying calibration, the LO leakage and IQ image performance of the RFDAC at 2.1 GHz are approximately 7.5 and 3.5 dB more superior than at 2.4 GHz. Nonetheless, utilizing the calibration makes them approximately the same. It should be reiterated that this calibration is carrier frequency dependent due to employing the passive power combiner.

The RFDAC linearity significantly improves by using either of the two DPD approaches that were discussed previously in Sections 10.9.2 and 10.9.3.

11.3.2 THE RFDAC'S LINEARITY USING AM-AM/AM-PM PROFILES

First, utilizing the AM-AM/AM-PM profiles of Section 10.9.2 and applying only a fourth-order memoryless polynomial approximation (Eqs. 10.9–10.12), the linearity of the RFDAC improves more than 25 dBc. Fig. 11.6A and B demonstrates the two-tone test measurement results before and after using the DPD of Section 10.9.2. In this measurement, the tone spacing is designated at 2.2 MHz, and the total power exceeds 16 dBm. The leakage level is below -55 dBm (-68 dBc) and the third-order intermodulation product (IM_3) is improved to more than -50.4 dBc. Since only the fourth-order polynomial is used, the nonlinearities of higher intermodulation products do not diminish as much as IM_3. Although the DPD improves the linearity of the lower-order odd intermodulation products (i.e., 3rd–7th), it deteriorates higher-order odd intermodulation products (bandwidth expansion). Comparing Fig. 11.6A and B, the 9th to 15th intermodulation products worsen.[4]

As stated in Sections 10.9.2, applying AM-AM/AM-PM DPD suppresses the higher intermodulation products from the envelope and phase profiles of the modulated RF signal. According to Fig. 11.6D, the phase deviation without applying DPD exceeds 10°. Applying DPD, however, makes the phase profile almost constant. Based on the simple expression in Appendix B, Section B.3, 10° produces a significant amount of third-order intermodulation (IM_3), which can be as high as -33 dBc. As a result, most of the linearity improvements originate in the phase correction. Note that, to justify the latter argument, a comparison is made between the frequency contents of the envelope profiles in Fig. 11.6C. Based on this analysis, the envelope profile is improved only about 3 dB when employing DPD. Thus, the linearity improvements are primarily related to the phase profile correction. It is worth noting that, when utilizing this DPD approach, a number of two-tone as well as multitone measurements have been performed in which their related signal bandwidth reaches up to 150 MHz. For example, for a 16-tone test signal with the bandwidth of 3.2 MHz, the spurious free dynamic range (SFDR) is 46 dB. By

---

[4]In Appendix B, Section B.4 a simple explanation is presented in order to disclose the supporting reason for bandwidth expansion of DPD.

FIG. 11.6

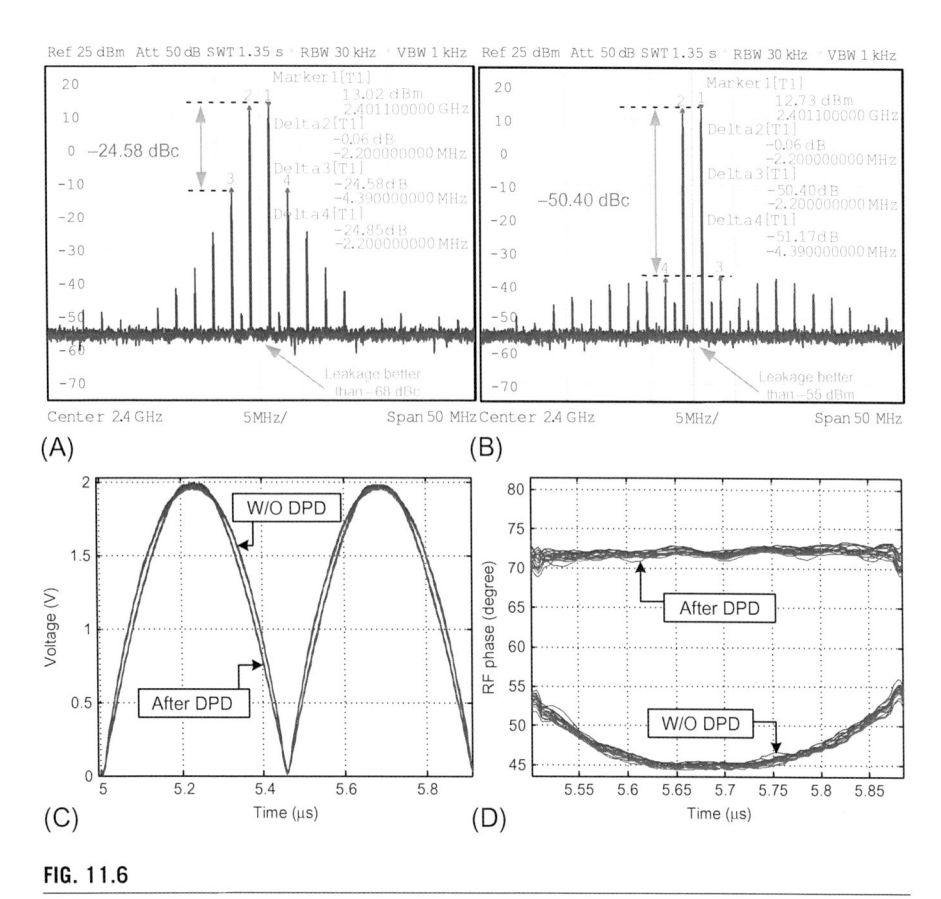

Two-tone linearity test: (A) without DPD; (B) with DPD. Its corresponding (C) envelope; and (D) phase.

doubling the bandwidth, the SFDR decreases by 3 dB which could be due to the fact that the "effective" oversampling rate becomes half.

Note that the AM-AM/AM-PM profiles are created utilizing the measurement results of the two-tone test. As stated in Ref. [155], there are three methods to perform the two-tone test. In this work, the second approach is employed whereby $I_{BB\text{-}up} = Q_{BB\text{-}up} = \cos(\omega_m t)$ are applied to the RFDAC chip. The related advantage of utilizing this approach lies in the fact that it is, indeed, a double-side band two-tone test scenario which more effectively employs the related bandwidth of the two-tone signal. Otherwise stated, measuring a single-side band two-tone signal in which its related tone spacing is 10 MHz requires 60 MHz bandwidth to measure IM_3 components whereas, in the double side band scenario, the required bandwidth is just 30 MHz.

11.3.3 THE RFDAC'S LINEARITY USING CONSTELLATION MAPPING

In addition, using the constellation-mapping DPD approach of Chapter 10, Section 10.9.3, a variety of I/Q signals have been tested. The QAM signals with different modulation complex points and bandwidths are applied to the transmitter.

First, a single-carrier "7.16 MHz QAM" signal is generated and applied to the chip. Fig. 11.7A exhibits the measured spectrum of a single-carrier "7.16 MHz 4-QAM" signal with and without applying the DPD. Utilizing the DPD improves the RFDAC linearity by more than 19 dB. The adjacent channel power ratio (ACPR) is better than −47.7 dBr while the alternate channel power ratio is better than −49 dBr. The I/Q trajectory and constellation diagram are depicted in Fig. 11.7B and C. The measured EVM is −38 dB. Moreover, Fig. 11.7D indicates the amplitude probability profile of the measured down-converted RF signal. Based on that, its mean and peak RF power are 18 dBm and 21.23 dBm, respectively, while its related drain efficiency is 24.9%.

Second, a single-carrier "7.16 MHz 64-QAM" signal is generated and applied to the chip. Fig. 11.8A shows the measured spectrum of a single-carrier "7.16 MHz 64-QAM" signal with and without application of the DPD. Utilizing the DPD improves the RFDAC linearity by more than 16 dB. The adjacent channel power ratio (ACPR) is better than −44 dBr while the alternate channel power ratio is better than −49 dBr. The I/Q trajectory and constellation diagram are depicted in Fig. 11.8B and C. The measured EVM is −33.5 dB. Note that, due to utilizing random in-phase and quadrature-phase codes as well as limited memory depth, 3 out of 64 codes are inaccessible in the original baseband I/Q codes. Thus, the resulting constellation diagram of Fig. 11.8C only comprises 61 baseband codes. Moreover, Fig. 11.8D illustrates the amplitude probability profile of the measured down-converted RF

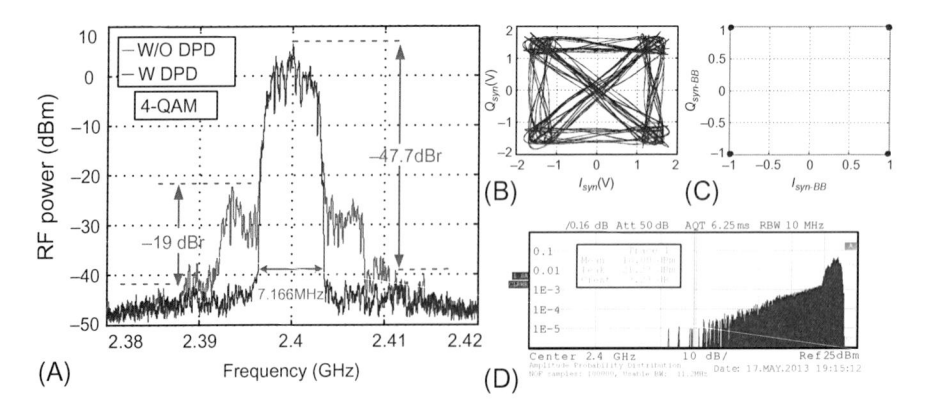

FIG. 11.7

"7.16 MHz 4-QAM" measurement results, (A) spectrum with and without DPD; (B) I_{syn}/Q_{syn} trajectories; (C) 4-QAM constellation; (D) amplitude probability distribution.

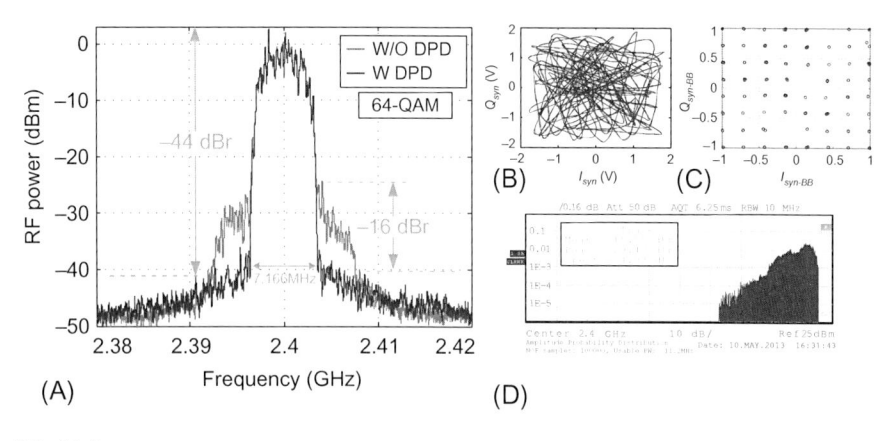

FIG. 11.8

"7.16 MHz 64-QAM" measurement results, (A) spectrum with and without DPD; (B) I_{syn}/Q_{syn} trajectories; (C) 64-QAM constellation; (D) amplitude probability distribution.

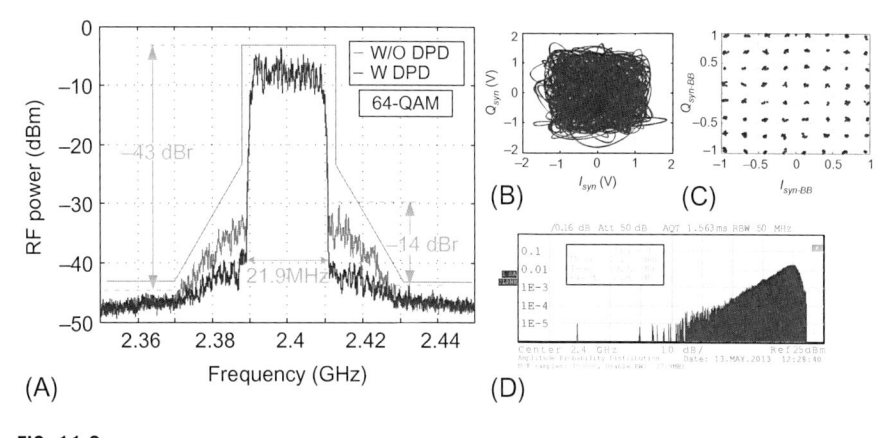

FIG. 11.9

"21.9 MHz 64-QAM" measurement results, (A) spectrum with and without DPD; (B) I_{syn}/Q_{syn} trajectories; (C) 64-QAM constellation; (D) amplitude probability distribution.

signal. According to Fig. 11.8D, its mean and peak RF power are 15.55 dBm and 20.48 dBm, respectively, while the related drain efficiency is 18%.

Third, a single-carrier "21.9 MHz 64-QAM signal" is generated and applied to the chip. Fig. 11.9A shows the measured spectrum of a single-carrier "21.9 MHz 64-QAM" signal with and without application of the DPD. Employing the DPD improves the RFDAC linearity by more than 14 dB. The adjacent channel power ratio (ACPR) is better than −40 dBr while the alternate channel power ratio is better than −43 dBr. The I/Q trajectory and constellation diagram are shown in Fig. 11.9B and C. The measured EVM is −28 dB. Moreover, Fig. 11.9D depicts the amplitude probability profile of the measured down-converted RF signal. Based on Fig. 11.9D, its mean and peak RF power are 13.30 dBm and 19.32 dBm, respectively, while the related drain efficiency is 14%.

Moreover, the chip is tested using a multicarrier "20-MHz, 256-QAM, OFDM" signal. The close-in and far-out spectrum measurement results are depicted in Fig. 11.10A and B, respectively. According to the measurement results of Fig. 11.10A, the close-in linearity exceeds 50 dB. Therefore, it can pass the close-in spectral mask by a large margin. Nonetheless, due to the zero-order-hold operation, its far-out spectrum contains replicas, which are distinguished in Fig. 11.10B. Additionally, according to an amplitude probability distribution measurement result, which is shown in Fig. 11.10C, the average power is 10.25 dBm, while its related PAPR is as high as 8.6 dB.

The chip performance is examined for other single-carrier complex QAM signals. The measured spectra of the RFDAC chip are demonstrated in Fig. 11.11A–C while employing a single-carrier "44 MHz 256-QAM," "88 MHz 256-QAM," and "154 MHz 1024-QAM," respectively. Since the operational bandwidth of our available VSA is limited to 7 MHz with the possibility of a 20 MHz extension, it was not feasible to measure EVM related to Fig. 11.11B and C. However, it is evidence that the simple DPD lookup table of Fig. 10.14B is still relevant up to 44 MHz.

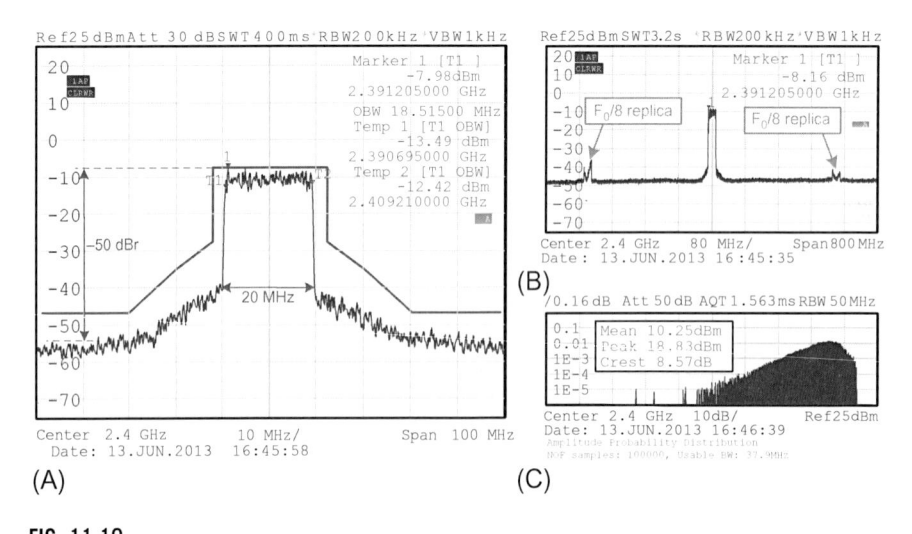

FIG. 11.10

"20 MHz 256-QAM OFDM" spectrum using DPD, (A) Close-in; (B) full span; (C) amplitude probability distribution.

11.3 Dynamic measurement results

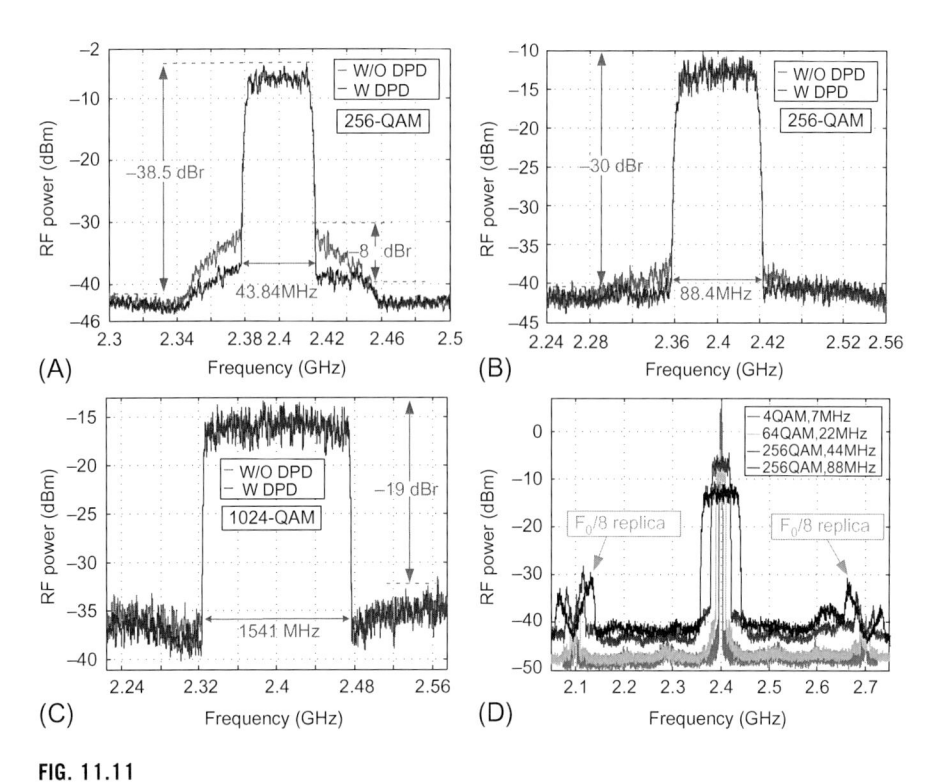

FIG. 11.11

Spectrum measurement results of different QAM signals with and without DPD, (A) "44 MHz 256-QAM," (B) "88 MHz 256-QAM," (C) "154 MHz 1024-QAM," (D) QAM spectrum including replicas.

The RFDAC indicates memory effects but only for higher frequencies and, as a result, the DPD lookup table should be amended.[5]

The memory effects are due to the limited amount of memory as well as on-chip ground sharing of the clock generation circuitry, digital baseband and the DRAC parts. Additionally, as discussed in Chapter 8, Section 8.1, signals with wider bandwidth exhibit higher out-of-band spectra (see Fig. 11.11D). The reason for such a higher noise floor lies in the fact that, due to the fixed upsampling clock rate ($f_{CKR} = f_0/8$), the "effective" oversampling rate of wider band signals is lower than for narrower band signals, therefore, the noise floor is higher. Moreover, in this chip, the SRAM memory size is limited to 8 kword which increases the noise floor for wideband signals. Note that for the "154 MHz 1024-QAM" scenario, the DPD improvement is very minimal. The reason lies in the fact that, since

---

[5] It should be pointed out that the measurement results in Figs. 11.8–11.11 have been obtained employing only a fixed, static, memoryless lookup table of Fig. 10.14B.

the upsampling clock, CK_R, is 300 MHz due to spectral regrowth, the out-of-band spectral images and the main signal spectrum, in this case, a 154 MHz 1024-QAM signal, are conjoined to one another. In addition, note that the DPD profiles are principally created considering only in-band nonlinearities. As a result, the DPD profiles must be updated to manage the aforementioned wideband signal. Fig. 11.11D also demonstrates the spectral replicas of the ZOH operation of RFDAC. Based on Section 8.1, one simple solution for decreasing the noise floor and spectral replicas of RFDAC would be increasing the upsampling clock rate, for example, $f_{CKR} = f_0/4$ or even higher.

Table 11.1 summarizes the performance of the proposed I/Q RFDAC. Table 11.2 compares this work against relevant I/Q publications [58–61, 64, 66]. The proposed RFDAC and the Mediatek work [66] are evidently more prominent than the rest for demonstrating the most superior performance. However, [66] exploits the 50% duty cycle, 40-nm CMOS technology, supply voltage of 1.8 V, upsampling clock rate of 804 MHz and, most importantly, requires a very sophisticated DPD algorithm. On the contrary, this work uses a very simple DPD lookup table due to the fact that the selected duty cycle is 25%, and the proper power combining network also improves isolation between I_{path} and Q_{path}. The drain efficiency of this work is higher than in Ref. [66] and, if our RFDAC were to be designed in a finer technology node, the drain efficiency would be even higher. Note that, in the Intel SoC work [64], they achieved high RF power with lower drain efficiency due to incorporating a conventional DAC, low pass filter, passive quadrature mixer and class-AB PA. In contrast, the proposed 2 × 13-bit RFDAC provides reasonable RF output power with higher efficiency using simpler architecture. In addition, Table 11.2 also

Table 11.1 Chip Performance Summary

Technology	65 nm CMOS
Transistor	Standard "Vt," baseline transistor
Power combining network	On-chip balun and capacitors
Input clock, single-ended	On-chip input transformer
Resolution	2 × 13-bit
Supply voltage	0.6–1.3 V
Operational carrier frequency	60 MHz to 3.5 GHz
Max RF power	22.8 dBm
Max drain/system efficiency	42/34%
LO generation current	33 mA @ 2.4 GHz
SRAM memory current	10 mA @ 2.4 GHz, 2 × 8 k-word
LO leakage/IQ image	> −70 dBc/−58 dBc
Baseband bandwidth	37 kHz…154 MHz @ 2.4 GHz
EVM 256-point/64-QAM	−32/−33.5 dB@15 dBm
Static noise floor full/low power	−160/−150 dBc/Hz
DPD	Simple lookup table

Table 11.2 Comparison Summary of I/Q, Polar, Out-Phasing Transmitter

Ref	Proc. (nm)	Resol. (bit)[a]	RF Power (dBm)	Drain Eff. (%)	EVM (dB)[b]	BB-BW (MHz)
Nokia [58]	30	11 I/Q	−2	0.04@1.2 V	−34	5
STMicro. [59]	65	12 I/Q	2.6[c]	1.2@1.2 V	−32.4	16
NXP [60, 61]	45	13 I/Q	5	10@1.8 V	−34	5
Intel [64]	32	11 I/Q	27.1	28.8@1.8 V	−25	20
Stanford [17]	180	10 Polar	20.7	23.2@1.7 V	−27	20
Berkeley [21]	65	9 Polar	23.3	43@1.2 V	−28	20
Intel [51]	32	8 Outphasing	26	35@2.0 V[d]	−31.5	40
Mediatek [66]	40	13 I/Q	24.7	37@1.8 V	−36	160
This work [119, 120]	65	13 I/Q	22.8	42@1.3 V	−28	154

[a]Bit-resolution including sign bit with its corresponding architecture of the transmitter.
[b]EVM is reported at maximum reported measurable bandwidth, which are either 5 or 20 MHz.
[c]The average power is reported. Perhaps the peak is 9 dBm with 7% drain efficiency (off-chip balun).
[d]They only reported their system efficiency. Note that their power combining network is off-chip.

presents the best performance numbers of recently published polar [17, 21, 49] and outphasing transmitters [51, 55, 56]. As evidenced, the I/Q transmitters can manage very wideband signals along with more effective EVM. Due to its versatility, high efficiency, wide bandwidth, and fine resolution while requiring only a small chip area, the implemented wideband 2×13-bit all-digital *orthogonal* I/Q RFDAC is a very promising candidate for future multimode/multiband RF CMOS transmitters.

11.4 CONCLUSION

In this chapter, the high-resolution wideband 2×13-bit all-digital I/Q transmitter, which was introduced in Chapter 10, is thoroughly measured. First, the chip is tested in continuous-wave mode operation. It is demonstrated that, with a 1.3 V supply and, of course, an on-chip power combiner, the RFDAC chip generates more than 21 dBm RF output power within a frequency range of 1.36–2.51 GHz. The peak RF output power, overall system, and drain energy efficiencies of the modulator are 22.8 dBm, 34%, and 42%, respectively. The measured static noise floor is below −160 dBc/Hz.

Employing a simple calibration algorithm, the digital I/Q RF modulator demonstrates an IQ image rejection and LO leakage of -65 dBc and -68 dBc, respectively. The RFDAC could be linearized employing either of the two DPD approaches: memoryless polynomial or a 2×1D lookup table. Its linearity is examined utilizing single-tone as well as multitone 4/16/64/256/1024-QAM baseband signals while their related modulation bandwidth can be as high as 154 MHz. Using AM-AM/AM-PM DPD improves the linearity by more than 25 dB. Moreover, the static, memoryless constellation-mapping, lookup table DPD is applied to the RFDAC which improves linearity by more than 14 dB while the measured EVM is better than -28 dB. These numbers indicate that this innovative concept is a viable option for the next generations of multiband/multistandard transmitters. It can perform as an energy-efficient RFDAC in a stand-alone digital transmitter directly (e.g., for WLAN) or as a predriver for high-power basestation PAs.

CHAPTER 12

Future of RFDAC

CHAPTER OUTLINE

12.1 The Outcome 245
12.2 Some Suggestions for Future Developments 247
12.3 Future Trends 249

12.1 THE OUTCOME

With the proliferation of wireless networks, there is a need for more compact, low-cost, power-efficient transceivers that are capable of supporting the comprehensive communication standards including GSM, WLAN, Bluetooth, GNSS, FM, and 4G of 3GPP cellular. To accommodate the ever-increasing appeal of higher data throughputs within the crowded frequency spectrum of 500 MHz to 6 GHz, these communication standards utilize extremely efficient complex-modulated baseband signals in which their related bandwidths can be as wide as 160 MHz. Moreover, due to the extreme cost pressures of consumer electronics, migration to a more advanced nanometer-scale CMOS process, which was primarily developed for fast and low-power digital circuits operating at low supply voltages, is necessary. Additionally, this fast-evolving CMOS technology unfolds the possibility of integrating the entire radio including the DSP as well as the RF transmitter and receiver with just one single chip. Conventionally, the RF transmitter could not be implemented as a fully integrated circuit like the other building blocks of the corresponding RF transceiver could be, as it consists of the bulky DACs, filters, mixers, and the power amplifier (PA). As a consequence of the existing previously mentioned components, especially the PA, the RF transmitter is the most power-consuming element of the RF transceivers. This part affects EVM, out-of-band emission, and, most significantly, the battery lifetime of the mobile devices.

Consequently, the objective of this book is to implement a novel, fully integrated RF transmitter, which should be power-efficient and, additionally, must support multimode and multiband communication standards. To achieve these ambitious goals, this work has extensively utilized more and more digital circuitry rather than the traditional analog counterparts. Otherwise stated, this book attempts to benefit from the ongoing CMOS technology properties, which tend to drive the design of cellular and wireless building blocks toward the digital domain where transistors are employed as switches rather than the current sources.

The concise review of the different transmitter architectures has revealed that the efficiency of the polar modulator is higher than the Cartesian (I/Q) counterpart. However, due to the bandwidth expansion of the polar transmitter resulting from employing the nonlinear CORDIC algorithm to convert in-phase and quadrature-phase baseband data into the envelope and phase information, the I/Q modulator is the preferred architecture as it is able to address very wideband complex-modulated baseband signals, while achieving a reasonable power efficiency. Likewise, to simplify the design, gain more benefits from the nanometer CMOS process, and to substantially eliminate the analog circuitry, the digital I/Q architecture is selected rather than the analog counterpart.

The most straightforward digital I/Q modulator, however, suffers from nonorthogonal operation due to overlap between their related 50% duty cycle differential quadrature clocks [66]. The nonorthogonal summation degrades in-band performance such as EVM, and it also deteriorates out-of-band spectrum due to generated spectral regrowth. To improve the foregoing dynamic performance of this modulator, therefore, employing sophisticated digital predistortion (DPD) is inevitable [66]. On the other hand, this book has proposed a novel time-multiplexing orthogonal summation, which has utilized the nonoverlapping 25% duty cycle differential quadrature clocks. Furthermore, it has been demonstrated that the orthogonal summation can still be obtained even when employing differential quadrature clocks, which have a related duty cycle of less than 25%. In addition, in contrast to previous publications [60, 61], the orthogonal summation is performed by digital power amplifier (DPA) unit cells that increase the RF output power as well as improve the power efficiency. Thus, the orthogonal digital I/Q modulator represents a radio-frequency digital-to-analog converter (RFDAC). To validate the orthogonal summing operation, a 2×3-bit all-digital I/Q RFDAC has been designed and implemented in a TSMC 65-nm bulk CMOS process. According to measurement results, the composite RF signal is the result of vectorial summation of in-phase (I) and quadrature-phase (Q) signals. Hence, the orthogonal summation has been substantiated.

It has been indicated that the direct digital modulation system produces quantization noise together with spectral replicas due to zero-order-hold operation. As stated, the noise performance of the RFDAC is related to the resolution of the digital power mixer which is, indeed, a digital-to-RF-amplitude converter (DRAC). Thus, the DRAC executes the orthogonal summing operation, frequency upconverting, and, most importantly, RF power amplification. The DRAC outputs are connected to a passive power combiner that facilitates the transformation of the upconverted digital signals into a "high-power" continuous-time RF output in an energy efficient manner. It has been revealed that employing upconverting clocks with the duty cycle of 25% is a necessary, but not sufficient, condition for the orthogonal operation. Consequently, this book has proposed an innovative class-E-based power combiner that properly realizes the orthogonal summing operation in a power-efficient approach. The power-combining network comprises a transformer balun together with primary and secondary switch-capacitor tuners. To achieve an efficient high-power RF output, the I/Q RFDAC is pushed to operate in the saturated power region, thus it must be digitally predistorted (DPD) in order to restore its in-band and out-of-band dynamic performance.

In contrast to Ref. [66] utilizing a very straightforward constellation mapping DPD, the out-of-band nonlinearity is improved by almost 19 dB. The proposed constellation mapping DPD is based on one-dimensional (1D) mapping of two individual in-phase and quadrature-phase upsampled baseband signals. In particular, since the in-phase and quadrature-phase paths are orthogonal, the DPD does not require a two-dimensional (2D) exhaustive search of the whole constellation diagram, which is required in Ref. [66].

To verify the concept of the proposed high-resolution direct digital RF transmitter, a wideband 2 × 13-bit all-digital I/Q RFDAC has been implemented in TSMC 65-nm bulk CMOS process. The maximum power and its related drain efficiency are 22.8 dBm and 42%, respectively. The digital I/Q RF modulator shows an IQ image rejection and LO leakage of −65 dBc and −68 dBc, respectively. Its linearity is examined using single-carrier, complex-modulated 4/16/64/256/1024-QAM as well as multicarrier 256-QAM OFDM baseband signals, while their related modulation bandwidth is as high as 154 MHz. The measured EVM for a "single-carrier 22 MHz 64-QAM" signal is better than −28 dB. The measurement results confirm that the proposed solution is a logical alternative for the future generations of multiband/multistandard transmitters.

12.2 SOME SUGGESTIONS FOR FUTURE DEVELOPMENTS

Although this book has introduced a novel 2 × 13-bit all-digital I/Q RFDAC that is able to manage very wideband baseband signals, it continues to require further steps to fully comply with the ever-evolving communication standards. Due to the lack of experience and knowledge regarding the digital hardware IC design flow, the digital synthesis, and the limited amount of time, the digital baseband processing (DSP) in this book was conducted off-chip. As indicated in previous chapters, the processed baseband I/Q data were transferred and stored in two on-chip SRAM memories. Since the SRAM memories are synchronized with the entire RFDAC exploiting an on-chip $f_0/8$ clock, the subsequent spectral replicas resulting from zero-order-hold operation are located at $f_n = f_0 \pm n \times f_0/8$ (see Figs. 8.3 and 8.4 as well as Eqs. 8.2–(8.3)). As a result, the following solutions are recommended in order to suppress them further.

- The simplest manner is to increase the clock sampling rate (f_{CKR}) from $f_0/8$ to possibly f_0 [59]. As stated in Chapter 8, its primary disadvantage is greater

power consumption, which will be less significant as a finer node process such as 28 nm CMOS technology will be used. Note that additional advantages of adopting a higher sampling rate include the improvement of the related dynamic range and the ability to manage wider modulation bandwidths.
- Instead of employing zero-order-hold interpolation, a first-order-hold interpolator will be incorporated [59, 156, 157]. This interpolator provides a double-notch filter, that is, $\text{sinc}^4((f-f_0)/f_{CKR})$, at $f_n = f_0 \pm n \times f_{CKR}$. Thus, it reduces the amplitude of the related spectral images.
- As an ultimate solution to eliminate even more of the corresponding spectral images, the Farrow interpolator [158] can be used. This interpolator generates a triple notch, that is, $\text{sinc}^6((f-f_0)/f_{CKR})$, at all multiples of f_{CKR} [156].

Consequently, the entire DSP, including the digital interpolators and their corresponding finite impulse response (FIR) filters, should be implemented in combination with the I/Q RFDAC. In addition, as stated in Chapters 10 and 11, the linearity of the RFDAC is improved using constellation mapping DPD so as to suppress the AM-AM and AM-PM nonlinearities of the RFDAC in order to meet stringent communication standards. Thus, the DPD lookup table must be stored in a designated on-chip SRAM. Note that this lookup table must be positioned before the interpolator blocks in order to minimize the amount of power employed by the low-speed synchronizing clock. As a result of this arrangement, the DPD lookup table must comprise not only the nonlinearity profiles of the RFDAC but also the nonideality of its corresponding interpolators and FIR filters. Hence, the DPD operation will be moderately complicated.

Note that, throughout this book, the general premise was that the RFDAC is always loaded by an ideal and fixed 50 Ω (*VSWR=1*), which resembles the impedance of an ideal antenna. Nonetheless, in reality, the antenna's impedance can be anything but 50 Ω. Although based on the brief discussion in Chapter 7, Section 7.1.5, Fig. 7.9C, the RFDAC is able to tolerate a small deviation from 50 Ω, yet it can definitely not manage large *VSWR* as it affects the output power, efficiency, EVM, and, most importantly, its linearity. The moderate *VSWR* such as (1.5:1) can be accomplished by employing coarse tuning switch-capacitor banks. The disadvantage would be larger area, layout complexity, and, most importantly, increased power loss due to the related losses of coarse tuning switches. However, in the case of *VSWR* larger than (2:1), an additional antenna tuning circuitry must be employed and positioned at the output node of the RFDAC. Using this technique, various corresponding load reflection coefficients (Γ_L), which are related to the antenna's impedance, will be transferred to the origin of the Smith-chart ($\Gamma_L = 0$).

Furthermore, in the proposed digital I/Q transmitter, the corresponding higher harmonic frequency contents of the related LO signal should be diminished. For example, if f_0 is 2.4 GHz, the second and third harmonic rejection ratios of the current RFDAC chip would be approximately −42 and −20 dBc, respectively. The second harmonic frequency content is primarily due to employing an imperfect balance-to-unbalance (balun) structure. As stated in Chapter 9, Section 9.6, the imbalanced property of the transformer balun is principally due to higher interwinding capacitance. Moreover, the balun should be laid out symmetrically to improve differential-to-single-ended operation. To legitimize these recommendations, based on the simulation as well as measurement results of the 2 × 3-bit I/Q modulator in Chapter 7, and due to employing a more balanced transformer structure, the second and third harmonic rejection ratios are −51 and −31 dBc, respectively. Since the modern communication standards require greater than 60 dBc harmonic rejections, the third harmonic frequency must be further filtered out. Thus, the third harmonic frequency suppression is a difficult task and, perhaps, requires novel breakthrough ideas and additional efforts. Note that, since the objective of the I/Q RFDAC is to operate as a software-defined transmitter, the desired third harmonic rejection filter must be tunable. This fact makes the third harmonic frequency suppression even more sophisticated. An additional way to mitigate this issue, since the third harmonic frequency component of the RF output signal is originated from the LO clock pulse (square waveform), would be to employ an LO clock that only contains fundamental and second harmonic frequency as well as exploiting more linear switches. This would dramatically reduce the subsequent third harmonic frequency content of the RF output signal.

Also, as stated previously, it is worth mentioning that the modern communication standards such as Wi-Fi and 4G of 3GPP cellular utilize the complex-modulated signals with the relative high peak-to-average power ratio (PAPR) to achieve higher data throughputs within the restricted bandwidth. As a result, the RFDAC should be efficiently operated not only at its maximum power but also at its power back-off region. For example, wireless communication standards such as WLAN, WiMAX, and LTE exploit the OFDM modulation in which the crest factor could be as high as 20.6 dB. The current RFDAC cannot effectively manage such a high PAPR (see Fig. 11.2D). Over the last several decades, three predominant power amplifier architectures have been proposed to efficiently operate in the power back-off region. They are as follows: outphasing [53], Doherty [159], and envelope tracking [160]. The latter approach is only applied to the polar transmitters whereby the PA supply voltage is modulated according to the envelope information while the phase signal drives the saturated PA. Moreover, it is also possible to manipulate its related matching network in such a way that it increasingly compensates the related power losses in the power back-off region [161]. The all-digital I/Q RFDAC, however, can be upgraded using either the mixed-mode N-way Doherty configuration [151, 162] or the mixed-mode N-way outphasing [154, 163].

12.3 **FUTURE TRENDS**

As mentioned in Chapter 1 as well as in this chapter, RF transceivers are directed toward the implementation of a universal radio which manages numerous communication standards. As a result, only very recently, has the introduction of combo chips occurred in order to perform Wi-Fi/Bluetooth/FM or Bluetooth/GNSS/FM communication standards employing only one single chip. These solutions as well as our proposed I/Q RFDAC, which can also serve as an SDT chip, will be the game changer and will reduce the cost, size, and, most probably, the weight of mobile devices. In addition, arrays of these devices can be integrated on just one IC chip to realize heterogeneous radios whose modes of operation are defined by the users. Furthermore, these wideband I/Q RFDACs will accelerate the adaptation of so-called 5G Wi-Fi (802.11ac) or the upcoming communication standards such as the subterahertz wireless communication as well as the fifth-generation mobile network (5G) into wireless portable devices.

Appendix for the polar transmitter A

CHAPTER OUTLINE

A.1 EDGE Modulation 251
    A.1.1 Symbol Mapping and Rotation 252
    A.1.2 Pulse Shaping Filter and Modulation 255
A.2 RF System Specifications for the EDGE Transmitter 256
A.3 Details of the Simulation Model 259
    A.3.1 Digital Amplitude and Phase Data Generation 260
    A.3.2 RF Front-End Model 260

A.1 EDGE MODULATION

EDGE is part of 3GPP standardization and hence can be classified as part of the 3G wireless radio technology but it is commonly referred to as 2.5G or 2.75G transitional cellular technology. EDGE provides three times the data rate compared to GSM—a maximum of 384 kbits/s data throughput under optimum conditions, while maintaining backward compatibility with GSM by using the same Time Division Multiple Access (TDMA) scheme along with the same frequency bandwidth utilization to allow the same frequency planning [113, 164]. This backward compatibility allows the deployment of EDGE networks without any modifications to the GSM backbone network but requires transceiver related modifications in the base stations. The basic GSM standard specifies use of the Gaussian minimum-shift keying (GMSK) modulation while the EDGE standard specifies $3\pi/8$-shifted 8-phase shift keying (8PSK) at the same symbol rate of 270.833 ksymbols/s enabling three times higher data rate in EDGE. GMSK is a frequency/phase-only modulation type where the amplitude is kept constant, which allows the use of a higher efficiency saturated PA and also simplifies the design of the transceiver. In contrast, 8PSK modulation is a complex modulation requiring both phase/frequency and amplitude modulation.

The symbol rate for GSM and EDGE is the same—270.833 ksymbols/s, but EDGE uses 3 bits per symbol modulation of 8PSK in contrast with the 1 bit per symbol GMSK of GSM. Using 200 kHz of modulation bandwidth (equal to the channel spacing), this translates into the bandwidth efficiency of 1.35 bits/s/Hz for GSM and 4.06 bits/s/Hz for EDGE.

A.1.1 SYMBOL MAPPING AND ROTATION

Every set of three modulation bits are Gray encoded and mapped to 8PSK symbols as per Eq. (A.1) [164]. Fig. A.1 shows the constellation of these symbols. Each symbol is given by,

$$S_i = e^{j2\pi l/8} \qquad (A.1)$$

where l is given in Table A.1.

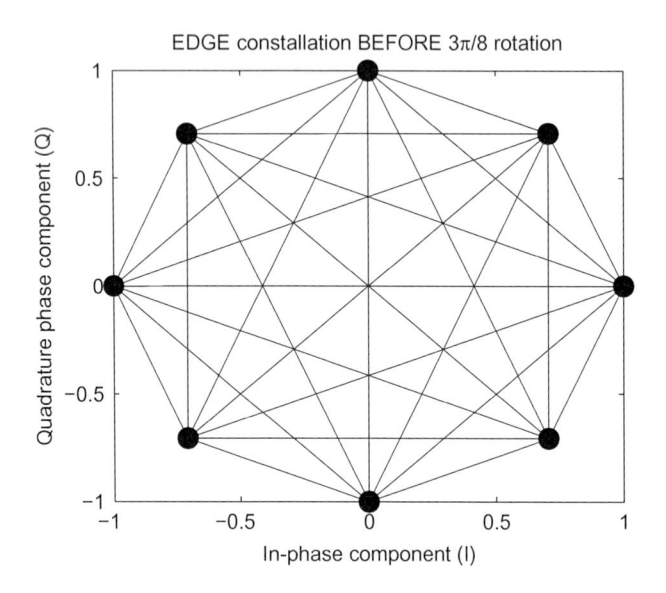

FIG. A.1

EDGE symbol mapper output represented by Eq. (A.1).

Table A.1 Mapping Between Modulating Bits and the 8PSK Symbol Parameter l [164]

Modulating Bits $d_{3i}, d_{3i+1}, d_{3i+2}$	Symbol Parameter l
(1,1,1)	0
(0,1,1)	1
(1,1,0)	2
(0,0,0)	3
(0,0,1)	4
(1,0,1)	5
(1,0,0)	6
(1,1,0)	7

A.1 EDGE modulation

After mapping, the symbols are rotated by $3\pi/8$ radians per symbol to avoid crossing through the origin. This stage of the modulation process ensures that the constellation does not have zero crossings and hence the spectrum is contained within 200 kHz modulation bandwidth making it more easily realizable in the power amplifier. It also results in bounded amplitude and phase modulation thus simplifying the transceiver and the power amplifier design. The spectrum of the phase signal alone is shown graphically in Fig. A.3. The zero crossing requires instantaneous change in the polarity of the signal, which requires infinite signal bandwidth. Thus Fig. A.3 shows that the spectrum without $3\pi/8$ has energy spread through the spectrum in contrast to the more contained spectrum resulting from $3\pi/8$ rotation. Similar comparison of the amplitude signal spectrum is shown in Fig. A.4. Although the combined spectrum of the complex signal with or without rotation is well contained, the individual amplitude/phase spectrum differences could result in higher level of design challenges of individual modules in the transmitter.

Even though the constellation in Fig. A.2 looks like a 16-PSK constellation, there are only 3 bits transmitted per symbol. And the receiver de-rotates first before making decisions on the received symbols. Thus the transmitter continues to function as 8-PSK constellation. The pulse shaping filter smoothes the transitions and adds extra data-points between these constellation points. It also creates finite amplitude modulation by controlling the rate of change of amplitude while going from one constellation point to another.

The rotation can be represented mathematically by Eq. (A.2), also graphically shown in Fig. A.2.

$$\hat{S}_i = S_i \cdot e^{ji3\pi/8} \quad (A.2)$$

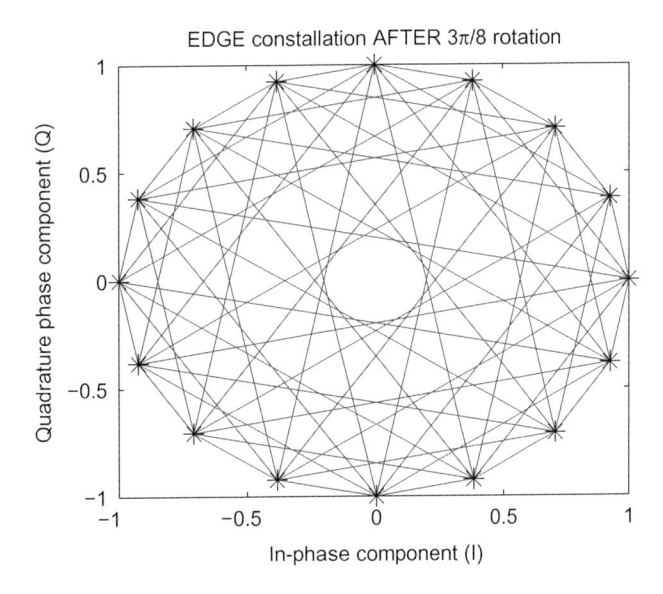

FIG. A.2

EDGE constellation represented by Eq. (A.2).

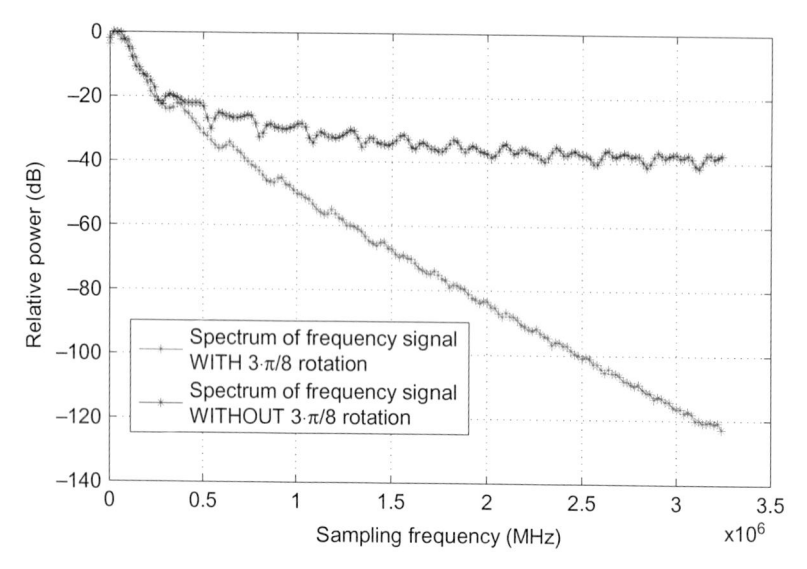

FIG. A.3

Comparison of frequency signal spectrum with and without $3\pi/8$ rotation.

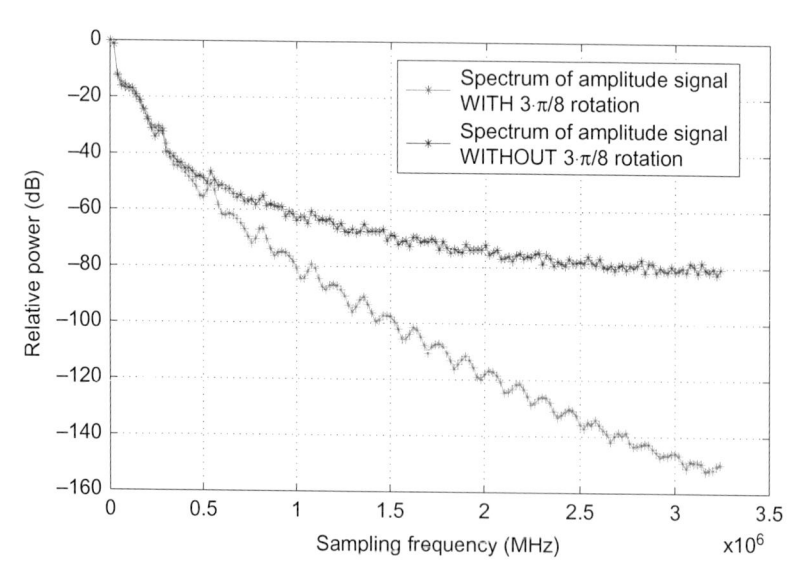

FIG. A.4

Comparison of amplitude signal spectrum with and without $3\pi/8$ rotation.

A.1.2 PULSE SHAPING FILTER AND MODULATION

A pulse shaping filter (PSF) is used to ensure that the transmit modulation spectrum is confined to a predetermined mask while causing insignificant degradation to the constellation. The measurement of the implemented modulation quality is defined by its error vector magnitude (EVM). The PSF provides the 8PSK spectrum with shaping that is almost identical to that of the GMSK spectrum. The impulse response of the Laurent pulse defined by the PSF is given by Eq. (A.3) [164].

$$c_0(t) = \begin{cases} \prod_{i=0}^{3} S(t+iT), & \text{for } 0 \leq t \leq 5T \\ 0, & \text{otherwise} \end{cases} \quad (A.3)$$

where T is the symbol duration given by

$$T = \frac{1}{270.833 \text{ kHz}}$$
$$= 3.6923 \text{ μs} \quad (A.4)$$

and,

$$S(t) = \begin{cases} \sin\left(\pi \int_0^t g(\alpha)d\alpha\right), & \text{for } 0 \leq t \leq 4T \\ \sin\left(\frac{\pi}{2} - \pi \int_0^{t-4T} g(\alpha)d\alpha\right), & \text{for } 4T < t \leq 8T \\ 0, & \text{otherwise} \end{cases}$$

$$g(t) = \frac{1}{2T}\left(Q\left(2\pi \cdot 0.3 \frac{t - \frac{5T}{2}}{T\sqrt{\ln(2)}}\right) - Q\left(2\pi \cdot 0.3 \frac{t - \frac{3T}{2}}{T\sqrt{\ln(2)}}\right)\right)$$

and,

$$Q(t) = \frac{1}{\sqrt{2\pi}} \int_t^{\infty} e^{-\frac{\tau^2}{2}} d\tau$$

The convolution of the impulse response with the individual symbols results in the baseband signal $y(t)$ as shown in Eq. (A.5).

$$y(t) = \sum_i \hat{S}_i \cdot c_0[t - iT + 2T] \quad (A.5)$$

This baseband signal $y(t)$ can be upconverted to RF carrier frequency by either Cartesian or polar scheme of modulation. Either way, the final RF modulation signal can be represented as $x(t)$ as given by Eq. (A.6).

$$x(t) = \sqrt{\frac{2E_s}{T}} Re[y(t) \cdot e^{j(2\pi f_0 t + \varphi_0)}] \quad (A.6)$$

where E_s is the energy per modulating symbol, f_0 is the carrier frequency, and φ_0 is a random phase, which is constant during a burst [164].

A.2 RF SYSTEM SPECIFICATIONS FOR THE EDGE TRANSMITTER

To achieve backward compatibility with the GSM RF system, the EDGE transmitter has to meet most of the stringent GSM specifications [113]. Critical RF system specifications at the antenna output are loosely divided into two categories: (a) near-carrier specification or close-in specification and (b) far-out spectral requirements. The modulation accuracy and frequency error-related European Telecommunications Standards Institute (ETSI) standardized EDGE parameters are summarized in Table A.2. An appropriate margin is added for individual modules to derive module-level specification for various blocks in the transmitter. The details for each are given in the corresponding subsections.

An output RF spectrum up to 6 MHz is typically considered near-carrier or close-in spectrum and the specifications are provided in dB relative to carrier power. These specifications are summarized in Fig. A.5. The shaded levels represent the fact that the spectral specifications change at these frequencies depending upon the output power control level. More details about the output power dependent specifications will be provided in the main text. Meeting the output RF spectrum mask given in Fig. A.5 through systems and circuits innovations is one of the key focus areas of this work.

Table A.2 Close-in Spectrum and Modulation Quality-Related ETSI RF Specifications for an EDGE Transmitter [113]

Mobile Specification	ETSI Requirement	Unit	Comments
RMS EVM	10	Percentage	The average distortion of the 8-PSK constellation based on Error Vector Magnitude (EVM) root mean square (RMS) value, calculated at the symbol rate.
Peak EVM	30	Percentage	Upper bound on the instantaneous constellation error, calculated at the symbol rate.
Frequency Error	40	Hertz	After synchronization between the mobile transmitter and base station receiver, frequency error represents the systematic signal processing error in the frequency, averaged over an EDGE burst.
Origin offset suppression	30	dB	Origin offset represents the accuracy of the constellation near zero point. It is strongly affected by the transmitter LO leakage (in Cartesian architectures), self-mixing and mismatches. For Polar transmitters, OOS is generally not a serious design challenge because the carrier is frequency modulated.

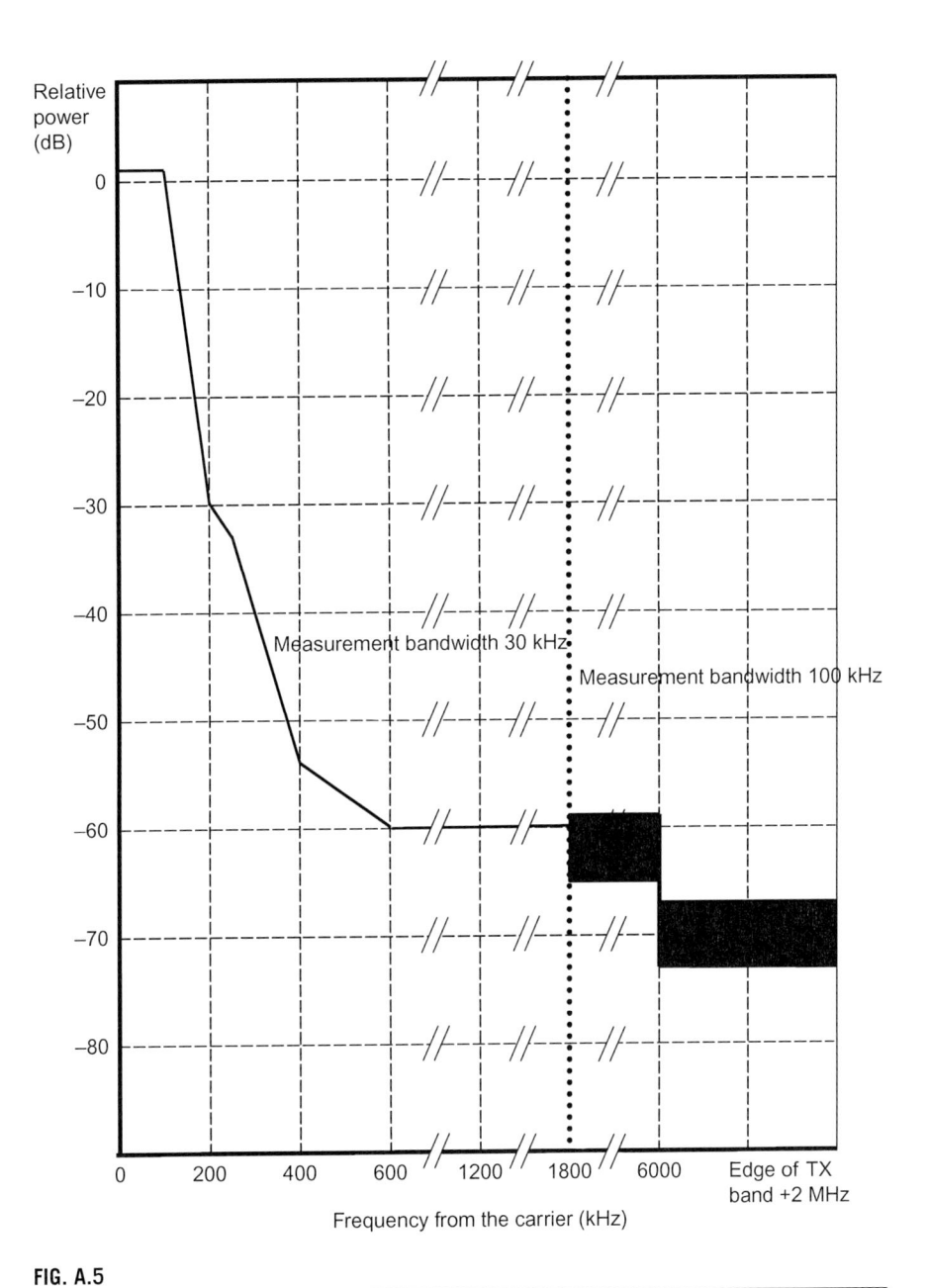

A.2 RF system specifications for the EDGE transmitter 257

FIG. A.5

EDGE RF spectrum at given frequency offset from carrier [113].

In addition to the above design parameters, switching transients and time masks are also typically considered part of the close-in parameters for a given design but they are beyond the scope of this research. More details about these parameters can be found in Ref. [113]. This work does, however, address the problem of power control over the required 30 dB dynamic range of EDGE modulation in Chapter 3. The power control requirements for one of the four bands (PCS 1900 band) are given in Table A.3. Only the power levels required to be supported by the presented transmitter are shown in the table. Similar specifications for additional frequency bands and power levels can be found in Ref. [113]. Transmitter SoC-level specifications targets ±1 dB power control to allow sufficient margin.

Far out spectrum and wide band noise (WBN) requirements typically focus on spectral emission within the adjacent receive band where they are most demanding, but also include other bands. For GSM/EDGE network frequency planning, typically the closest receive band starts from 20 MHz offset and hence, for most transceivers, 20 MHz offset spectral noise is one of the key specifications. A considerable part of this work is dedicated to designing the necessary system and circuit-level solutions required to meet the 20 MHz offset specification. The transmitter is required to operate in the four frequency bands given in Table A.4, which include bands used in Europe and America.

For Low Band (LB)—GSM 850 and GSM 900 frequency band—the spectral emissions have to be lower than −79 dBm when measured in 100 kHz bandwidth at 20 MHz offset. For case of High Band (HB)—DCS 1800 and DCS 1900 frequency

Table A.3 Power-Level Requirements for PCS 1900 Band in EDGE Modulation [113]

Power Control Level	Output Power (dBm)	Required Tolerance (dB)
0	30	±3
1	28	±3
2	26	±3
3	24	±3
4	22	±3
5	20	±3
6	18	±3
7	16	±3
8	14	±3
9	12	±4
10	10	±4
11	8	±4
12	6	±4
13	4	±4
14	2	±5
15	0	±5

Table A.4 Frequency Bands of Operation

Band Type	Frequency Band	Frequency Range (MHz)
Low Band (LB)	GSM 850	824.2–848.8
	GSM 900	880.2–914.8
High Band (HB)	DCS 1800	1710.2–1784.8
	PCS 1900	1850.2–1909.8

band—the spectral emissions have be to be lower than −71 dBm when measured in 100 kHz bandwidth at 20 MHz offset. These specifications have to be met over the power levels, as given in Table A.3. In addition, there are other WBN specifications given in Ref. [113] which are considered relatively easy to meet and are evaluated thoroughly as part of this research.

A.3 DETAILS OF THE SIMULATION MODEL

Matlab language [165] based model is used to accurately model the nonidealities of the transmitter. The model is used to design many critical design parameters in the proposed transmitter system. The structure of the model follows the transmitter block diagram shown in Fig. A.6. Some of the key features of the model are covered in this appendix.

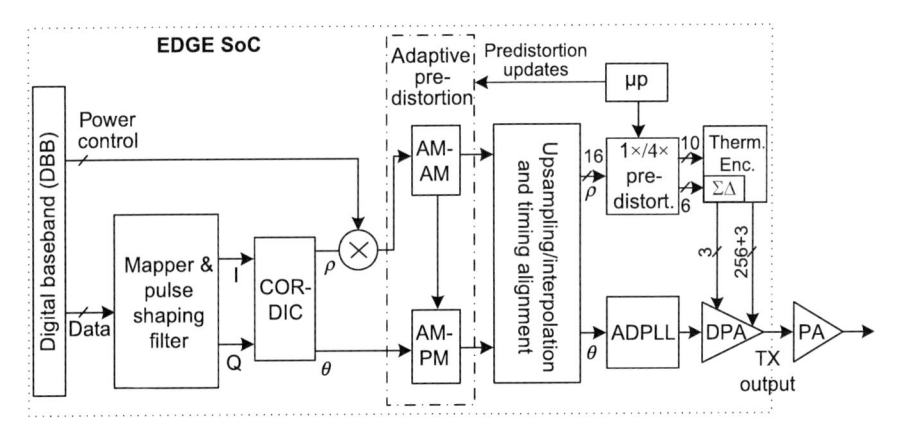

FIG. A.6

Structure of the simulation model for the EDGE Transmitter.

A.3.1 DIGITAL AMPLITUDE AND PHASE DATA GENERATION

Digital baseband is used to supply the binary data to the transmitter. This is modeled in Matlab language as a random data source generator with the capability of reproducing the same data using a fixed seed for the random number generator. The mapper and pulse shaping filter are represented by the fixed point implementation of the equations highlighted in Section A.1. The bit-widths are various points in the datapath are provisioned to be programmable to allow for fine tuning the design.

Similarly CORDIC and power control block are represented by their equivalent fixed point representation. The predistortion block is used in conjunction with the AM-AM and AM-PM distortion model covered later in this section. The model has the capability of incorporating quantization noise as well as the measurement noise for the distortion values stored in the LUTs. The fixed point effects are modeled appropriately with the distortion data stored in the LUTs as covered in Section 3.2. The use of dynamic inversion and adaptive interpolation is modeled accurately to facilitate the calculation of various quantization parameters highlighted in Section 3.2.3. Along the same lines, interpolative filter is modeled as per the block diagram shown in Fig. 3.22. The sampling rates are adjusted in the model based on the interpolative factors used at various points in the model. The clock signal is not directly modeled and hence the phase noise of the clock is modeled as an additive phenomenon at the final RF spectrum. These subsections of the model create digital amplitude and phase information sampled at CKV/16 frequency.

A.3.2 RF FRONT-END MODEL

The high-speed digital and analog components of the transmitter are modeled in this section. The digital sections are easier to model and the measured performance matches almost identically with the model performance. But the analog component modeling involves certain approximations, which create uncertainty between measured and simulation performance. With multiple iterations of such modeling, the correlation between the simulation model performance and measured performance was found to be within 1–2 dB of each other as shown in Fig. 5.1.

First-order phase modulation approximation is used for the ADPLL modeling. A separate model has been used to evaluate ADPLL in closed-loop operation but the runtime for the closed-loop ADPLL model is extremely high which makes the design difficult. Hence for most of the simulations, a basic phase modulator model is used which introduces minimal distortion to the phase signal. In contrast, the Σ-Δ and DPA models have been extremely detailed containing following components:

- DPA transistor mismatch between various transistors ($1\times$ fractional, $1\times$ integer, $4\times$)—both amplitude and phase mismatch (Section 4.4)
- DPA AM-AM and AM-PM response (Section 2.4.3)
- Clock skew effects in the DPA (Section 4.6)
- Data-clock misalignment in the DPA (Section 4.6)
- Coupling between fractional and integer transistors (Section 4.7)

- Σ-Δ modulator—first and second order (Section 4.2)
- 1×-4× predistortion (Section 4.5.1)
- Dynamic element matching (Section 4.5.3)

The final RF spectrum is obtained by combining amplitude and phase information signals at baseband. It is not possible to run the model directly at the RF carrier frequency because that would require more than 10 GHz oversampling frequency which will result in prohibitive simulation run time and computer memory consumption. The spectral measurements are performed at the desired offset from baseband (0 Hz) frequency. Measurements of EVM and close-in spectrum are performed in accordance with the ETSI specifications [113].

Appendix for I/Q RFDAC

CHAPTER OUTLINE

B.1 Universal Asynchronous Receiver/Transmitter 263
B.2 Matching Network Equations 263
B.3 AM-AM/AM-PM Relationship 266
B.4 DPD Bandwidth Expansion 267

B.1 UNIVERSAL ASYNCHRONOUS RECEIVER/TRANSMITTER

The Universal Asynchronous Receiver/Transmitter (UART) controller takes a byte of data and transmits the individual bits in a sequential style. In other words, the basic job of UART is to convert data from parallel to serial format for the transmission as well as converting data from serial to parallel format during reception. Fig B.1 depicts the block diagram of the implemented UART controller in combination with the two on-chip SRAM memories in the proposed 2×13-bit all-digital I/Q RFDAC. The upsampled baseband signals are generated in PC (MATLAB) and they are subsequently transferred to the chip via UART protocol. The I/Q data are stored into two on-chip SRAMs. Note that in order to preserve continuity of the baseband input data, two auto-loop memory readout units are exploited.

B.2 MATCHING NETWORK EQUATIONS

Based on Refs. [23, 143], $f(D)$ and $gt(D)$ in Chapter 9 are two functions that depend on D_1, φ, and g. For the sake of clarity, these parameters are redefined here. "D_1" is related to the duty cycle of the upconverting clock. In Raab's paper, it is defined as y and is located in Ref. [143, Fig. 2] and expressed as:

$$D_1 = \pi \times (1 - D) \qquad (B.1)$$

APPENDIX B Appendix for I/Q RFDAC

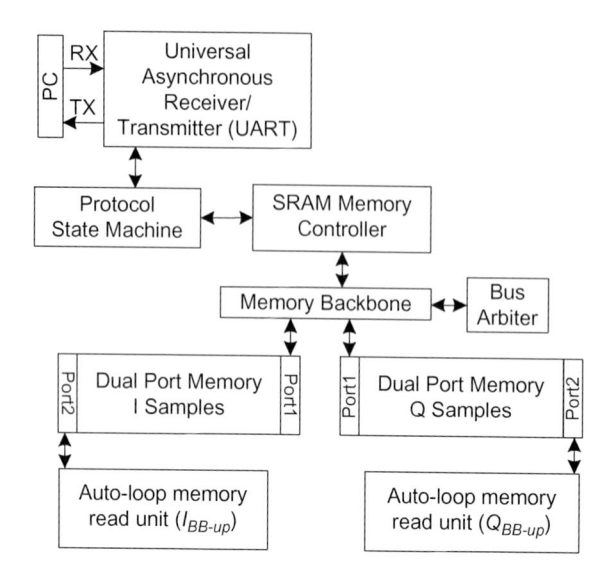

FIG. B.1

The schematic of the implemented UART controller along with on-chip SRAMs.

"φ" is the phase difference between the load output voltage and the drain voltage of the switching power amplifier. It is located in Ref. [143, Figs. 1 and 2, and eq. 3.8]

$$\varphi(D_1) = \tan^{-1}\left(\left(\frac{-1}{D_1}\right) + \cot(D_1)\right) \tag{B.2}$$

Note that here, ς, the slope of the waveform at the time of turn-on [143, eqs. 3.3 and 3.4], is considered zero. "$g(D_1)$" is a proportionality function, which relates the load output voltage (c) with the choke DC current (I_{DC}) and the output load resistor (R_{load}). It can be expressed as [143, eq. 2.19]:

$$g(D_1) = \frac{c}{I_{DC} \times R_{load}} \tag{B.3}$$

It is located in Ref. [143, eq. 3.9]:

$$g(D_1) = \frac{D_1}{\cos(\varphi) \times \sin(D_1)} \tag{B.4}$$

Based on Ref. [143, eq. 3.11], the capacitor's susceptance is defined as:

$$B(D_1) = \frac{f(D_1)}{R_{load}} \tag{B.5}$$

Consequently, $f(D)$ can be expressed as:

$$f(D) = f(D_1) = \frac{2}{\pi g^2} \times \left(D_1^2 + D_1 g \sin(\varphi - D_1) - g \sin(\varphi) \sin(D_1)\right) \quad \text{(B.6)}$$

Moreover, $gt(D)$ is a proportionality function that relates the reactance of the L_{add} inductance ("jX" in Ref. [143, Fig. 1]) with the output load resistor (R_{load}), and it is defined in Ref. [143, eq. 3.28]:

$$X(D_1) = gt(D_1) \times R_{load} \quad \text{(B.7)}$$

According to Ref. [143, eqs. 3.13–3.15 and 3.18–3.19] "$gt(D)$" is expressed as:

$$gt(D) = \frac{q_n(D)}{q_d(D)} = \frac{q_n(D_1)}{q_d(D_1)} \quad \text{(B.8)}$$

where q_n and q_d are determined in Ref. [143, eq. 3.18]. Therefore, for $D = 25\%$, $g(D), f(D)$, and $gt(D)$ are as follows:

$$g(D = 25\%) = 5.79925 \quad \text{(B.9)}$$
$$f(D = 25\%) = 0.21322 \quad \text{(B.10)}$$
$$gt(D = 25\%) = 3.5619 \quad \text{(B.11)}$$

Moreover, DET_2 and DET_1, which are employed in Eqs. (9.30)–(9.32) are defined to simplify the equations, and these two parameters also depend on B_{Cs}, X_{Ladd}, Z_{RX}, G_{sw}, and R_L:

$$DET_2 = 0.0625 + 4 \times \left(\frac{B_{Cs}}{G_{sw}}\right)^2 \quad \text{(B.12)}$$

$$DET_1 = \left(\frac{1}{4} + \frac{R_L}{|Z_{RX}|^2 \times G_{sw}}\right)^2 - \left(\frac{1}{4} + \frac{R_L}{|Z_{RX}|^2 \times G_{sw}}\right) \times \left(\frac{0.028}{DET_2}\right) + \frac{0.0027}{DET_2}$$
$$+ \left(\frac{0.011}{DET_2}\right)^2 - 0.02533 + \left\{\left(\frac{B_{Cs}}{G_{sw}}\right) \times \left(1 + \left(\frac{0.09}{DET_2}\right)\right) - \frac{X_{Ladd}}{|Z_{RX}|^2 \times G_{sw}}\right\}^2 \quad \text{(B.13)}$$

Consequently, H, K, and T can be expressed as:

$$H = \left(-2 + \frac{0.106}{DET_2}\right) \times \left(0.0908 - \left(\frac{0.0056}{DET_2}\right) + \frac{R_L}{|Z_{RX}|^2 \times G_{sw}}\right) - \left(\frac{0.4244}{DET_2}\right)$$
$$\times \left(\frac{B_{Cs}}{G_{sw}}\right) \times \left(\left(\frac{B_{Cs}}{G_{sw}}\right) \times \left(1 + \left(\frac{0.09}{DET_2}\right)\right) - \frac{X_{Ladd}}{|Z_{RX}|^2 \times G_{sw}}\right) \quad \text{(B.14)}$$

$$K = -(V_{DD} \times 0.0796) - a_1 \times (0.075) - b_1 \times (0.3) \times \left(\frac{B_{Cs}}{G_{sw}}\right) \quad \text{(B.15)}$$

$$T = \left(\frac{-0.4244}{DET_2}\right) \times \left(\frac{B_{Cs}}{G_{sw}}\right) \times \left(0.41 - \left(\frac{0.0225}{DET_2}\right) + \frac{R_L}{|Z_{RX}|^2 \times G_{sw}}\right)$$
$$- \left(-2 + \frac{0.106}{DET_2}\right) \times \left(\left(\frac{B_{Cs}}{G_{sw}}\right) \times \left(1 + \left(\frac{0.09}{DET_2}\right)\right) - \frac{X_{Ladd}}{|Z_{RX}|^2 \times G_{sw}}\right) \quad (B.16)$$

B.3 AM-AM/AM-PM RELATIONSHIP

For a memoryless nonlinear power amplifier described in Chapter 10, the input-output relationship can be expressed with a polynomial approximation:

$$V_{out} = f(V_{in})$$
$$= a_1 V_{in} + a_2 V_{in}^2 + a_3 V_{in}^3 + a_4 V_{in}^4 + \cdots \quad (B.17)$$

In the two-tone test scenario, the input signal is:

$$V_{in} = \cos\omega_1 t + \cos\omega_2 t \quad (B.18)$$

Applying Eq. (B.18) as an input signal in Eq. (B.17), the following expression is obtained:

$$V_{out} = f(\cos\omega_1 t + \cos\omega_2 t)$$
$$= a_1 (\cos\omega_1 t + \cos\omega_2 t) + a_2 (\cos\omega_1 t + \cos\omega_2 t)^2 + a_3 (\cos\omega_1 t + \cos\omega_2 t)^3 + \cdots$$
$$= \left(a_1 + \frac{9}{4} a_3\right) (\cos\omega_1 t + \cos\omega_2 t) + \frac{3}{4} a_3 (\cos(2\omega_1 - \omega_2)t + \cos(2\omega_2 - \omega_1)t) + \cdots$$
$$= \mu_1 V(f_m) + \mu_3 V(3f_m) + \cdots \quad (B.19)$$

where $\mu_3 = \frac{3}{4} a_3$ is the IM_3 product, and

$$V(f_m) = V_{in} = \cos\omega_1 t + \cos\omega_2 t \quad (B.20)$$

is the main two-tone signal and

$$V(3f_m) = \cos(2\omega_1 - \omega_2)t + \cos(2\omega_2 - \omega_1)t \quad (B.21)$$

is the third-order intermodulation signal. The foregoing expressions were obtained assuming only amplitude modulation. In fact, Eq. (B.19) indicates the AM-AM relationship between the input and the output signal. If the input is a phase-modulated signal, however, the effect of AM-PM must be considered. If the output phase can simply be approximated as [22]:

$$\Phi_{out} = \frac{\Phi}{2}(1 + \cos 2\omega_m t) \quad (B.22)$$

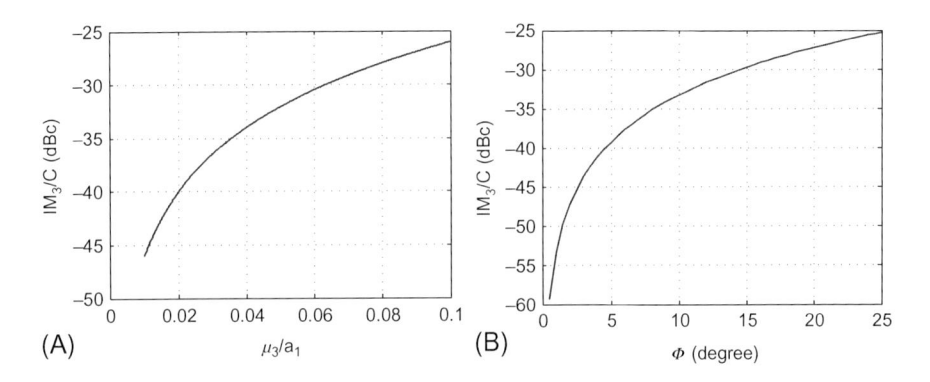

FIG. B.2

IM_3/C versus (A) μ_3 (B) Φ.

where Φ is the maximum phase deviation of the phase profile (Eq. B.19). The overall RF output can be simply modeled as:

$$V_{out} = (AM - AM)\exp(j(\omega_0 + AM - PM))$$
$$= (\mu_1 V(f_m) + \mu_3 V(3f_m) + \cdots)\exp\left(j\left(\omega_0 + \frac{\Phi}{2}(1 + \cos 2\omega_m t)\right)\right) \quad (B.23)$$

Now, the real part of IM_3 could be simply expressed as follows [22]:

$$v_{im3} = \mu_3 \cos(3\omega_m t) \times \cos\left\{\omega t + \frac{\Phi}{2}(1 + \cos 2\omega_m t)\right\}$$
$$\cong \frac{\mu_3}{2}\cos\left\{((\omega_0 \pm 3\omega_m) \times t) - \frac{\Phi}{2}\right\} + \frac{\Phi}{8}\sin((\omega_0 \pm 3\omega_m) \times t) \quad (B.24)$$

where $\omega_m = 2\pi f_m$ is baseband natural frequency. Based on Eq. (B.24), the AM-AM and AM-PM effects are almost orthogonal and could not cancel each other [22]. Fig. B.2 illustrates the IM_3-to-carrier (IM_3/C) based on envelope and phase variations. According to Fig. B.2 and Eq. (B.24), if $\mu_3 = 0.04$, then it produces the IM_3/C of -33 dBc. Furthermore, for 1 and 10 degree of phase deviation, the generated IM_3/C would be -53 dBc and -33 dBc, respectively.

B.4 DPD BANDWIDTH EXPANSION

Note that, as was demonstrated in Section 11.3.2, DPD deteriorates higher-order odd intermodulation products. Otherwise stated, it entails a bandwidth expansion and it is, indeed, common for DPD operations. The simple explanation is as follows. Suppose the nonlinearity profile of a PA or RFDAC can be expressed as:

$$Y_{out} = X_{in} + X_{in}^2 + X_{in}^3 \quad (B.25)$$

As a result, the predistortion profiles would contain $\sqrt{X}$ term in which its related Taylor series can be approximately expressed as follows:

$$Y_{out} = \left(\frac{1}{2}\right) X_{in} - \left(\frac{1}{8}\right) X_{in}^2 + \left(\frac{1}{16}\right) X_{in}^3 \tag{B.26}$$

Replacing Eq. (B.26) into Eq. (B.25) leads to the generation of up to sixth harmonic. Thus, DPD causes at least twofold bandwidth expansion.

References

[1] Wireless LAN Medium Access Control (MAC) and Physical Layer (PHY) specifications, 2012, http://standards.ieee.org/getieee802/download/802.11-2012.pdf (online).

[2] Wireless Medium Access Control (MAC) and Physical Layer (PHY) specifications for Wireless Personal Area Networks (WPANs), 2005, http://ieeexplore.ieee.org/stamp/stamp.jsp?tp=&arnumber=1490827 (online).

[3] User Equipment (UE) conformance specification, 3GPP TS 34.121-1 V8.7.0, Tech. Specification Group Radio Access Network, 2009.

[4] K. Fazel, S. Kaiser, Multi-Carrier and Spread Spectrum Systems: From OFDM and MC-CDMA to LTE and WiMAX, Wiley, New York, 2008, http://books.google.nl/books?id=RzUJsPqe-dgC (online).

[5] M. Ergen, Mobile Broadband: Including WiMAX and LTE, Springer, Berlin, 2009, http://books.google.nl/books?id=v1nItLFLgx0C (online).

[6] B. Razavi, RF Microelectronics, Prentice Hall PTR, Upper Saddle River, NJ, 1998, http://books.google.nl/books?id=TQZTAAAAMAAJ (online).

[7] T.H. Lee, The Design of CMOS Radio-Frequency Integrated Circuits, Cambridge University Press, Cambridge, 1998, http://books.google.nl/books?id=zLeWQgAACAAJ (online).

[8] I. Aoki, S.D. Kee, D.B. Rutledge, A. Hajimiri, Distributed active transformer—a new power-combining and impedance-transformation technique, IEEE Trans. Microwave Theory Tech. 50 (1) (2002) 316–331.

[9] I. Aoki, S. Kee, R. Magoon, R. Aparicio, F. Bohn, J. Zachan, G. Hatcher, D. McClymont, A. Hajimiri, A fully-integrated quad-band GSM/GPRS CMOS power amplifier, IEEE J. Solid-State Circuits 43 (12) (2008) 2747–2758.

[10] I. Aoki, S.D. Kee, D.B. Rutledge, A. Hajimiri, Fully integrated CMOS power amplifier design using the distributed active-transformer architecture, IEEE J. Solid-State Circuits 37 (3) (2002) 371–383.

[11] R.B. Staszewski, P.T. Balsara, All-Digital Frequency Synthesizer in Deep-Submicron CMOS, Wiley, New York, 2006, http://books.google.nl/books?id=2VHFD-7LgAwC.

[12] G. Hueber, R.B. Staszewski, Multi-Mode/Multi-Band RF Transceivers for Wireless Communications: Advanced Techniques, Architectures, and Trends, Wiley, New York, 2011, http://books.google.nl/books?id=kc1eCPzuMbgC.

[13] R.B. Staszewski, K. Muhammad, D. Leipold, C.-M. Hung, Y.-C. Ho, J.L. Wallberg, C. Fernando, K. Maggio, R. Staszewski, T. Jung, J. Koh, S. John, I.Y. Deng, V. Sarda, O. Moreira-Tamayo, V. Mayega, R. Katz, O. Friedman, O.E. Eliezer, E. de Obaldia, P.T. Balsara, All-digital TX frequency synthesizer and discrete-time receiver for Bluetooth radio in 130-nm CMOS, IEEE J. Solid-State Circuits 39 (12) (2004) 2278–2291.

[14] R.B. Staszewski, J.L. Wallberg, S. Rezeq, C.-M. Hung, O.E. Eliezer, S.K. Vemulapalli, C. Fernando, K. Maggio, R. Staszewski, N. Barton, M.-C. Lee, P. Cruise, M. Entezari, K. Muhammad, D. Leipold, All-digital PLL and transmitter for mobile phones, IEEE J. Solid-State Circuits 40 (12) (2005) 2469–2482.

[15] S. Kousai, A. Hajimiri, An octave-range, watt-level, fully-integrated CMOS switching power mixer array for linearization and back-off-efficiency improvement, IEEE J. Solid-State Circuits 44 (12) (2009) 3376–3392.

[16] P. Eloranta, P. Seppinen, Direct-digital RF modulator IC in 0.13 μm CMOS for wideband multi-radio applications, in: IEEE International Solid-State Circuits Conference Digest of Technical Papers (ISSCC), vol. 1, 2005, pp. 532–615.

[17] A. Kavousian, D.K. Su, M. Hekmat, A. Shirvani, B. Wooley, A digitally modulated polar CMOS power amplifier with a 20-MHz channel bandwidth, IEEE J. Solid-State Circuits 43 (10) (2008) 2251–2258.

[18] S.-M. Yoo, J.S. Walling, E.-C. Woo, B. Jann, D.J. Allstot, A switched-capacitor RF power amplifier, IEEE J. Solid-State Circuits 46 (12) (2011) 2977–2987.

[19] D. Chowdhury, L. Ye, E. Alon, A.M. Niknejad, An efficient mixed-signal 2.4-GHz polar power amplifier in 65-nm CMOS technology, IEEE J. Solid-State Circuits 46 (8) (2011) 1796–1809.

[20] S. Zheng, H.C. Luong, A CMOS WCDMA/WLAN digital polar transmitter with AM replica feedback linearization, IEEE J. Solid-State Circuits 48 (7) (2013) 1701–1709.

[21] L. Ye, J. Chen, L. Kong, E. Alon, A.M. Niknejad, Design considerations for a direct digitally modulated WLAN transmitter with integrated phase path and dynamic impedance modulation, IEEE J. Solid-State Circuits 48 (12) (2013) 3160–3177.

[22] S.C. Cripps, RF Power Amplifiers For Wireless Communications, second ed., Artech House, Incorporated, 2006, http://books.google.nl/books?id=JklyQgAACAAJ.

[23] A. Grebennikov, N.O. Sokal, M.J. Franco, Switchmode RF Power Amplifiers, Elsevier Science, Amsterdam, 2011, http://books.google.nl/books?id=kulWnlH4e4AC.

[24] B. Razavi, RF Microelectronics, Prentice-Hall PTR, Englewood Cliffs, NJ, 2011, http://books.google.nl/books?id=_TccKQEACAAJ&hl.

[25] T.H. Lee, The Design of CMOS Radio-Frequency Integrated Circuits, Cambridge University Press, 2004, http://books.google.nl/books?id=io1hL48OqBsC.

[26] R.B. Staszewski, M.S. Alavi, Digital I/Q RF transmitter using time-division duplexing, in: Proceedings of IEEE Radio-Frequency Integration Technology (RFIT) Symposium, 2011, pp. 165–168.

[27] A.A. Abidi, RF CMOS comes of age, IEEE J. Solid-State Circuits 39 (4) (2004) 549–561.

[28] M. Zargari, D.K. Su, C.P. Yue, S. Rabii, D. Weber, B.J. Kaczynski, S.S. Mehta, K. Singh, S. Mendis, B.A. Wooley, A 5-GHz CMOS transceiver for IEEE 802.11a wireless LAN systems, IEEE J. Solid-State Circuits 37 (12) (2002) 1688–1694.

[29] A.R. Behzad, Z.M. Shi, S.B. Anand, Li Lin, K.A. Carter, M.S. Kappes, T.-H. Lin, T. Nguyen, D. Yuan, S. Wu, Y.C. Wong, V. Fong, A. Rofougaran, A 5-GHz direct-conversion CMOS transceiver utilizing automatic frequency control for the IEEE 802.11a wireless LAN standard, IEEE J. Solid-State Circuits 38 (12) (2003) 2209–2220.

[30] M. Zargari, M. Terrovitis, S.H.-M. Jen, B.J. Kaczynski, M. Lee, M.P. Mack, S.S. Mehta, S. Mendis, K. Onodera, H. Samavati, W.W. Si, K. Singh, A. Tabatabaei, D. Weber, D.K. Su, B.A. Wooley, A single-chip dual-band tri-mode CMOS transceiver for IEEE 802.11a/b/g wireless LAN, IEEE J. Solid-State Circuits 39 (12) (2004) 2239–2249.

[31] M. Zargari, S. Mehta, D. Su, Challenges in the design of CMOS transceivers for the IEEE 802.11 wireless LANs: past, present and future, in: International Conference On ASIC, vol. 1, 2005, pp. 1153–1156.

[32] A. Behzad, K.A. Carter, H.-M. Chien, S. Wu, M.-A. Pan, C.P. Lee, Q. Li, J.C. Leete, S. Au, M.S. Kappes, Z. Zhou, D. Ojo, L. Zhang, A. Zolfaghari, J. Castanada, H. Darabi, B. Yeung, A. Rofougaran, M. Rofougaran, J. Trachewsky, T. Moorti,

R. Gaikwad, A. Bagchi, J.S. Hammerschmidt, J. Pattin, J.J. Rael, B. Marholev, A fully integrated MIMO multiband direct conversion CMOS transceiver for WLAN applications (802.11n), IEEE J. Solid-State Circuits 42 (12) (2007) 2795–2808.

[33] A. Afsahi, J.J. Rael, A. Behzad, H.-M. Chien, M. Pan, S. Au, A. Ojo, C.P. Lee, S.B. Anand, K. Chien, S. Wu, R. Roufoogaran, A. Zolfaghari, J.C. Leete, L. Tran, K.A. Carter, M. Nariman, K.W.-K. Yeung, W. Morton, M. Gonikberg, M. Seth, M. Forbes, J. Pattin, L. Gutierrez, S. Ranganathan, N. Li, E. Blecker, J. Lin, T. Kwan, R. Zhu, M. Chambers, M. Rofougaran, A. Rofougaran, J. Trachewsky, P. Van Rooyen, A low-power single-weight-combiner 802.11abg SoC in 0.13 μm CMOS for embedded applications utilizing an area and power efficient cartesian phase shifter and mixer circuit, IEEE J. Solid-State Circuits 43 (5) (2008) 1101–1118.

[34] M. Zargari, L.Y. Nathawad, H. Samavati, S.S. Mehta, A. Kheirkhahi, P. Chen, K. Gong, B. Vakili-Amini, J. Hwang, S.-W.M. Chen, M. Terrovitis, B.J. Kaczynski, S. Limotyrakis, M.P. Mack, H. Gan, M. Lee, R.T. Chang, H. Dogan, S. Abdollahi-Alibeik, B. Baytekin, K. Onodera, S. Mendis, A. Chang, Y. Rajavi, S.H.-M. Jen, D.K. Su, B.A. Wooley, A dual-band CMOS MIMO radio SoC for IEEE 802.11n wireless LAN, IEEE J. Solid-State Circuits 43 (12) (2008) 2882–2895.

[35] A. Matsuzawa, Digital-centric RF CMOS technology, in: IEEE International Workshop on Radio-Frequency Integration Technology (RFIT), 2007, pp. 122–126.

[36] P. Nagle, P. Burton, E. Heaney, F. McGrath, A wide-band linear amplitude modulator for polar transmitters based on the concept of interleaving delta modulation, IEEE J. Solid-State Circuits 37 (12) (2002) 1748–1756.

[37] E. McCune, W. Sander, EDGE transmitter alternative using nonlinear polar modulation, in: Proceedings of International Symposium on Circuits and Systems (ISCAS), vol. 3, 2003, pp. 594–597.

[38] W.B. Sander, S.V. Schell, B.L. Sander, Polar modulator for multi-mode cell phones, in: Custom Integrated Circuits Conference, 2003. Proceedings of the IEEE 2003, 2003, pp. 439–445.

[39] T. Sowlati, D. Rozenblit, E. MacCarthy, M. Damgaard, R. Pullela, D. Koh, D. Ripley, Quad-band GSM/GPRS/EDGE polar loop transmitter, in: IEEE International Solid-State Circuits Conference Digest of Technical Papers (ISSCC), vol. 1, 2004, pp. 186–521.

[40] M.R. Elliott, T. Montalvo, B.P. Jeffries, F. Murden, J. Strange, A. Hill, S. Nandipaku, J. Harrebek, A polar modulator transmitter for GSM/EDGE, IEEE J. Solid-State Circuits 39 (12) (2004) 2190–2199.

[41] P. Reynaert, M. Steyaert, A 1.75 GHz GSM/EDGE polar modulated CMOS RF power amplifier, in: IEEE International Solid-State Circuits Conference Digest of Technical Papers (ISSCC), vol. 1, 2005, pp. 312–600.

[42] A.W. Hietala, A quad-band 8PSK/GMSK polar transceiver, IEEE J. Solid-State Circuits 41 (5) (2006) 1133–1141.

[43] M. Youssef, A. Zolfaghari, B. Mohammadi, H. Darabi, A.A. Abidi, A low-power GSM/EDGE/WCDMA polar transmitter in 65-nm CMOS, IEEE J. Solid-State Circuits 46 (12) (2011) 3061–3074.

[44] Y. Huang, J.H. Mikkelsen, T. Larsen, Investigation of polar transmitters for WCDMA handset applications, in: Norchip Conference, 2006, pp. 155–158.

[45] S. Akhtar, P. Litmanen, M. Ipek, J. Lin, S. Pennisi, F.-J. Huang, R.B. Staszewski, Analog path for triple band WCDMA polar modulated transmitter in 90 nm CMOS,

in: Proceedings of IEEE Radio Frequency Integrated Circuits (RFIC) Symposium, 2007, pp. 185–188.

[46] R.B. Staszewski, D. Leipold, O. Eliezer, M. Entezari, K. Muhammad, I. Bashir, C.-M. Hung, J. Wallberg, R. Staszewski, P. Cruise, S. Rezeq, S. Vemulapalli, K. Waheed, N. Barton, M.-C. Lee, C. Fernando, K. Maggio, T. Jung, I. Elahi, S. Larson, T. Murphy, G. Feygin, I. Deng, T. Mayhugh, Y.-C. Ho, K.-M. Low, C. Lin, J. Jaehnig, J. Kerr, J. Mehta, S. Glock, T. Almholt, S. Bhatara, A 24 mm^2 quad-band single-chip GSM radio with transmitter calibration in 90 nm digital CMOS, in: IEEE International Solid-State Circuits Conference Digest of Technical Papers (ISSCC), 2008, pp. 208–607.

[47] J. Mehta, R.B. Staszewski, O. Eliezer, S. Rezeq, K. Waheed, M. Entezari, G. Feygin, S. Vemulapalli, V. Zoicas, C.-M. Hung, N. Barton, I. Bashir, K. Maggio, M. Frechette, M.-C. Lee, J. Wallberg, P. Cruise, N. Yanduru, A 0.8 mm^2 all-digital SAW-less polar transmitter in 65 nm EDGE SoC, in: IEEE International Solid-State Circuits Conference Digest of Technical Papers (ISSCC), 2010, pp. 58–59.

[48] D. Chowdhury, S.V. Thyagarajan, L. Ye, E. Alon, A.M. Niknejad, A fully-integrated efficient CMOS inverse class-D power amplifier for digital polar transmitters, IEEE J. Solid-State Circuits 47 (5) (2012) 1113–1122.

[49] L. Ye, J. Chen, L. Kong, P. Cathelin, E. Alon, A.M. Niknejad, A digitally modulated 2.4 GHz WLAN transmitter with integrated phase path and dynamic load modulation in 65 nm CMOS, in: IEEE International Solid-State Circuits Conference Digest of Technical Papers (ISSCC), 2013, pp. 330–331.

[50] P. Cruise, C.-M. Hung, R.B. Staszewski, O. Eliezer, S. Rezeq, K. Maggio, D. Leipold, A digital-to-RF-amplitude converter for GSM/GPRS/EDGE in 90-nm digital CMOS, in: Proceedings of IEEE Radio Frequency Integrated Circuits (RFIC) Symposium, 2005, pp. 21–24.

[51] A. Ravi, P. Madoglio, H. Xu, K. Chandrashekar, M. Verhelst, S. Pellerano, L. Cuellar, M. Aguirre-Hernandez, M. Sajadieh, J.E. Zarate-Roldan, O. Bochobza-Degani, H. Lakdawala, Y. Palaskas, A 2.4-GHz 20-40-MHz channel WLAN digital outphasing transmitter utilizing a delay-based wideband phase modulator in 32-nm CMOS, IEEE J. Solid-State Circuits 47 (12) (2012) 3184–3196.

[52] S.-M. Yoo, J.S. Walling, O. Degani, B. Jann, R. Sadhwani, J.C. Rudell, D.J. Allstot, A class-G switched-capacitor RF power amplifier, IEEE J. Solid-State Circuits 48 (5) (2013) 1212–1224.

[53] H. Chireix, High power outphasing modulation, Proc. Inst. Radio Eng. 23 (11) (1935) 1370–1392.

[54] P.A. Godoy, S. Chung, T.W. Barton, D.J. Perreault, J.L. Dawson, A 2.4-GHz, 27-dBm asymmetric multilevel outphasing power amplifier in 65-nm CMOS, IEEE J. Solid-State Circuits 47 (10) (2012) 2372–2384.

[55] H. Xu, Y. Palaskas, A. Ravi, M. Sajadieh, M.A. El-Tanani, K. Soumyanath, A flip-chip-packaged 25.3 dBm class-D outphasing power amplifier in 32 nm CMOS for WLAN application, IEEE J. Solid-State Circuits 46 (7) (2011) 1596–1605.

[56] P. Madoglio, A. Ravi, H. Xu, K. Chandrashekar, M. Verhelst, S. Pellerano, L. Cuellar, M. Aguirre, M. Sajadieh, O. Degani, H. Lakdawala, Y. Palaskas, A 20 dBm 2.4 GHz digital outphasing transmitter for WLAN application in 32 nm CMOS, in: IEEE International Solid-State Circuits Conference Digest of Technical Papers (ISSCC), 2012, pp. 168–170.

[57] A. Jerng, C.G. Sodini, A wideband $\Delta\Sigma$ digital-RF modulator for high data rate transmitters, IEEE J. Solid-State Circuits 42 (8) (2007) 1710–1722.

[58] P. Eloranta, P. Seppinen, S. Kallioinen, T. Saarela, A. Parssinen, A multimode transmitter in 0.13 μm CMOS using direct-digital RF modulator, IEEE J. Solid-State Circuits 42 (12) (2007) 2774–2784.

[59] A. Pozsgay, T. Zounes, R. Hossain, M. Boulemnakher, V. Knopik, S. Grange, A fully digital 65 nm CMOS transmitter for the 2.4-to-2.7 GHz WiFi/WiMAX bands using 5.4 GHz $\Delta\Sigma$ RF DACs, in: IEEE International Solid-State Circuits Conference Digest of Technical Papers (ISSCC), 2008, pp. 360–619.

[60] X. He, J. van Sinderen, A low-power, low-EVM, SAW-less WCDMA transmitter using direct quadrature voltage modulation, IEEE J. Solid-State Circuits 44 (12) (2009) 3448–3458.

[61] X. He, J. van Sinderen, R. Rutten, A 45 nm WCDMA transmitter using direct quadrature voltage modulator with high oversampling digital front-end, in: IEEE International Solid-State Circuits Conference Digest of Technical Papers (ISSCC), 2010, pp. 62–63.

[62] T.W. Barton, S. Chung, P.A. Godoy, J.L. Dawson, A 12-bit resolution, 200-M Sample/second phase modulator for a 2.5 GHz carrier with discrete carrier pre-rotation in 65 nm CMOS, in: Proceedings of IEEE Radio Frequency Integrated Circuits (RFIC) Symposium, 2011, pp. 1–4.

[63] W.M. Gaber, P. Wambacq, J. Craninckx, M. Ingels, A CMOS IQ direct digital RF modulator with embedded RF FIR-based quantization noise filter, in: Proceedings of European Solid-state Circuits Conference (ESSCIRC), 2011, pp. 139–142.

[64] H. Lakdawala, M. Schaecher, C. Fu, R. Limaye, J. Duster, Y. Tan, A. Balankutty, E. Alpman, C.C. Lee, K.M. Nguyen, H. Lee, A. Ravi, S. Suzuki, B.R. Carlton, H.S. Kim, M. Verhelst, S. Pellerano, T. Kim, S. Venkatesan, D. Srivastava, P. Vandervoorn, J. Rizk, C. Jan, S. Ramamurthy, R. Yavatkar, K. Soumyanath, A 32 nm SoC with dual core ATOM processor and RF WiFi transceiver, IEEE J. Solid-State Circuits 48 (1) (2013) 91–103.

[65] W.M. Gaber, P. Wambacq, J. Craninckx, M. Ingels, A CMOS IQ digital doherty transmitter using modulated tuning capacitors, in: Proceedings of European Solid-state Circuits Conference (ESSCIRC), 2012, pp. 341–344.

[66] C. Lu, H. Wang, C.H. Peng, A. Goel, S. Son, P. Liang, A.M. Niknejad, H.C. Hwang, G. Chien, A 24.7 dBm all-digital RF transmitter for multimode broadband applications in 40 nm CMOS, in: IEEE International Solid-State Circuits Conference Digest of Technical Papers (ISSCC), 2013, pp. 332–333.

[67] GSM World, Market Data Summary, 2010, http://www.gsacom.com/news/statistics.php4.

[68] TriQuint Semiconductor, TQM7M5005H Data Sheet, Hillsboro, Oregon, USA, 2009.

[69] TriQuint Semiconductor, TQM7M5012H Data Sheet, Hillsboro, Oregon, USA, 2010.

[70] S. Modi, S. Kanigere, O. Eliezer, P. Balsara, A limited bandwidth envelope follower for efficiency enhancement in a linear power amplifier in broadband transmitters, in: Circuits and Systems Workshop: System-on-Chip-Design, Applications, Integration, and Software, 2008 IEEE Dallas, IEEE, pp. 1–4.

[71] M. Elliott, T. Montalvo, F. Murden, B. Jeffries, J. Strange, S. Atkinson, A. Hill, S. Nandipaku, J. Harrebek, A polar modulator transmitter for EDGE, Solid-State Circuits Conference, 2004. Digest of Technical Papers. ISSCC. 2004 IEEE International 1 (2004) 190–522.

[72] R. Staszewski, P. Balsara, All-Digital Frequency Synthesizer in Deep-Submicron CMOS, Wiley, New York, 2005.

[73] R. Staszewski, J. Wallberg, S. Rezeq, et al., All-digital PLL and GSM/EDGE transmitter in 90 nm CMOS, in: IEEE Solid-State Circuits Conf., San Francisco, CA, 2005, pp. 316–317.

[74] Texas Instruments Inc., OMAP Processor Platform for Wireless Solution Development, 2011, http://en.wikipedia.org/wiki/Texas_Instruments_OMAP.

[75] J. Lin, A low-phase-noise 0.004-ppm/step DCXO with guaranteed monotonicity in the 90-nm CMOS process, IEEE J. Solid-State Circuits 40 (12) (2005) 2726–2734.

[76] D. Griffith, F. Dulger, G. Feygin, A.N. Mohieldin, P. Vallur, A 65 nm CMOS DCXO system for generating 38.4 MHz and a real time clock from a single crystal in $0.09\,mm^2$, in: Radio Frequency Integrated Circuits Symposium (RFIC), 2010 IEEE, pp. 321–324.

[77] R. Andraka, A survey of CORDIC algorithms for FPGA based computers, in: FPGA '98: Proceedings of the 1998 ACM/SIGDA Sixth International Symposium on Field Programmable Gate Arrays, ACM, New York, NY, 1998, pp. 191–200.

[78] J. Mehta, V. Zoicas, O. Eliezer, R.B. Staszewski, S. Rezeq, M. Entezari, P. Balsara, An efficient linearization scheme for a digital polar EDGE transmitter, IEEE Trans. Circuits Syst. II, Express Briefs 57 (3) (2010) 193–197.

[79] J. Mehta, I. Bashir, V. Zoicas, Y. Wang, O. Eliezer, K. Waheed, M. Entezari, S. Larson, D. Shrestha, S. Rezeq, B. Staszewski, P. Balsara, Self-calibration of a digital pre-power amplifier in a polar transmitter, in: Circuits and Systems Workshop: System-on-Chip—Design, Applications, Integration, and Software, 2010 IEEE Dallas, 2010, pp. 1–4.

[80] E. Hogenauer, An economical class of digital filters for decimation and interpolation, IEEE Trans. Acoust. Speech Signal Process. 29 (2) (1981) 155–162.

[81] V. Parikh, P. Balsara, O. Eliezer, J. Mehta, A low power and low quantization noise digital sigma-delta modulator for wireless transmitters, in: IEEE ISCAS, New Orleans, 2007.

[82] B. Razavi, A study of injection locking and pulling in oscillators, IEEE J. Solid-State Circuits 39 (9) (2004) 1415–1424.

[83] I. Bashir, R.B. Staszewski, O. Eliezer, K. Waheed, V. Zoicas, N. Tal, J. Mehta, M.-C. Lee, P.T. Balsara, B. Banerjee, An EDGE transmitter with mitigation of oscillator pulling, in: Radio Frequency Integrated Circuits Symposium (RFIC), 2010 IEEE, 2010, pp. 13–16.

[84] O.E. Eliezer, A Phase Domain Approach for Mitigation of Self-Interference in a Transmitter, Ph.D. Dissertation, 2008.

[85] I.L. Syllaios, P.T. Balsara, R.B. Staszewski, Recombination of envelope and phase paths in wideband polar transmitters, vol. 57, 2010, pp. 1891–1904.

[86] T. Hentschel, G. Fettweis, Sample rate conversion for software radio, IEEE Commun. Mag. 38 (8) (2000) 142–150.

[87] I. Elahi, K. Muhammad, P.T. Balsara, I/Q mismatch compensation using adaptive decorrelation in a low-IF receiver in 90-nm CMOS process, IEEE J. Solid-State Circuits 41 (2) (2006) 395–404.

[88] J.E. Volder, The CORDIC trigonometric computing technique, IRE Trans. Electron. Comput. (3) (1959) 330–334.

[89] J.L. Dawson, T.H. Lee, Feedback Linearization of RF Power Amplifiers, Springer, Berlin, 2004.

[90] T. Sowlati, D. Rozenblit, R. Pullela, M. Damgaard, E. McCarthy, D. Koh, D. Ripley, F. Balteanu, I. Gheorghe, Quad-band GSM/GPRS/EDGE polar loop transmitter, IEEE J. Solid-State Circuits 39 (12) (2004) 2179–2189.

[91] J. Vuolevi, T. Rahkonen, Distortion in RF Power Amplifiers, Artech House Publishers, Norwood, MA, 2003.

[92] G. Seegerer, G. Ulbricht, EDGE transmitter with commercial GSM power amplifier using polar modulation with memory predistortion, in: Microwave Symp. Digest, 2005 IEEE MTT-S, 2005, pp. 1553–1556.

[93] K.J. Muhonen, M. Kavehrad, R. Krishnamoorthy, Look-up table techniques for adaptive digital predistortion: a development and comparison, IEEE Trans. Veh. Technol. 49 (5) (2000) 1995–2002.

[94] J.K. Cavers, A linearizing predistorter with fast adaptation, in: Vehicular Technology Conference, 1990 IEEE 40th, 1990, pp. 41–47.

[95] N. Ceylan, J.-E. Mueller, R. Weigel, Optimization of EDGE terminal power amplifiers using memoryless digital predistortion, in: Radio Frequency Integrated Circuits (RFIC) Symposium, 2004. Digest of Papers. 2004 IEEE, 2004, pp. 373–376.

[96] L. Sundstrom, M. Faulkner, M. Johansson, Quantization analysis and design of a digital predistortion linearizer for RF power amplifiers, IEEE Trans. Veh. Technol. 45 (4) (1996) 707–719.

[97] J.K. Cavers, Optimum table spacing in predistorting amplifier linearizers, IEEE Trans. Veh. Technol. 48 (5) (1999) 1699–1705.

[98] K.J. Muhonen, M. Kavehrad, R. Krishnamoorthy, Adaptive baseband predistortion techniques for amplifier linearization, in: Conference Record of the Thirty-Third Asilomar Conference on Signals, Systems, and Computers, 1999, vol. 2, IEEE, 1999, pp. 888–892.

[99] R.B. Staszewski, I. Bashir, O. Eliezer, RF built-in self test of a wireless transmitter, IEEE Trans. Circuits Syst. II, Express Briefs 54 (2) (2007).

[100] K. Waheed, S.N. Ba, Adaptive digital linearization of a DRP based EDGE transmitter for cellular handsets, in: 50th Midwest Symposium on Circuits and Systems, 2007. MWSCAS, pp. 706–709.

[101] Wikipedia, Linear Feedback Shift Register, 2011, http://en.wikipedia.org/wiki/Linear_feedback_shift_register.

[102] R.M. Gray, Spectral analysis of quantization noise in a single-loop sigma-delta modulator with DC input, IEEE Trans. Commun. 37 (6) (1989) 588–599.

[103] I. Galton, Delta-sigma data conversion in wireless transceivers, IEEE Trans. Microwave Theory Tech. 50 (1) (2002) 302–315.

[104] S. Luschas, R. Schreier, H.-S. Lee, Radio frequency digital-to-analog converter, IEEE J. Solid-State Circuits 39 (9) (2004) 1462–1467.

[105] K. Bernstein, D.J. Frank, A.E. Gattiker, W. Haensch, B.L. Ji, S.R. Nassif, E.J. Nowak, D.J. Pearson, N.J. Rohrer, High-performance CMOS variability in the 65-nm regime and beyond, IBM J. Res. Dev. 50 (4.5) (2010) 433–449.

[106] P.G. Drennan, C.C. McAndrew, Understanding MOSFET mismatch for analog design, IEEE J. Solid-State Circuits 38 (3) (2003) 450–456.

[107] M.J.M. Pelgrom, A.C.J. Duinmaijer, A.P.G. Welbers, Matching properties of MOS transistors, IEEE J. Solid-State Circuits 24 (5) (1989) 1433–1439.

[108] D.S. Boning, S. Nassif, Models of process variations in device and interconnect, in: Design of High Performance Microprocessor Circuits, Citeseer, 1999 (Chapter 6).

[109] T. Shui, R. Schreier, F. Hudson, Mismatch shaping for a current-mode multibit delta-sigma DAC, IEEE J. Solid-State Circuits 34 (3) (1999) 331–338.

[110] I. Galton, Why dynamic-element-matching DACs work, IEEE Trans. Circuits Syst. II, Express Briefs 57 (2) (2010) 69–74.

[111] J. Welz, I. Galton, Necessary and sufficient conditions for mismatch shaping in a general class of multibit DACs, IEEE Trans. Circuits Syst. II, Analog Digit. Signal Process. 49 (12) (2002) 748–759.

[112] R.M. Gray, Quantization noise spectra, IEEE Trans. Inf. Theory 36 (6) (1990) 1220–1244.

[113] 3rd Generation Partnership Project, Technical Specification Group GSM/EDGE Radio Access Network—Radio Transmission and Reception, 2005, 3GPP TS 45.005 V5.12.0 Release 5, 2005-04.

[114] S.F. Chen, Y.B. Lee, B. Tzeng, C.C. Tang, C. Chiu, R. Yu, O. Lin, L.W. Ke, C.P. Wu, C.W. Yeh, P.Y. Chen, G.K. Dehng, A GSM/EDGE transmitter in 0.13 μm CMOS using offset phase locked loop and direct conversion architecture, in: Radio Frequency Integrated Circuits Symposium, 2008. RFIC 2008. IEEE, 2008, pp. 581–584.

[115] Aeroflex, Project P25, 2011, http://www.p25.com.

[116] M.S. Alavi, A. Visweswaran, R.B. Staszewski, L.C.N. de Vreede, J.R. Long, A. Akhnoukh, A 2-GHz digital I/Q modulator in 65-nm CMOS, in: Proceedings of IEEE Asian Solid-State Circuit Conference (A-SSCC), 2011, pp. 277–280.

[117] M.S. Alavi, R.B. Staszewski, L.C.N. de Vreede, J.R. Long, Orthogonal summing and power combining network in a 65-nm all-digital RF I/Q modulator, in: Proceedings of IEEE Radio-Frequency Integration Technology (RFIT) Symposium, 2011, pp. 21–24.

[118] M.S. Alavi, R.B. Staszewski, L.C.N. de Vreede, A. Visweswaran, J.R. Long, All-digital RF I/Q modulator, IEEE Trans. Microwave Theory Tech. 60 (11) (2012) 3513–3526.

[119] M.S. Alavi, G. Voicu, R.B. Staszewski, L.C.N. de Vreede, J.R. Long, A 2×13-bit all-digital I/Q RF-DAC in 65-nm CMOS, in: Proceedings of IEEE Radio Frequency Integrated Circuits (RFIC) Symposium, 2013, pp. 167–170.

[120] M.S. Alavi, R.B. Staszewski, L.C.N. de Vreede, J.R. Long, A widband 2×13-bit all-digital I/Q RF-DAC, IEEE Trans. Microwave Theory Tech. 62 (4) (2014) 732–752.

[121] B. Razavi, Design of Analog CMOS Integrated Circuits, McGraw-Hill Education, New York, 2000, http://books.google.nl/books?id=X_rAQgAACAAJ.

[122] H.R. Rategh, T.H. Lee, Superharmonic injection-locked frequency dividers, IEEE J. Solid-State Circuits 34 (6) (1999) 813–821.

[123] J.M. Rabaey, A.P. Chandrakasan, B. Nikolic, Digital Integrated Circuits: A Design Perspective, Pearson Education, Upper Saddle River, NJ, 2003, http://books.google.nl/books?id=_7daAAAAYAAJ.

[124] B. Razavi, Design of Integrated Circuits for Optical Communications, McGraw-Hill, New York, 2003, http://books.google.nl/books?id=Pl9TAAAAMAAJ.

[125] B. Razavi, K.F. Lee, R.H. Yan, Design of high-speed, low-power frequency dividers and phase-locked loops in deep submicron CMOS, IEEE J. Solid-State Circuits 30 (2) (1995) 101–109.

[126] J.R. Long, Monolithic transformers for silicon RF IC design, IEEE J. Solid-State Circuits 35 (9) (2000) 1368–1382.

[127] B. Razavi, Principles of data conversion system design, IEEE Press, New York, 1995, http://books.google.nl/books?id=mKYoAQAAMAAJ.

[128] W.A. Kester, Data Conversion Handbook, Elsevier, Amsterdam, 2005, http://books.google.nl/books?id=0aeBS6SgtR4C.

[129] M. Gustavsson, J.J. Wikner, N. Tan, CMOS Data Converters for Communications, Springer, Berlin, 2000, http://books.google.nl/books?id=D_I2XvNOc4wC.

[130] P. Chatzimisios, C. Verikoukis, I. Santamaria, M. Laddomada, O. Hoffmann, Mobile Lightweight Wireless Systems: Second International ICST Conference, in: Lecture Notes of the Institute for Computer Sciences, Social Informatics and Telecommunications Engineering, Springer, Berlin, 2010, http://books.google.nl/books?id=tSwKZxtx82gC.

[131] N. Dinur, D. Wulich, Peak-to-average power ratio in high-order OFDM, IEEE Trans. Commun. 49 (6) (2001) 1063–1072.

[132] M. Helaoui, S. Boumaiza, A. Ghazel, F.M. Ghannouchi, On the RF/DSP design for efficiency of OFDM transmitters, IEEE Trans. Microwave Theory Tech. 53 (7) (2005) 2355–2361.

[133] T.H. Lee, Planar Microwave Engineering: A Practical Guide to Theory, Measurement, and Circuits, Cambridge University Press, Cambridge, 2004, http://books.google.nl/books?id=uoj3IWFxbVYC.

[134] H.A. Mahmoud, H. Arslan, Error vector magnitude to SNR conversion for nondata-aided receivers, IEEE Trans. Commun. Wirel. 8 (5) (2009) 2694–2704.

[135] T. Pollet, M. Van Bladel, M. Moeneclaey, BER sensitivity of OFDM systems to carrier frequency offset and Wiener phase noise, IEEE Trans. Commun. 43 (234) (1995) 191–193.

[136] R. Liu, Y. Li, H. Chen, Z. Wang, EVM estimation by analyzing transmitter imperfections mathematically and graphically, Analog Integr. Circ. Sig. Process 48 (3) (2006) 257–262, doi:10.1007/s10470-006-7701-0.

[137] G. Liu, P. Haldi, T.-J.K. Liu, A.M. Niknejad, Fully integrated CMOS power amplifier with efficiency enhancement at power back-off, IEEE J. Solid-State Circuits 43 (3) (2008) 600–609.

[138] P. Haldi, D. Chowdhury, P. Reynaert, G. Liu, A.M. Niknejad, A 5.8 GHz 1 V linear power amplifier using a novel on-chip transformer power combiner in standard 90 nm CMOS, IEEE J. Solid-State Circuits 43 (5) (2008) 1054–1063.

[139] D. Chowdhury, C.D. Hull, O.B. Degani, Y. Wang, A.M. Niknejad, A fully integrated dual-mode highly linear 2.4 GHz CMOS power amplifier for 4G WiMax applications, IEEE J. Solid-State Circuits 44 (12) (2009) 3393–3402.

[140] H. Wang, C. Sideris, A. Hajimiri, A CMOS broadband power amplifier with a transformer-based high-order output matching network, IEEE J. Solid-State Circuits 45 (12) (2010) 2709–2722.

[141] G.D. Ewing, High-Efficiency Radio-Frequency Power Amplifiers, Ph.D. thesis, Dept. Elect. Eng., Oregon State University, 1964, http://ir.library.oregonstate.edu/xmlui/handle/1957/20196 (online).

[142] N.O. Sokal, A.D. Sokal, Class E—a new class of high-efficiency tuned single-ended switching power amplifiers, IEEE J. Solid-State Circuits 10 (3) (1975) 168–176.

[143] F.H. Raab, Idealized operation of the class E tuned power amplifier, IEEE Trans. Circuits Syst. I, Regul. Pap. 24 (12) (1977) 725–735.

[144] S.D. Kee, I. Aoki, A. Hajimiri, D. Rutledge, The class-E/F family of ZVS switching amplifiers, IEEE Trans. Microwave Theory Tech. 51 (6) (2003) 1677–1690.

[145] Y. Yoon, J. Kim, H. Kim, K.H. An, O. Lee, C.-H. Lee, J.S. Kenney, A dual-mode CMOS RF power amplifier with integrated tunable matching network, IEEE Trans. Microwave Theory Tech. 60 (1) (2012) 77–88.

[146] M. Babaie, R.B. Staszewski, A study of RF oscillator reliability in nanoscale CMOS, in: European Conference on Circuit Theory and Design (ECCTD), 2013, pp. 1–4.

[147] A. Hajimiri, T.H. Lee, Design issues in CMOS differential LC oscillators, IEEE J. Solid-State Circuits 34 (5) (1999) 717–724.

[148] C.-H. Lin, K. Bult, A 10 bit 250 M Sample/s CMOS DAC in 1 mm^2, in: IEEE International Solid-State Circuits Conference Digest of Technical Papers (ISSCC), 1998, pp. 214–215.

[149] C.-H. Lin, K. Bult, A 10-b, 500-M Sample/s CMOS DAC in 0.6 mm^2, IEEE J. Solid-State Circuits 33 (12) (1998) 1948–1958.

[150] C.-H. Lin, F.M.I. van der Goes, J.R. Westra, J. Mulder, Y. Lin, E. Arslan, E. Ayranci, X. Liu, K. Bult, A 12 bit 2.9 GS/s DAC with IM3 $\ll$ −60 dBc beyond 1 GHz in 65 nm CMOS, IEEE J. Solid-State Circuits 44 (12) (2009) 3285–3293.

[151] W.C.E. Neo, J. Qureshi, M.J. Pelk, J.R. Gajadharsing, L.C.N. de Vreede, A mixed-signal approach towards linear and efficient N-way Doherty amplifiers, IEEE Trans. Microwave Theory Tech. 55 (5) (2007) 866–879.

[152] V. Petrovic, Reduction Of spurious emission from radio transmitters by means of modulation feedback, in: Proceeding of IEE Conference on Radio Spectrum Conservation Techniques, 1983, pp. 44–49.

[153] G. Karam, H. Sari, A data predistortion technique with memory for QAM radio systems, IEEE Trans. Commun. 39 (2) (1991) 336–344.

[154] J.H. Qureshi, M.J. Pelk, M. Marchetti, W.C.E. Neo, J.R. Gajadharsing, M.P. van der Heijden, L.C.N. de Vreede, A 90-W peak power GaN outphasing amplifier with optimum input signal conditioning, IEEE Trans. Microwave Theory Tech. 57 (8) (2009) 1925–1935.

[155] H. Granberg, Measuring the intermodulation distortion of linear amplifier, in: Motorola Semiconductor Engineering Bulletin, 1993, pp. 1–4, http://cache.freescale.com/files/rf_if/doc/eng_bulletin/EB38.pdf (online).

[156] F. Op't Eynde, Filtering aspects of all-digital RF transceivers for mobile applications, in: Workshop on Pushing the Ultimate Performance Limits of RF CMOS, Radio Frequency Integrated Circuits Symposium (RFIC), 2013.

[157] J. Chen, L. Ye, D. Titz, F. Gianesello, R. Pilard, A. Cathelin, F. Ferrero, C. Luxey, A.M. Niknejad, A digitally modulated mm-wave cartesian beamforming transmitter with quadrature spatial combining, in: IEEE International Solid-State Circuits Conference Digest of Technical Papers (ISSCC), 2013, pp. 232–233.

[158] R.E. Crochiere, L.R. Rabiner, Multirate Digital Signal Processing, Prentice Hall PTR, Upper Saddle River, NJ, 1983, http://books.google.nl/books?id=X_NSAAAAMAAJ.

[159] W.H. Doherty, A new high efficiency power amplifier for modulated waves, Proc. Inst. Radio Eng. 24 (9) (1936) 1163–1182.

[160] L.R. Kahn, Single-sideband transmission by envelope elimination and restoration, Proc. IRE 40 (7) (1952) 803–806.

[161] M.S. Alavi, F. van Rijs, M. Marchetti, M. Squillante, T. Zhang, S.J.C.H. Theeuwen, Y. Volokhine, H.F.F. Jos, M.P. Heijden, M. Acar, L.C.N. de Vreede, Efficient LDMOS device operation for envelope tracking amplifiers through second harmonic manipulation, in: IEEE Microwave Symposium Digest (MTT), 2011, pp. 1–4.

[162] M.J. Pelk, W.C. Neo, J.R. Gajadharsing, R.S. Pengelly, L.C.N. de Vreede, A high-efficiency 100-W GaN three-way Doherty amplifier for base-station applications, IEEE Trans. Microwave Theory Tech. 56 (7) (2008) 1582–1591.

[163] D.A. Calvillo-Cortes, M.P. van der Heijden, M. Acar, M. de Langen, R. Wesson, F. van Rijs, L.C.N. de Vreede, A package-integrated Chireix outphasing RF switch-mode high-power amplifier, IEEE Trans. Microwave Theory Tech. 61 (10) (2013) 3721–3732.
[164] 3rd Generation Partnership Project, Technical Specification Group GSM/EDGE Radio Access Network—Modulation, 2003, 3GPP TS 45.004 V5.1.1 Release 5, 2003-09.
[165] Mathworks, Mathworks Matlab—The Language Of Technical Computing, 2011, http://www.mathworks.com/products/matlab/.

Index

Note: Page number followed by *f* indicate figure *t* indicate tables *np* indicate footnotes.

A

ACPR. *See* Adjacent channel power ratio (ACPR)
Adaptive interpolation, 57–59, 58*t*, 60*f*, 63*f*, 260
Adaptive predistortion, 52
Adjacent channel power ratio (ACPR), 8–9, 238–240
ADPLL. *See* All-digital phase-locked loop (ADPLL)
All-digital high resolution I/Q RFDAC, 174
All-digital I/Q modulator
  concept of, 129–131, 130*f*
  idealized block diagram, 130–131, 130*f*
  RFDAC, orthogonal summing operation, 131–141
  SPICE simulated constellation diagram, 131–132, 133*f*
All-digital I/Q RFDAC, 162–163, 171*f*
All-digital I/Q transmitter, 16–18, 17*f*
All-digital *orthogonal* I/Q modulator, 161
All-digital phase-locked loop (ADPLL), 24, 41–42, 47
All-digital polar transmitter, 14–16, 15*f*
AM-AM. *See* Amplitude distortion (AM-AM)
Amplitude control word (ACW), 42, 53*f*
Amplitude-dependant-phase distortion (AM-PM), 34, 53–55, 94–95, 219–220
  DPA distortion, 64
  dynamic measurement, 236–237
  LUT-based predistortion scheme, 55, 56*f*
  RFDAC's linearity, 236–237
Amplitude distortion (AM-AM), 34, 53–55, 67, 99–100
  digital I/Q transmitter, 232–233
  dynamic measurement, 236–237
  LUT-based predistortion scheme, 55, 56*f*
  nonlinearity, 216–217, 219–220
  predistortion module, 60, 62*f*
  RFDAC's linearity, 236–237
  table inversion and adaptive interpolation, 62–63, 63*f*
Amplitude mismatch, 92–93
Amplitude modulation, 42
  digital design, 81–82, 81*f*, 82*f*
  overview, 80–81, 81*f*
  transfer function and spectrum, 82–83, 83*f*
AM-PM. *See* Amplitude-dependant-phase distortion (AM-PM)
Analog-intensive RF transmitter, 9–12, 10*f*, 11*f*
Analog *vs.* digital transmitter, 9
Antenna's impedance, 248

B

Back-off power levels, 191–193
Balun converter, 181–182, 193–200, 194*f*, 194*t*, 196*f*
Balun transformer, 248–249
Baseband data, 139–141
Baseband digital transmitter, 42, 43*f*
Baseband information, 131
Baseband signal, 255
Binary-coded system, 91
Binary-to-thermometer encoders, 209, 214–215
Built-in self-calibration (BISC), 67
Built-in self-test (BIST), 52, 66, 69

C

Calibration algorithm, 110
Calibration mechanism
  HB WBN performance, 110, 111*f*
  LB WBN performance, 110–112, 111*f*
Cartesian modulators, 129
Cartesian-to-polar conversion, 75–76
Cascaded integrator-comb (CIC) structure, 69–70, 70*f*, 71*f*
Circuit building blocks
  digital I/Q modulator, 144–154
Circuit-level SPICE simulations, 99
CKV. *See* Variable clock (CKV)
Class-B power amplifier, 191–192
Class-D power combiner
  I/Q RFDAC, 173
  switched-capacitor digital polar transmitter, 173
Class-E based power combiner, 179–187, 180*f*, 181*f*, 182*f*, 197–198
  back-off levels, 191–193
  design parameter, 188–189, 189*t*
Clock delay alignment
  problem explanation, 106–108, 107*f*
  self-calibration and compensation mechanism, 108–112, 109*f*, 111*f*
Clock ÷2 dividers, 203–204, 203*f*

281

Clock frequency, 110
Clock input transformer, 202–203
Clock sampling rate, 247–248
Clock skew alignment, 125
Closed form equations, 113–114
Closed-loop ADPLL model, 260–261
Closed-loop continuous feedback predistortion, 53
Close-in spectrum, 256, 256t
CMOS technology, 4–5, 22
Combo chips, 249–250
Complementary quadrature sign bit, 204–205, 205f
Complex-modulated baseband signals, 166–167, 245
Composite IQ RF vectors, 161
Constant integer, 113
Continuous traverse, 208f, 209
Conventional class-E network, 179–180, 180f
Conventional RF transceiver performance, 2–3
COordinate Rotation DIgital Computer (CORDIC), 26, 44, 49–51, 50f, 51f, 57
Current limited regime, 198–200
Current sources
  controlling signals of, 134–135
  FIR arrays summing, 136, 137f

D

DAT topology, 173
DEM. *See* Dynamic element matching (DEM)
Demodulator, 3
Differential I/Q DPA
  back-off levels, 191–193
  balun design, 193–200
  class-E based power combiner, 179–187
  duty cycle, 190–191
  idealized power combiner, 174–179
  I/Q RFDAC efficiency, 187–190
  rise/fall time, 190–191
Differential quadrature 25% duty cycle generator, 205–207, 206f
Digital amplitude, 260
Digital baseband (DBB), 42, 42f, 47, 55
Digital carrier signals, 131–132
Digital communications theory, 131
Digital I/Q calibration. *See* Digital predistortion (DPD)
Digital I/Q modulator, 161–162
  digitally controlled oscillator, 145–146, 145f
  divide-by-two circuit, 146, 146f
  duty cycle generator, 146–147, 147f
  implicit mixer circuit, 148, 149f
  sign bit circuit, 147–148, 148f

2 × 3-bit I/Q switch array circuits, 148–154, 149f, 159t
Digital I/Q transmitter, 232–233
Digitally controlled oscillator (DCO), 28
  core circuit, 145–146, 145f
  8GHz DCO resonating frequency, 145
  frequency pulling, 33–34, 34f
  frequency pushing, 32–33
  operating frequency, 29
  three-bit switched-capacitor, 145–146
Digitally intensive amplitude path, 23
Digitally intensive RF transmitter, 12–13
Digital power amplifier (DPA), 24, 169
Digital predistortion (DPD), 51–52, 99, 216–228, 236–237, 246
  AM-AM and AM-PM profiles, 219–220
  bandwidth expansion, 267–268
  constellation-mapping approach, 238
  I/Q code mapping, 220–224, 226–228
  IQ image and leakage suppression, 217–218
  memory and time, 224–225
  spectrum measurement results, 240–241, 241f
  temperature and aging, 225–226
  two-tone linearity test, 236–237, 237f
Digital PRE-PA
  mismatches, 92–95
  overview, 83–84, 85f
  quantization noise, 84–90, 87f, 88f, 89f, 90f
  structural design, 91, 92f
Digital signal processing, 162–163
Digital-to-analog converter (DAC) functionality, 78–79
Digital-to-RF-amplitude converter (DRAC), 130–131, 210–214, 211f, 246–247
  mixer and switch array, 140–141, 140f
Digital transmitter (DTX), 41–42, 42f, 43f
DPD. *See* Digital predistortion (DPD)
DRAC. *See* Digital-to-RF-amplitude converter (DRAC)
Drain capacitance *vs.* drain voltage, 176–177, 176f
Drain efficiency, 176–177, 176f
Drain voltage, 138
D-to-Q delay, 204
DTX. *See* Digital transmitter (DTX)
Duty cycle, 131–132, 190–191
  generator, 146–147, 147f
Dynamic element matching (DEM), 91, 93, 102–104
Dynamic inversion, 260
Dynamic LUT inversion, 59–64, 61f, 63f
Dynamic measurement, 234–243

I/Q RFDAC, LO leakage and IQ image
    suppression of, 234–236
  RFDAC's linearity
    AM-AM/AM-PM profiles, 236–237
    constellation mapping, 238–243
Dynamic nonlinearity (DNL), 93
Dynamic range, 161–162
  modulated/multitone signal, 167
  of RFDAC, 166–167

E

Efficiency contour, 178–179, 178*f*
8PSK symbol parameter, 252, 252*f*
Enhanced Data Rates for GSM Evolution (EDGE),
    21, 41–42, 43*f*, 58–59
  constellation, 253, 253*f*
  pulse shaping filter (PSF), 255
  RF system specifications, 256–259, 256*t*
  simulation model, 259–261, 259*f*
    digital amplitude and phase data generation,
      260
    RF front-end model, 260–261
  symbol mapping and rotation, 252–253, 252*f*
Error vector magnitude (EVM), 255
  distortion and predistortion, 65–66
  DPD algorithm, 169
European Telecommunications Standards Institute
    (ETSI), 256, 256*t*

F

Farrow interpolator, 248
Figures-of-merit, RF transmitter, 8–9
Finite impulse response (FIR) filters, 247–248
First- and second-order $\Sigma\Delta$, 81, 82*f*
First-order digital $\Sigma\Delta$ modulator, 80–81, 81*f*
First-order-hold interpolator, 248
First order phase modulation, 260–261
First-order $\Sigma\Delta$ ADC, 80, 81*f*
5GWi-Fi, 249–250
Flip-flop based frequency divider, 203–204
Floorplanning
  of 2×13-bit DRAC, 207–209, 208*f*
Fourier representation, 187–188
Frequency bands, 258–259, 259*t*
Frequency control word (FCW), 28, 41
Frequency down conversion, 146

G

Gate/drain capacitance *vs.* on-switches,
    176–177, 176*f*
Gaussian minimum-shift keying (GMSK), 41, 251

GetFlipFlopOutput procedure, 109–110
Gilbert cell mixer, 134–135
Global System for Mobile communication (GSM),
    21, 41
Gradient mismatch, 215*np*

H

Harmonic frequency, 248–249
  components, 134, 134*t*
High-resolution RFDAC, 162–169
High speed digital logic, 112
High-speed rail-to-rail differential dividers,
    203–204

I

Idealized power combining network, 174*f*, 175, 179
Idle-tones, 112–113, 112*f*
Implicit mixer circuit, 148, 149*f*
IM3-to-carrier *(IM3/C)*, 266, 267*f*
In-phase and quadrature digital RF vectors, 161
In-phase and quadrature-phase paths, 200
In-phase/quadrature (I/Q) RFDAC, 184–185, 185*t*
  baseband interpolations, 162–163
  calibration technique, 234–236
  chip performance, 242–243, 242*t*
  code mapping, 220–224, 226–228
  conceptual diagram of, 131–132, 132*f*
  duty cycle, 190–191
  efficiency of, 187–190, 187*f*
  leakage and image suppression, 217–218, 219*f*,
      234–236, 235*f*
  mismatch and gain calibrations, 67
  modulation accuracy, 193
  noise performance, 191–192, 192*f*
  rise/fall time effect, 190–191, 191*f*
  segmented thermometer code, 207–209
  summation, 136
Integer bits, 79
Integer transistor, 106
Interdigitated transformer balun, 195–196
Interpolative filter, 59, 69–75, 70*f*, 71*f*, 72*f*, 74*t*
  highpass noise addition, 73, 73*f*
Intra-die variation, 96–97
Inverter based buffer delay, 110–112
Invertor delay code, 108

K

KDCO gain, 31
Kirchoff's voltage and current law, 187–188

L

Lagrange polynomial, 47–48
Linear interpolation, 63, 71
Local oscillator (LO), 234–236
Look-up-table (LUT), 49, 54–55, 100
  AM-PM characteristics, 65–66
  quantization effects, 64–66, 65f
Lossless power combining network, 193–194
Low-dropout (LDO) linear regulators, 68, 68f
Low noise amplifier (LNA), 3, 67
LUT. *See* Look-up-table (LUT)

M

MASH $\Sigma\Delta$ architecture, 82
Matching network, 136, 174f, 175–176
  equations, 263–266
MATLAB, 162–165
  language based model, 259
Mismatches and DEM, 102–104, 104f
  measurement results, 104–105, 105f, 106f
  MSB (4×) transistor mismatch, 96–97
  simulation-based specifications, 101–102, 102f, 103f
  systematic mismatch between 1× and 4× transistors, 98–101, 98f, 100f
  unit (1×) transistor mismatch, 97
Modulating bits, 252, 252f
Modulation error, 177, 178f
Multicarrier M-QAM OFDM signals, 166–167
Multimode/multiband transmitters, 161–162

N

Nanometer-scale CMOS process, 245
  RF transmitter, 13–14
Narrowband polar transmitters
  conventional analog approaches, 24–25
  motivation, 21–23
  phase modulation, 30–32
  RFDAC-based polar transmitter architecture, 25–29
  small-signal polar transmitter, challenges, 32–40
Noise shaping, 82–83, 117–118, 123–125
Noncontinuous traverse, 208f, 209
Nonmonotonous amplitude response, 114–115
Nyquist sampling theorem, 75–76

O

On-chip differential power combining network, 174
On-chip microprocessor, 108–109
On-chip receiver-based approach, 104–105
On-chip SRAM memories, 263, 264f
One-dimensional (1D) mapping, 247
Open loop feedforward predistortion, 52
Optimum delay relationship, 108
Orthogonal power combining network, 179
Orthogonal summation, 133, 141
  digital I/Q modulator, circuit building blocks of, 144–154
  measurement results, 154–159
  RFDAC, 131–141
Oversampling ratio (OSR), 83

P

Parasitic coupling, 125
  idle-tones, 113–115, 115f, 116f
  possible coupling paths, 112–113, 112f
Peak-to-average power ratio (PAPR), 249
Pelgrom's law, 93
Phase data generation, 260
Phase distortion, 95
Phase mismatch, 93–95, 94f, 96f
Polar bandwidth expansion, 75–76, 75f
Polar transmitter
  measurement results, 118–125
  simulation results, 117–118, 118f
Polynomial approximation, 266
Power amplifier (PA), 4, 169, 245
Power back-off levels, 173
Power combining network, 144, 150–151
  back-off levels, 191–193
  balun design, 193–200
  class-E based, 179–187
  duty cycle, 190–191
  idealized, 174–179
  I/Q RFDAC efficiency, 187–190
  rise/fall time, 190–191
  simulation results, 185–187, 186f
Power consumption, 125
Power efficiency, 134–135
Power-efficient RF transmitter, 173
Power spectral density, 114
Predistortion module, 118, 119f
  operation principle, 55–64
  overview, 51–55
  quantization effects, 64–66
  self-calibration, 66–69
  temperature variations, 69
Predistortion self-calibration, 66–69, 67f, 68f
Printed circuit board (PCB), 229, 230f
Pseudo-differential quadrature structure, 175, 175f
Pull-down NMOS transistor, 84
Pulse shaping filter (PSF), 43–47, 45f, 46f, 253, 255

Q

Quadrature modulation, 131–132, 134
Quadrature passive mixer, 136, 137f
Quadrature phase shift keying (QPSK) modes, 156–157, 157f
Quadrature-phase sweeping time, 225
Quantization effects, 64–66, 65f
Quantization noise, 166–167

R

Radio frequency (RF)
    front-end model, 260–261
    system specifications, 256–259, 256t
Radio-frequency digital-to-analog converter (RFDAC)
    ACPR, 239–240
    AM-AM profiles, 236–237
    AM-PM distortion, 232–233, 236–237
    block diagram of, 78, 78f
    clock delay alignment, 106–112
    constellation mapping, 238–243
    digital PRE-PA, 83–91
    DPA transistor mismatches, 92–95
    efficiency vs. RF output power, 231–232, 232f
    future developments, suggestions, 247–249
    future trends, 249–250
    mismatches and DEM, 96–105
    outcome, 245–247
    overview, 78–80, 78f
    parasitic coupling analysis, 112–115
    $\Sigma\Delta$ amplitude modulation, 80–83, 83f
    static phase noise, 233–234
Radio frequency (RF) modulator, 2–3
Radio frequency (RF) transmitter, 3–4
    adjacent channel power ratio, 8–9
    advantages, 5–6
    all-digital I/Q transmitter, 16–18, 17f
    all-digital polar transmitter, 14–16, 15f
    analog-intensive, 9–12, 10f, 11f
    analog vs. digital, 9
    digitally intensive, 12–13
    drain efficiency, 8–9
    motivation, 4–6
    in nanometer-scale CMOS, 13–14
    system efficiency, 8–9
    WCDMA baseband signal, 6–8, 6f, 7f
Random mismatch, 215np
Resampler
    implementation, 47–48, 48f
    need for, 47
Resolution bandwidth (RBW), 118, 165
RFDAC-based polar transmitter architecture
    advantages, 26–28
    CKV clock, 29
    CORDIC module, 26
    DCO operating frequency, 29, 30t
    DPA, 28–29, 29f
    EDGE transmitter system level block diagram, 25–26, 26f
    interpolation filter, 28
    SoC containing the EDGE transceiver, 25, 26f
    transmitter connectivity, 25, 27f
Rise/fall time effect
    I/Q RFDAC, 190–191, 191f
Root mean square (RMS) power, 79

S

Sampling rate selection, 43–44
Sandard deviation (STDEV), 105
Second divider, 146–147, 147f
Self-calibration, 108–109. See also Predistortion self-calibration
$\Sigma\Delta$ amplitude modulator (SAM), 41, 71–72
Sign bit circuit, 147–148, 148f
Sinc function, 163–164, 166–167
Single-carrier M-QAM signals, 166–167
Single-switch DRAC structure, 175
64-point constellation, 140–141, 140f
Small-signal analog polar transmitter, 24–25
Small-signal polar transmitter, 22–23
    amplitude path DAC mismatches, 39
    amplitude path DC offset, 37–38, 38f
    amplitude-phase delay misalignment, 35–36, 36f
    amplitude-phase path transfer function mismatch, 37
    DCO frequency pulling, 33–34, 34f
    DCO frequency pushing, 32–33
    DCO phase noise, 32
    DPA, additional distortions in, 40
    dynamic range limitations, 38–39, 39f
SoC. See System on a Chip (SoC)
Spurious free dynamic range (SFDR), 236–237
Static measurement
    I/Q RFDAC, 231
    output power, 231–232
    power efficiency, 231–232, 232f
    RF measurements, 231–232, 232f
Sub-DRAC cell, 210–212, 213f
Switchable cascode DRAC structure, 175, 175f
Switched-capacitor digital polar transmitter, 173
Switched-mode behavior, 136
Symbol mapping and rotation, 252–253, 252f
Systematic gradient-based variations, 97
Systematic mismatch, 215np

System level analysis, 96
System level simulations, 84–86, 167f
　signal bandwidth, 164–165, 165f
　SRAM memory length, 164–165, 165f
　test-bench, 162–163, 162f
System on a Chip (SoC), 25, 26f

T

Temperature sweeping, 225–226, 226f
Thermometer encoders, 209, 210f
3π/8 rotation
　amplitude signal spectrum comparison, 253, 254f
　frequency signal spectrum comparison, 253, 254f
Time division multiple access (TDMA) scheme, 251
Time domain filter impulse response, 43, 44f
Transformer-based power combining techniques, 173
Transistor threshold voltage, 92
Transmitter close-in performance, 118–123
　EDGE
　　burst transmission, 118–120, 119f
　　EVM and constellation, 120, 121f
　　400 kHz offset spectral performance, highband frequencies, 121–123, 123f
　　400 kHz offset spectral performance, lowband frequencies, 121–123, 122f
　　highband frequencies, 120–121, 122f
　　lowband frequencies, 120–121, 121f
　　spectral performance, 120, 120f
Transmitter (TX) digital baseband, 41–51
Transmitter wideband noise performance, 123–125
　for highband channel, 123–125, 124f
　for lowband channel, 123–125, 124f
25% duty cycle generator. *See* Differential quadrature 25% duty cycle generator
2 × 3-bit I/Q switch array circuits
　balun transformer, 150–152, 152f
　disadvantages, 148–150
　drain current, 150, 151f
　harmonic distortion, 148
　implementational block diagram, 144, 144f
　NMOS switches, 150–151
　output power, 150
　passive current source, 150
　primary and secondary inductance, 151–152, 153f
　simulations, 150
　single-ended version, 148–150
　S-parameter model, 151–152
　supply and ground bond-wire, 150
　T-section model, 151–152
Two-tone signal, 266
2 × 13-bit all-digital I/Q RFDAC, 202f
　chip micrograph, 229, 230f
　clock input transformer, 202–203
　complementary quadrature sign bit, 204–205
　differential quadrature 25% duty cycle generator, 205–207
　digital I/Q calibration, 216–228
　DPD techniques, 216–228
　　AM-AM and AM-PM profiles, 219–220
　　I/Q code mapping, 220–224, 226–228
　　IQ image and leakage suppression, 217–218
　　memory and time, 224–225
　　temperature and aging, 225–226
　DRAC unit cell, 210–214
　dynamic measurement results, 234–243
　floorplanning, 207–209
　high-speed rail-to-rail differential dividers, 203–204
　measurement setup, 230–231, 230f
　MSB/LSB selection choices, 210–215
　multimode/multiband RF CMOS transmitters, 242–243
　static measurement results, 231–234
　3-to-7 and 4-to-15, thermometer encoders of, 209

U

Unit-weighted digital switches, 137–138, 137f
Universal asynchronous receiver/transmitter (UART), 263, 264f
Upconverted clock, 134, 134f, 136
　duty cycle, 131–133, 137f
　IQ signal, 134

V

Variable clock (CKV), 29, 31
Vectorial summing, 129–130
VLSI design flow, 78
Voltage limited regime, 198–200

W

Watt-level power generation, 173
WBN. *See* Wide band noise (WBN)
Wideband CDMA (WCDMA), 127
　close-in range spectrum, 7, 7f
　far-out spectrum, 7, 7f
　pulse-shaped oversampling rate, 6–7
　upsampling and interpolation process, 6–7, 6f
Wide band noise (WBN), 88, 125–127, 258
Wider band signals, 164, 165f
Wireless communication standards, 249
Wireless LAN (WLAN), 5–6

Z

Zero crossings, 253
Zero-order-hold, 90